China's e-Science Blue Book 2023

Chinese Academy of Sciences · Ministry of Education of the PRC · Ministry of Science and Technology of the PRC · China Association for Science and Technology · Chinese Academy of Social Sciences · Chinese Academy of Engineering · National Natural Science Foundation of China · Chinese Academy of Agricultural Sciences
Editors

China's e-Science Blue Book 2023

中国工信出版集团

電子工業出版社
PUBLISHING HOUSE OF ELECTRONICS INDUSTRY
http://www.phei.com.cn

Springer

Editors
Chinese Academy of Sciences
Beijing, China

Ministry of Education of the PRC
Beijing, China

Ministry of Science and Technology of the PRC
Beijing, China

China Association for Science and Technology
Beijing, China

Chinese Academy of Social Sciences
Beijing, China

Chinese Academy of Engineering
Beijing, China

National Natural Science Foundation of China
Beijing, China

Chinese Academy of Agricultural Sciences
Beijing, China

ISBN 978-981-99-8272-1   ISBN 978-981-99-8270-7 (eBook)
https://doi.org/10.1007/978-981-99-8270-7

Jointly published with Publishing House of Electronics Industry
The print edition is not for sale in China (Mainland). Customers from China (Mainland) please order the print book from: Publishing House of Electronics Industry.

This Springer imprint is published by the registered company Springer Nature Singapore Pte Ltd.
The registered company address is: 152 Beach Road, #21-01/04 Gateway East, Singapore 189721, Singapore

Paper in this product is recyclable.

Preface

According to the report of the 20th National Congress of the Communist Party of China, momentous changes of a like not seen in a century are accelerating across the world. A new round of scientific and technological revolution and industrial transformation is well under way, and a significant shift is taking place in the international balance of power, presenting China with new strategic opportunities in pursuing development. In the new stage of development, scientific and technological innovation plays a key role in the global development pattern and competition pattern. In recent years, with a high-speed development trend, the new generation of information technology, as a key area of technological innovation, intensively and extensively integrates with various industries and fields, driving the transformation of production models and highlighting its guiding and supporting role in the high-quality development of the economy, society, and technological innovation. Sizing up the trend of events with great foresight, the CPC Central Committee, with Comrade Xi Jinping at its core, has set up a national strategic development goal of "building a strong network country and a digital China". It attaches great importance to informationization work and has repeatedly emphasized that "informationization has brought a golden opportunity once in a blue moon to the Chinese nation" and "we must keenly seize the historical opportunity of informationization development."

The e-Science is typical of research environment and activities in the information age. It is an important part of China's informationization construction and a powerful means to improve research productivity, promote technological innovation, and produce significant results. Presently, e-Science plays an important role in promoting the convergence and sharing of scientific and technological resources, triggering changes in research organization and modes, and promoting technological transformation.

The e-Science is an accelerator for significant technological breakthroughs. As scientific research advances towards the super-macro and super-micro directions, information technology is increasingly closely linked with scientific research activities, gradually becoming an indispensable driving force for technological achievements. For example, the research on "complex physical systems" which won the 2021 Nobel Prize in Physics successfully predicted the trend of global warming. The achievement was made with the method of physics simulation, which relied heavily on the application of the e-Science.

The e-Science is the driving force behind the paradigm shift in scientific research. As human society enters the era of big data, scientific research has also entered the fourth paradigm characterized by "data-intensive" and "AI + big data". The application of information technology represented by big data and artificial intelligence can not only help scientists improve the productivity of scientific research, but also overturn traditional scientific research modes, leading to paradigm shifts and driving advancements in the field. For example, in 2021, the DeepMind introduced AlphaFold2, an artificial intelligence program capable of accurately predicting the 3D structure of proteins, ushering in an era in which biologists use computational structure prediction as a research tool.

The e-Science is a crucial pillar in the pursuit of becoming a technologically advanced nation. Currently, all countries and regions around the world attach great importance to the construction of research informationization and consider research informationization as a strategic measure to enhance innovation capacity and international competitiveness. The e-Science is the foundation for the interconnection of major scientific and technological infrastructure, supercomputing centers, scientific data centers, field scientific stations, as well as research institutes, universities, enterprises, and other research elements in China. The national-level research informationization infrastructure platform, known as the "China Science & Technology Cloud" integrates and converges fundamental resources like networks, computing, storage, and software. It promotes the open sharing of scientific and technological resources, provides integrated cloud services of discovery, access, use, and delivery for scientific and technological professionals, and greatly enhances the ability and efficiency of converting scientific data into knowledge discovery. Moreover, this platform also offers critical support for major scientific and technological innovation activities.

Under the strong leadership of the Central Committee of the Communist Party of China, China's e-Science has made great strides. Recent years have witnessed a series of new progress, new achievements, and new breakthroughs in the field of research informationization supporting scientific and technological innovation and promoting output of achievements. Research informationization technology and infrastructures have provided storage, computing, and transmission solutions for massive data generated by major scientific and technological infrastructures such as the Five-hundred-meter Aperture Spherical Radio Telescope (FAST), the Large High Altitude Air Shower Observatory (LHAASO), and the High Energy Photon Source (HEPS), and supported the international cooperation between China Spallation Neutron Source (CSNS) and the European Organization for Nuclear Research (CERN). The e-VLBI technology based on high-speed networks guarantees the tracking and positioning of China's lunar and planetary exploration projects. The multi-star and multi-task parallel information system based on "one network and two platforms" supports the massive data processing of space science series satellites such as the "DAMPE", "Practice 10", "QUESS", and "HXMT". The Scientific Observing Network of the

Chinese Academy of Sciences (CASSON) in the western Pacific Ocean has overcome the world's difficult problem of real-time transmission of deep-water and long-time-sequence subsurface buoy data. The application of research informationization in the fields of industry, agriculture, energy, and environment continues to deepen, constantly promoting the development of the national economy and industrialization, supporting disaster prevention and reduction research under the "Belt and Road" initiative, driving the digital and intelligent transformation of agriculture, and the study on the possibility of China's economic low-carbon transformation. The deep integration and application of information technologies, especially big data, artificial intelligence, and the Internet of Things in the fields of medicine, biology, and genomics have made positive explorations in and contributions to the research and development of the COVID-19 vaccine and epidemic prevention and control.

The world is currently experiencing a new round of deepening technological revolution and industrial transformation; fueling the need for the reform of the new scientific research paradigm and the restructuring of the innovation model. Information technology is at the forefront, driving innovation and redefining our core competitiveness. In order to summarize and showcase China's experience and achievements in e-Science in the past two years, and to further promote the development of e-Science in China, the Chinese Academy of Sciences, together with the Ministry of Education of the PRC, the Ministry of Science and Technology of PRC, the China Association for Science and Technology, the Chinese Academy of Social Sciences, the Chinese Academy of Engineering, the National Natural Science Foundation of China, and the Chinese Academy of Agricultural Sciences jointly compiled and published the "China's e-Science Blue Book 2023". The book has been highly valued and strongly supported by the leadership of participating parties. The book includes 29 articles written by authoritative experts in the field of e-Science in China, covering the trend strategy, application practice, and infrastructure construction of scientific research informationization. It presents the development state, application achievements, and construction status of e-Science in China in the past two years from different perspectives. The book is rich in content, with detailed cases, and boasts both theoretical significance and practical reference value, which can be read and referred to by the leadership, administrators, and frontline scientific researchers engaged in informationization or scientific research work in scientific research institutions, universities, and related enterprises.

Finally, due to a brief working cycle and limited access to complete information, the content of this book may not fully and comprehensively reflect the work and achievements of scientific research informationization construction at all levels in China. We apologize for any shortcomings in this regard. However, we highly value feedback and suggestions from readers from all walks of life, and we are committed to continuous improvement in our future work. Your valuable opinions and suggestions are most welcome as we strive to enhance our content and deliver better results in the future.

Beijing, China
March 2023

The Committee for the Compilation of
China's e-Science Blue Book 2023

Contents

Situational Strategies

E-Science—A Key Driver for Scientific Research Paradigm

Jiulin Sun, Xuehai Hong, Yang Wang, and Yan Ban

Abstract Scientific research paradigm is an important driving force to promote the transformation of scientific paradigm. At present, it is of great significance to actively explore the transformation of scientific research paradigm for improving the efficiency of scientific research, promoting the output of major achievements and promoting the reform of scientific and technological mechanism and system. This paper expounds the connotation and characteristics of scientific research paradigm, and explains the types of transformation of scientific research paradigm, its manifestations in different scientific fields and its multi-dimensional expression. It is pointed out that scientific research informatization promotes the transformation of scientific research paradigm in different scientific fields, and that scientific research informatization is an important driving force to promote the transformation of scientific research paradigm. Finally, the paper proposes that the construction of scientific research informatization should adapt to and promote the transformation of scientific research paradigm.

Keyword Scientific paradigm · Scientific research paradigm · Scientific research informatization · Paradigm elements · Transformation

On May 28, 2021, General Secretary Xi Jinping pointed out in his speech at the 20th Congress of the Chinese Academy of Sciences, the 15th Congress of the Chinese Academy of Engineering and the 10th National Congress of the Chinese Association for Science and Technology that the current new round of scientific and technological

J. Sun (✉)
Institute of Geographic Sciences and Natural Resources Research, Chinese Academy of Sciences, Beijing, China
e-mail: sunjl@igsnrr.ac.cn

X. Hong
Institute of Computing Technology, Chinese Academy of Sciences, Beijing, China

X. Hong · Y. Wang · Y. Ban
Computer Network Information Centre, Chinese Academy of Sciences, Beijing, China
e-mail: banyan@cnic.cn

China's e-Science Blue Book 2023, https://doi.org/10.1007/978-981-99-8270-7_1

revolution and industrial change is advancing rapidly, scientific research paradigms are undergoing profound changes, the interdisciplinary integration of disciplines is developing, and the integration of science and technology and economic society development are accelerating [1]. This shows that the paradigm shift in scientific research has become a new situation for the scientific community, which requires a concerted effort to adapt to and achieve a major paradigm shift in the mindset, methods and organization of research activities.

Scientific research informatization is the informatization of scientific research elements in research management and research activities, and is an important support and driving force for achieving changes in the scientific research paradigm. Since mankind entered the age of informatization, the methods, paths and evaluation systems of scientific research have undergone significant changes, and these changes are very different from the previous paradigm of scientific research. Some old establishments, methods, tools and research paths are no longer adopted, a new set of rules is rapidly emerging and old scientific research paradigms are being replaced. Therefore, in the Blue Book on China's Research Informatization 2015, which was jointly prepared by CAS and the State Internet Information Office, the Ministry of Science and Technology, the Ministry of Education, the Ministry of Industry and Information Technology, the Academy of Social Sciences and the National Natural Science Foundation of China, CAS academician Tan Tieniu has clearly pointed out that research informatization will further promote the change of research paradigm. He also pointed out that scientific research is currently facing more profound and complex challenges in both micro and macro directions, and that several data-intensive scientific fields such as astronomy, meteorology, geophysics, earth system science, biology, genomics, etc. need new and more advanced interdisciplinary tools and informatization platforms to manage and analyze massive amounts of scientific data and solve complex scientific problems [2].

In the current environment of increasingly fierce global competition in science and technology, it is of great significance to the development of science and technology innovation in China as to how to make full use of the platform, environment, data and mechanism of research informatization to further improve the efficiency of scientific research, how to achieve rapid scientific discoveries and breakthroughs in major scientific and technological problems, and to promote the changes of paradigm in scientific research in China.

1 The Meaning and Characteristics of the Scientific Research Paradigm

1.1 Connotations of the Scientific Research Paradigm

The term Paradigm is an abstract concept introduced by the American scientist Thomas Kuhn in his book *The Structure of Scientific Revolutions* [3]. It is primarily concerned with scientific paradigms. According to Thomas Kuhn, one of the hallmarks of a scientific paradigm (or disciplinary paradigm) is that if a discipline relies primarily on monographs to advance knowledge, it has no paradigm, suggesting that everyone needs to start with basic questions; disciplines with paradigms rely primarily on journal articles to advance knowledge production. Disciplines that have a consensus on paradigms rarely discuss paradigms, while disciplines that do not have a consensus on paradigms often have paradigm debates in which people prefer to use the term 'paradigm' when describing themselves or the camp they support. Those paradigms that do not yet reach the level of normal science expand into the term 'research paradigm'. This can easily lead to confusion in understanding and expressing the terms 'scientific paradigm' and 'research paradigm'.

In a narrow sense, the paradigm proposed by Thomas Kuhn should be a scientific paradigm. A scientific paradigm is a theoretical system that is commonly believed and accepted by the contemporary scientific community as a theoretical presupposition for conventional scientific work in the disciplines. For example, the scientific research activity of mankind today is still generally confined to the basic theoretical framework established by the revolution in physics at the beginning of the twentieth century, and is far from being in a period of theoretical decline. In this sense, the scientific paradigm in today's world has not changed much.

A scientific research paradigm, also known as a scientific research paradigm or research paradigm, is how scientific research is conducted by the scientific community in a particular historical period, in line with the requirements of the inherent laws of science and technology innovation. We believe that a scientific research paradigm is a collection of rules and systems that the academic community relies on and generally adopts to make its daily scientific research activities efficient and orderly, including the establishment environment, research paths, evaluation systems, research methods, research tools, technical routes and research models. Although this definition is not a strict conceptual definition, the connotation of a research paradigm is clear: it is a collection of rules that are recognized or accepted by the academic community and are influenced by social, economic, cultural, and international environments and individual preferences. Accordingly, we believe that a research paradigm is a system of rules that contains a mixture of elements such as research methods, research approaches and paths, research resources, research organization, research evaluation and research systems for research activities. This system of rules does not necessarily have to be a hierarchical structure between the various mixed elements, but can also be a mesh structure. Different people can give their understanding and knowledge of the research paradigm in terms of different rule element dimensions, and define the process characteristics of the relevant research activities accordingly. Therefore, at present, research paradigms are more often summarized and described in terms of one element, such as data-intensive fourth research paradigm and national institutional research paradigm, in terms of research tools (e.g. research informatics), research resources (e.g. big data) and ways of organizing research activities (e.g. individual scientists or teams) and research management systems. Although these summaries and generalizations are useful for guiding the organization and management of current research activities, they can also lead to a one-sided understanding of research paradigms. Therefore, there is a need to systematically summarize the common features of research paradigms and the different characteristics of research paradigms in each discipline, to help guide the research activities in specific fields.

1.2 Characteristics of the Scientific Research Paradigm

A scientific paradigm has a degree of acceptance within an academic community and is a whole consisting of its fundamental laws, theories, applications, etc. Its existence provides scientists with an agenda to follow in their research. A breakthrough in a scientific paradigm can lead to a scientific revolution, which can give science a whole new look. For example, the energy science paradigm is a scientific paradigm change based on energy conversion, which triggered the transformation of traditional fossil energy sources to nuclear energy and led to the development from human-powered.

carriages to internal combustion engines. The breakthrough in the research paradigm will lead to changes in the means of research, improvements in the efficiency of research and changes in the organization and management of research. Unlike the scientific paradigm inherent in a discipline, a paradigm change in research does not lead to a change in the scientific mechanisms and laws inherent in a discipline. Research paradigms are characterized by several features.

(1) The scientific paradigm has characteristics that cannot be described by a scientific paradigm. A scientific paradigm cannot be characterized by the fact that research activities in all disciplines follow the same code of conduct. For example, a common code of conduct for research activities in the natural sciences cannot be applied to research and development activities in engineering and technology, nor to research activities in the humanities and social sciences.
(2) Research paradigms are individual in character. Research paradigms are different in different disciplines and even in different stages of development of the same discipline. For example, the scientific research paradigm of natural science research is not of the same type as that of engineering science and technology. Scientific research in the natural sciences lies in the creation or discovery of new knowledge and takes the form of papers and monographs that are carriers of knowledge. Engineering science and technology is usually technological

innovation and development research based on the basic principles and knowledge of the natural sciences, which takes the form of technology patents and products. For example, protein structure analysis in the life sciences, which was previously based on experimental equipment (e.g. cryo-electron microscopy), has now developed into a DATA + AI based research model.

(3) Scientific research paradigms are characterised by constant change and evolution. According to the characteristics of the different stages, scientific research paradigms are usually divided into the first, second, third, fourth and even fifth paradigms. These paradigms are a category of common characteristics that evolve in stages as scientific research activities progress, the complexity of the scientific problems faced grows, and the technological advances in research tools and instruments evolve.

As can be seen from the above description, a scientific paradigm is a theory, the knowledge, that a scientific community follows. Scientific Paradigm is a different approach or model of scientific research developed for a specific discipline, based on the scientific paradigm of a discipline. A scientific paradigm is the unification of the content and methods of a discipline, and a scientific research paradigm is the methodological part of a paradigm discipline. A scientific paradigm and a research paradigm are not the same concept. In general, a scientific paradigm provides successful precedents for scientific research that can be imitated, as a tool that relies on its successful demonstration, a method for solving problems, an artificial paradigm, or a constructive paradigm. For example, the current paradigm of relying on "big data + artificial intelligence" to drive research activities in various disciplines is a paradigm that has been derived from the scientific paradigm of artificial intelligence and the use of big data technology to adapt to the needs of research activities in many disciplines and is a paradigm shift caused by a change in the research resources element of the research paradigm. It is a paradigm shift in research caused by changes in the resources of the research paradigm. Therefore, changes in some elements of this type of research paradigm give rise to new research paradigms, which is the theoretical basis for why research informatics can contribute to contemporary paradigm shifts in research. The development of research informatics, for example, has the potential to provide the elemental changes needed to facilitate paradigm change in research.

2 Paradigm Shift in Scientific Research and Its Multidimensional Manifestations

2.1 Paradigm Shift in Scientific Research

The paradigm shift in scientific research is based on two main drivers. One is the intrinsic drive for developments in the field of science. The other is the extrinsic drive to improve research methods as a result of technological advances, typically the computerization of research. In the history of human scientific research, four paradigm shifts have taken place. From the first paradigm, which used simple experimental methods, to the second paradigm, based on theoretical deduction, to the third paradigm, based on computer simulations, and the fourth paradigm, driven by massive amounts of data. Currently, the concept of a fifth paradigm is being proposed in the scientific community. These scientific paradigms are not substitutes for each other. Each research paradigm shift inherits the strengths of the previous generation of research paradigms and is parallel and may co-exist in various subject areas. Table 1 shows a comparison of the paradigm shifts at each stage, including the attributes of the scientific research problem, the research tools, and typical examples.

In general, the third paradigm is a combination of "human brain + computation", where the human brain is the main character and computational methods are auxiliary. The fourth paradigm is a combination of "computer + human brain + data", where the computer is the main character, the human brain is the dominant character and the data is the auxiliary material. The "fifth paradigm", as pointed out by Xueqi Cheng et al. in their article[4], is based on an ontological understanding of data science; the "fifth paradigm" emphasizes an ontological perspective on data, considering that data itself contains the laws of natural intelligence, and is also the carrier and product of new types of intelligence. It is expected that while data drives intelligence, it will break through the boundaries of the existing capabilities of computational intelligence and construct a new paradigm of intelligence with the help of natural intelligence.

The evolution of the scientific research paradigm as described above shows that the characteristics of the various stages of the scientific research paradigm actually blend the concepts of scientific and research paradigms. The first and second paradigms, which essentially follow the general consensus of the scientific paradigm, are of an acceptable nature, i.e. there is a research paradigm of experimental observation and theory in any scientific study. The third and fourth paradigms, on the other hand, are actually paradigm shifts in scientific research, formed by changes in the paradigm elements of scientific tools and data, and are paradigmatic (Exemplar), existing only in certain fields of scientific research. The fifth paradigm is a paradigm shift in research across a wide range of disciplines, triggered by the great development of computing power and the breakthroughs in AI, which is also essentially a paradigm shift in the elements of research. As can be seen from the evolution of the paradigm described above, computing has played a role in the paradigm shift in research by providing elements of tools and resources, which lays a solid basis for the theory that research information technology promotes the transformation of the research paradigm.

2.2 *Paradigm Shifts in Different Scientific Fields*

From a macro perspective, research paradigms vary and shift in different disciplinary fields as the elements of the research paradigm change in various research activities. Research paradigms in the natural sciences, in engineering and technology, in humanities and social sciences all vary, and there are even different stages of

Table 1 Comparison of paradigm shifts in research by phase

	Properties of scientific research questions	Research tools and typical cases
First paradigm	Experimental science: the discovery of natural laws. Such as heat from fire, gravitational acceleration, etc., characterized by the recording and description of natural phenomena. Natural phenomena are characterized by Modern discoveries of natural physical mechanisms. The Higgs boson, the "God particle", is mainly through experimental recording of data and analysis of the intrinsic Mechanisms	Early experiments such as drilling for fire and Galileo's Tower of Pisa ball tossing. Modern large scientific instruments and large scientific installations, such as the Large Hadron Collision Accelerator
Second paradigm	Theoretical science: limited by experimental conditions, it is difficult to achieve a more precise and complex understanding of natural phenomena. Theoretical science. Scientists have tried to remove minor distractions and leave only the key factors to simplify the experimental model as much as possible and then Theoretical science This paradigm of research continues to this day. This paradigm has continued to this day	Based on mathematical equations for calculation and reasoning. For example, Newton's three classical laws of mechanics, Maxwell's theory of electromagnetism, and quantum mechanics and relativity
Third paradigm	Computational simulations: As it became increasingly difficult and economically expensive to verify classical and modern physical theories (requiring large installations), scientific research began to enter a phase of intractable problems. The emergence of computers with the von Neumann architecture led to the rapid spread of the use of computers to simulate scientific experiments, allowing complex phenomena to be simulated and simulated, and complex problems to be clearly explained	Based on physical theory, computational theory and computational techniques. Typical examples are simulated nuclear tests, weather forecasting, etc. As the use of computer simulations increasingly Replaces experiments, they are becoming a regular mode of modern scientific research

(continued)

Table 1 (continued)

	Properties of scientific research questions	Research tools and typical cases
Fourth paradigm	Data Intelligence: With the rapid development of modern information technology and the explosive growth of data, it is possible to discover scientific knowledge and the laws of nature and physical society from the vast amount of data, i.e. to mine the data for correlations, cause-effect relationships and knowledge of various things and events and to make predictions. This paradigm is a unique "data-intensive knowledge discovery research paradigm" separated from the third paradigm	Based on big data theory and technology and computing technology, we use algorithms such as data mining and machine learning. Typical examples are the discovery of potential human diseases based on high-throughput genomic test data, the discovery of gravitational waves based on cosmic observation data, the discovery of interpersonal relationships based on online social data, etc.
Fifth paradigm	AI for Science: To overcome the "black box" problem of machine learning in the fourth paradigm, which relies solely on data-driven approaches, and to overcome the problems of modeling complex systems in the real world, understanding high-dimensional spaces and making effective predictions in engineering practice, by integrating physical knowledge, using algorithms related to artificial intelligence and advanced computing techniques, to solve problems describing complex. This paradigm is a new development in the intersection and synthesis of the second, third and fourth paradigms, and is a research paradigm under exploration	Pre-training and inference techniques using machine learning neural networks based on big data and physical knowledge and laws (including deep learning networks, convolutional networks and physical information networks, etc.). Typical examples are DeepMind's AlphaFold2 for protein de novo structure analysis and prediction, large-scale equation prediction, large-scale equation set compression, and physics-based information neural networks for the application of Navier–Stokes equations for solving Navier–Stokes equations, etc.

research paradigms in different disciplines. At each stage in the history of human research activity, the research paradigm has changed in line with its own scientific and technological progress. This means that the organisation and management of research activities cannot be driven by the same research paradigm, nor by the same type of research elements. This is the motivation and basis for the paradigm shift that research informatics can facilitate. However, regardless of the subject area, the objectives of the research paradigm are: to achieve rapid scientific.

discoveries, to solve major economic and social development problems, to effectively improve the efficiency of research and to optimise the objectives of research tasks.

In the field of natural science research, the transformation of research paths andmodels is one of the most common paradigm changes in scientific research. After the SecondWorld War, the paths and paradigms of scientific research changed fundamentally as the shape of scientific existence changed from small science to big science. In the era of small science, the paradigm of scientific research embraced individual heroism, which led to the achievement of numerous scientific wizards, such as Albert Einstein and Madame Curie, who, through their individual efforts, achieved outstanding scientific achievements. But in the era of big science, it is far beyond the power of a single individual to achieve major scientific achievements, such as the Manhattan Project, the Apollo moon landing, gravitational wave detection and China's Shenzhou spacecraft, etc. These costly scientific projects are all organically integrated by huge research teams through complex modern management techniques, and no individual is capable of completing them. It is easy to see that this is a shift in the paradigm of scientific research, triggered by changes in the human element and in the elements of scientific organisation and management models. Research activities in the era of big science are not "islands" of individual "solitary" and categorical research. The paradigm of "solo" and "all-inclusive" full-spectrum innovation is no longer adapted to the requirements of scientific innovation in the era of big science.

In the field of natural science, major natural science discoveries are increasingly dependent on major science and technology infrastructures, both at present and in the future. Major science and technology infrastructure is the material and technical basis for breaking through the frontiers of science and solving major scientific and technological problems for economic and social development and the country. This is a paradigm shift and an upgrade of scientific research triggered by changes in the tools and means of scientific research. The paradigm upgrade in scientific research shows that major scientific discoveries and technological changes increasingly rely on major scientific and technological infrastructure, innovation platforms and extreme experimental conditions and large scientific teams to collectively tackle them, which is also the policy foothold for the coordoinated use of national strategic scientific and technological forces and the realization of major scientific and technological innovations. In the future, we should further promote the upgrading of the scientific research paradigm through national strategic science and technology forces.

In recent decades, with the rapid development of information technology, high-speed high-capacity communications, the Internet of Things, big data, artificial intelligence, high-performance computing and other technological advances, a paradigm shift in data-driven natural science research has taken place in the field of natural science research, giving rise to a new paradigm of scientific discovery based on intensive data, the Fourth Scientific Research Paradigm. This is one of the elements of scientific research, the data element, which has changed the traditional scientific research paradigm of relying solely on experimental observation and theoretical deduction in natural science research.

Thus, the paradigm shift in scientific research in the natural sciences is essentially a change in the elements of the relevant scientific paradigm, which has triggered a paradigm shift in natural science research. The paradigm of scientific research is different from the traditional natural science paradigm in the past, such as the collection, management and exploitation of massive scientific data generated by major scientific facilities, and knowledge discovery. These paradigm shifts cannot be achieved without the support of tools, platforms, data and environments for research informatics.

In the field of engineering science and technology, engineering science is the study of various disciplines of engineering technology, which is a comprehensive body of knowledge used in engineering. The focus of engineering technology research does not lie in new scientific discoveries, nor scientific principle innovation, but mainly presents forms relying on scientific principles and knowledge in terms of innovation and progress in technological implementation. For example, in the field of information technology, people's knowledge of theories and knowledge of semiconductors belongs to knowledge discovery (such as the energy band theory of quantum mechanics), but relying on such theories and knowledge to make semiconductor devices requires innovation in engineering technology. In recent times, engineering technology has more directly linked scientific discoveries to industrial development and has become a major driver of economic and social development. Engineering science and technology has transformed knowledge into products, bringing unprecedented convenience to human production and life. Several major achievements in engineering science and technology, such as the "Two Bombs, One Satellite", manned spaceflight and lunar exploration projects, have greatly enhanced China's comprehensive national power and international status. The successful construction of a large number of major projects such as the Three Gorges Project, the Qinghai-Tibet Railway and high-speed railways has significantly enhanced the innovation capacity and level of China's basic industries, manufacturing industries and emerging industries. The innovation drive of engineering science and technology has been instrumental in this.

The distinctive feature of engineering science and technology is "engineering", which is the creation of "physical objects". From the perspective of whether there is a paradigm shift in scientific research, to observe whether engineering science and technology has achieved a paradigm shift, we need to pay attention to whether there is a shift in the relevant elements of the scientific research paradigm in "engineering", such as scientists, scientific research resources, scientific research tools, scientific research organization and management, research paths, research models and mechanisms, etc.

From the perspective of the human element, a distinctive feature of contemporary engineering science and technology is "large", which means that the realization of "engineering" is not achieved by a single scientist (technical expert), but requires a large scientific research team to implement it. From the perspective of the means, tools and resources for the realization of engineering science and technology, the realization of a project in modern engineering science and technology no longer relies only on a certain scientific research instrument or equipment, nor computer tools or data resources, but on various scientific research instruments, computer tools and various data resources, etc., whether in the pre-proof of engineering design, or various tools and resources used in the process of engineering realization, all are so. As a result, the paradigm elements of engineering science and technology have changed from a single element to a multi-element synthesis. From the perspective of the organization and management elements of engineering science and technology, which used to rely on the organization and management of a single person or unit, modern engineering science and technology are generally large projects that require complex organization and management, and have evolved into large team organization and multi-unit collaborative organization and management, or even transnational organization and collaboration. Even in contemporary large-scale natural science research, such as gravitational wave detection, celestial science discovery, human genome project, etc., it is still necessary to rely on large team building to complete the implementation of major scientific research basis of big science projects, on which the major research objectives of natural science can be achieved, which is essential for engineering science and technology.

From the above analysis, it is clear that with the advancement of modern technology, elements of the scientific research paradigm in engineering science and technology are achieving a shift. Each link and each stage of engineering science and technology may have its scientific research paradigm. As a result, the paradigm shift of scientific research in engineering science and technology, regardless of the stage of R&D in engineering science and technology, cannot be simply summarized as the first paradigm. In fact, in engineering science and technology, the first and second paradigms provide the immediate observational and theoretical basis, while the third and fourth paradigms and even the fifth paradigm provide the technology and methods for engineering realization. Each of these scientific research paradigms has a role to play in engineering. Therefore, each scientific research paradigm element may achieve a transformation in engineering realization, thus contributing to the scientific research paradigm shift in engineering science and technology. Therefore, we cannot simply use a single research organization and management model for the management of engineering science and technology. This requires us to change the "one-size-fits-all" approach to research management and policy formulation. At the same time, the distinctive feature of engineering science and technology is "engineering", and the intersection and integration of multiple disciplines within engineering science and technology is intricate and complex, which presents many opportunities for the realization of the value of research informatics.

At present, there are many "neck" problems in China, which essentially belong to engineering science and technology problems. What foreign countries are able to "neck" us are mature engineering technologies that have been developed for decades. The problem is not that we do not understand the scientific principles and knowledge, but that we have fallen behind in the "continuous improvement" stage. The improvement phase is the accumulation of small advances, each of which may not be at a very high theoretical level, but can improve the system. What needs to be changed is the organization, management and evaluation of the research paradigm and how to use research information technology to accelerate this paradigm shift.

In the field of social science and economic research, the American sociologist Riltz defines three different research paradigms in sociology[5]: the social fact paradigm, the social definition paradigm and the social behavior paradigm, a division that mainly indicates the different ways sociologists look at social phenomena or different perspectives of observation. The social fact paradigm (structural–functional school, conflict school, neo- Marxist school) emphasizes the objectivity of social phenomena. The social definition paradigm (symbolic interactionism, phenomenology, folklore methodology) emphasizes the subjective nature of social phenomena and the study of how people build society and act in it at a micro level. The social behavior paradigm (behavioral theory, exchange theory) emphasizes the objective and precise analysis of the social behavior of individuals. In our view, the scientific research paradigm in the social sciences is mainly a research model in which social science research scholars observe and summarize social development phenomena and trends and then abstract theoretical perspectives. On the whole, this belongs to the first paradigm. Of course, modern social science research emphasizes refinement, causality and interpretability of social phenomena, which inevitably leads to a trend of data-driven research, which will also promote the emergence of a data-intensive research paradigm for social science research (the fourth paradigm). At present, the scientific research activities of social science research based on internet big data are taking place in the field of social science and information science cross research.

Similarly, there are two dominant economic paradigms (scientific paradigms) in modern economics research: the "neoclassical" economic paradigm and the "neo-Keynesian" paradigm. These two paradigms are, on the one hand, observational and statistical, and, on the other hand, based on basic assumptions, followed by theoretical and explanatory studies of economic development using the respective theories. From the perspective of the scientific research paradigm, this is typical of the first and second paradigms. In macroeconomic research activities, more use is made of the development of theoretical mathematical models followed by computational and analytical research, which is the third paradigm. In microeconomic research activities, there is an increasing reliance on mathematical modeling and data-driven research, which is part of the fourth paradigm. It can be seen that these four research paradigms co-exist in economic research activities and that different research paradigms may exist at different research stages of the same research activity.

2.3 *Multidimensional Manifestations of the Paradigm Shift in Scientific Research*

As mentioned earlier, the current description and summary of the research paradigm are more often summarized in terms of one of the elements such as research tools (e.g. research informatics), research resources (e.g. big data), the way research activities are organized (e.g. individual scientists or teams) and the research management system. In the following, we summarize the paradigm shift in research in terms of several elements of the paradigm shift.

2.3.1 From the Organizational Dimension of the Researcher, the Paradigm Shift from Individuals to Small Teams to Large Teams in Research

In the beginning, human scientific research activities were driven by the interest of individual scientists and were mainly conducted by individual scientists using observation and experimental means, which belonged to the small scientific research paradigm. However, the increasing complexity and depth of scientific research problems have made it impossible to solve complex and in-depth problems by relying solely on the small scientific research paradigm of an individual scientist. Scientific problems that require the strength of a team (not just a small team, perhaps a team from a research institution, or even teams organized together across units and countries) to work together collaboratively to solve research problems. In the field of engineering science and technology, the large research model of teams is even more prominent. Thus, while the dynamic role of the individual scientist and the implementation of an interest-driven research model advocates the release of individual scientist energy, the collaborative research model of the research team is also emphasized, emphasizing team energy to achieve big scientific research goals. The 'elite-centric' research paradigm has recently been proposed, pointing out that the traditional research system has created a scientific paradigm centered on the scientific elite, citing then Harvard University President Conant In the words of Dr. James B. Conant: Ten second-rate people are not worth one first-rate person". This "elite-centric" paradigm emphasizes the role of the "scientific elite". This 'scientific elite' can be a single outstanding scientist or a 'group of innovators'. In modern research organizations, the paradigm of relying on a single 'scientific elite' still exists in some disciplines, particularly in purely theoretical and fundamental scientific research, such as mathematics, as well as in the social sciences and economics. However, more and more research areas, such as engineering and science, and large-scale natural science research, rely on the paradigm of 'team' organization.

2.3.2 From the Dimension of Tools (Platforms) and Resources on Which Research Activities Depend, there is a Paradigm Shift in Research From Using Simple Research Tools and Simple Small Data to Relying on Individual Large Devices and Big Data to Relying on Networked Devices

Earlier scientific research activities carried out by individual scientists and even by scientists in some current disciplines, used tools that may have been simple tools such as rulers, microscopes and one or a limited number of other scientific tools and small data. However, relying on major research facilities (installations) and big data for scientific research is now becoming a way to get ahead of the current level of research and to quickly contribute to the production of scientific results. For example, in the field of life science research, scientists rely on tens of millions of dollars a piece for cryo-electron microscopes, nuclear magnetic resonance equipment, etc. to conduct research in their disciplines. In the field of earth sciences, scientists rely on large supercomputing platforms and big data resources to carry out simulations and calculations of geophysical processes. In addition, in modern international transnational scientific research, as scientific research is organized in a networked manner, the equipment used is often also distributed in different countries, resulting in a networked scientific research organization model of shared equipment and data-intensive knowledge discovery with big data integration. Examples include global seismic observation networks for earthquake science research, and global ocean observation networks for marine environmental research.

2.3.3 The Change from a Single-Disciplinary to a Multidisciplinary Research Paradigm, in Terms of the Dimension of the Change in the Technical Lines of Research

The interdisciplinary research and mission-oriented research that has emerged since the Second World War has gradually condensed and evolved into fusion science in the early twenty-first century. The research of many major scientific problems is now faced with highly complex large systems and problems that cannot be solved by the research of a single discipline, for example, the research of major engineering science and technology problems, more often presents a scientific research paradigm of multidisciplinary integration, that is, fusion science is a new scientific research paradigm based on multidisciplinary integration to solve major problems, and is actively represented by the United States as a typical representative of the developed countries of science and technology Initiative[4].

The reform of China's science and technology sector in recent years has also been in line with the international trend towards a new paradigm of "convergent science". For example, the Chinese Academy of Sciences has proposed eight major areas of innovation in its 13th Five-Year Plan, based on the strengths of multidisciplinary integration, and the National Natural Science Foundation of China has promoted the establishment of the Department of Interdisciplinary Sciences and the reform of its support program. These strategic plans and the reform of the support program show that the paradigm of convergent scientific research has become the consensus of the academic community and continues to play a role in the scientific research activities in the field of engineering science and technology, as well as being systematically supported in the research activities in the field of natural science research.

In the history of science, the convergence of scientific paradigms is a phase in the evolution of scientific research paradigms. The history of science and technology in mankind is a process of the creation and even integration of ideas and branches of the discipline. When science is fragmented, there is an instinctive urge to integrate. For example, the history of scientific research began with a limited number of observations and experimental means of obtaining a limited amount of "small data" for scientific research. Today, scientific research activities have evolved to analyze big data, condensing it into small intelligence, deep intelligence, and precise knowledge for research. In the field of synthetic biology, the intersection of quantitative biology and synthetic biology in a complementary research approach will drive the change of synthetic biology from qualitative, descriptive and localized research to quantitative, theoretical and systematic. In the field of artificial intelligence research, its fusion of scientific thinking is an internal fusion and unification of research and thinking within multiple disciplines such as computer science, mathematics and physics, and is the key to spawning a new research paradigm.

2.3.4 From the Dimension of Change in Research Methodology, the Shift From a Single Research Paradigm in The Early Days to a Combination of Multiple Research Paradigms Alongside a Single Research Paradigm, Promoting the Integration of Reductionist and Holistic Research Paradigms

Scientific research methods are the tools and means of discovering new phenomena and things, proposing new theories and ideas, and revealing the inner laws of things in research. Due to the perspective of people's understanding of the problem, the complexity of the research object and other factors, it is difficult to have a completely unified understanding of the classification of research methods. However, they can generally be classified as empirical methods, theoretical research methods and systematic scientific methods.

Currently, in the scientific research activities of many disciplines, the model of a single scientific research paradigm still exists, as well as a combination of multiple research paradigms, depending on the different characteristics of the scientific research activities of different disciplinary attributes. According to Professor Rao Yi, the main scientific research paradigm in the life sciences is still at the stage of describing the phenomena of life, while further work is done to understand the basic description of the phenomena, processes and functions of biology, and then to promote life science research through a cross of mainly chemistry and physics. In the field of mathematics, scientists have instead taken the Keplerian paradigm of observational data-driven research more (the first paradigm) and the fundamentals-driven Newtonian paradigm (the second paradigm) are effectively combined for scientific research activities. According to Academician Weinan E, scientists currently strive to combine data-driven machine learning methods, fundamental principles, and fundamental principles in physics such as quantum mechanics and molecular dynamics to solve previously unsolvable scientific problems, achieving an upgrade from a small farmer's workshop model to an integrated large platform model, completing the combination of multiple research paradigms. By effectively combining the reductionist and holistic scientific paradigms through a large platform, the whole is first reduced to its components for study, and then complex system problems are studied at a high level in themselves and as a whole, so that the two scientific paradigms are no longer opposed to each other. Thus, the evolution from a combination of multiple research paradigms to a research paradigm that relies on a large platform promotes a combination of reductionist and holistic research paradigms, giving today's research paradigm a multi-paradigm fusion.

2.3.5 From the Dimension of Research Organisation and Scientific Evaluation, the Research Paradigm Has Changed from a Single Model to a Multi-modal

Different research systems will affect different models of research organization and, consequently, different paradigms of research. For our scientists, the more familiar system is that of the state (and now the new state system). The overall backwardness of China's science and technology in recent times has led to a shift from 'interest-driven' scientific research to the organization of major scientific and technological research activities around national goals, from 'what can be done'(interest-driven) before and after the reform and opening up to 'what should be done' (current). From "what can be done" (interest-driven) before and after the reform and opening up to "what should be done" (national goal-driven). From the "two bombs and one star" major science and technology projects in the past, to the current efforts to solve the "neck" problem, this is the paradigm of the national system of scientific research. Of course, under China's current research system, the state also encourages scientists to engage in a "research interest" driven research paradigm. Many of the surface research projects supported by the NSFC in China are driven by "research interests", but of course, such support falls within one of the "four types of attribute problems" supported by the NSFC.

In addition, the ongoing reform of the research system will cause a shift from a 'centralized' to a 'decentralized' research paradigm. The literature [6] suggests that the 'elite-centric' paradigm of scientific research, which revolves around a scientific elite, will be transformed into a 'decentric' paradigm. The research paradigm has dominated the international scientific community since the.

middle of the twentieth century. Its main features are the professionalization of scientific research, the hierarchy of researchers and the specialization of scientific communication. Although this "elite-centric" research paradigm has been successful to some extent, it has gradually evolved into a highly "involuted" and closed system, leading to a tendency to allocate research resources in a "four-only phenomenon. This has led to a tendency to allocate research resources in a "four-only phenomenon"(" paper-only, title-only, education-only and award-only tendency"). In a report, academician Xu Kuangdi said that "disruptive technologies, such innovations, are difficult to achieve under the current administrative approval and assessment system", referring to this "inward scroll". This is contrary to the spirit of encouraging scientific "freedom of inquiry". At present, the state is implementing a reform of the science and technology evaluation system - "breaking the four only" and implementing a multi-dimensional evaluation system of representative works and typical achievements, which also requires the exploration and establishment of a "decentralized" new paradigm of scientific research with a win–win effect. This also requires the exploration and establishment of a new paradigm of "decentralized" scientific research with win–win effects, unleashing the vitality of scientists and the tension of science. However, at present, this scientific research paradigm has not formed effectively, and the "four-only" scientific research paradigm and the "four-only" scientific research paradigm will continue to exist for some time.

3 Research Informatics for Paradigm Shift in Research

The term "informatization of scientific research" refers to the infiltration of information technology into scientific research activities, resulting in the informatization of the organization and management of scientific research and the informatization of the process of specific scientific research activities of individual scientists, teams and organizations. Both types of informatization rely on software and hardware systems, data resources, computing platforms, and supporting policies and systems for data and computing, networks and communications, information and security, and other information technology attributes. Research informatization has developed new research tools, research models, research platforms and research environments, which have effectively promoted the progress of various research work in contemporary times, whether in the field of natural science research, engineering science and technology research, and social science researchand become an important driving force for the paradigm shift in research in various disciplinary research fields.

3.1 Research Informatization Promotes a Paradigm Shift in the Research Paradigm of Research Organization and Management

Research organization and management are key elements of the research paradigm, and changes in any of these elements can bring about changes in the research paradigm. As mentioned earlier, in the early days of human research, research activities were mainly driven by scientists' interests, i.e. they existed mainly in the first and second paradigms, and research organization and management had little impact on scientists' research activities. In modern research activities, however, the organization and management of research have a huge impact on the efficiency and orientation of research. This is because the organization and management of contemporary research activities must be adapted to the country's scientific and technological institutions and mechanisms, which have a global impact and will directly lead to different models of research.

The development of management informatization in scientific research has effectively promoted the transformation of the research organization model, thus enabling the transformation of the research paradigm. Through various types of management information systems, the research organization model has broken through the boundaries of individual and small team research organization and management, and can efficiently support the formation of large, established and distributed research organization and management models, and in major research tasks can even form cross-organization and cross-national (international) research organization and management models. For example, the ARP (Academia Resource Planning) system, which has been built by CAS over the past 20 years, has effectively achieved full coverage of research management activities of over 100 research units and related management bodies across the CAS system. The system covers the management and sharing mechanism of information related to research management, such as projects, personnel, equipment and funding, and realizes the dynamic supervision of the implementation process of research projects and research results. The ARP system has effectively supported the organization and management of a series of major scientific research projects of CAS, such as the CAS Pilot Project, and has effectively improved the efficiency of scientific research management, accelerated the output of major scientific research results, and effectively ensured the role of CAS as an established team (large team) of national strategic forces in science and technology. At the national level, the Ministry of Science and Technology has built the National Science and Technology Management Information System Service Platform, which has greatly enhanced the efficiency of national science and technology organization and management, and provided solid support for the implementation of a nationwide large research organization model. The ARP system.

and the National Science and Technology Management Information System (NSTMS) service platform have successfully ensured and realized the transformation from the previous model of a small team research organization to the "national system" research paradigm of large teams.

3.2 *Research Informatics for a Paradigm Shift in Natural Science Research*

In the natural sciences, a scientific paradigm shift, also known as a 'scientific revolution', occurs when existing research paradigms fail to effectively address the challenges faced and a new paradigm approach is needed to address these challenges. The paradigm shift from the first 'experimental science' paradigm to the second 'theoretical science' paradigm took two to three centuries, but since the invention of the computer and the rapid development of information technology, the scientific paradigm has undergone a third paradigm shift in the last few decades. The "computational science" paradigm has evolved over the past few decades to the current fourth "data-intensive" paradigm and even the fifth "AI for Science" paradigm. This shows that the development of information technology and the construction and application of information technology in research have efficiently facilitated and ensured the paradigm shift in research.

The paradigm shift is most evident in the natural sciences, which are the main fields of study in mathematics, science and chemistry, as well as in heaven and earth and life. This is because the development of information technology has led to a paradigm shift in scientific research, with information technology in research penetrating all disciplines and its techniques, methods and tools being used for scientific innovation and knowledge discovery in the disciplines. The paradigm shift in research has even had a profound impact on the development of disciplines, which have become intertwined with information technology in a dichotomous development—"computing X" and "X informatics The "X" represents a specific discipline, such as biology, physics, astronomy, etc.).

The 2013 Nobel Prize in Chemistry was awarded to three American scientists who "created multiscale models for complex chemical systems". Scientists, who began their research in the 1970s to understand and predict chemical processes using computer programs that enabled computational models to translate basic understanding at the microscopic scale into predictive power at the macroscopic scale, gave rise to the discipline of computational chemistry. Changing the paradigm of research where chemists once used balls and sticks made of plastic to build molecular models and test tubes and beakers for experimental analysis, computational chemistry now allows scientists to use computers to build models, make experimental predictions and study the mechanisms of chemical reactions. With the current rise of artificial intelligence and big data technologies, chemists are increasingly trying to use the creation of a robust database of basic chemical reactions combined with AI machine learning to accelerate the discovery of chemicals. 2020 saw the development of the Mobile Robotic Chemist, an intelligent mobile AI chemist developed by a team of researchers at the University of Liverpool that not only works around the clock, but can also analyze from The Mobile Robotic Chemist will not only be able to work around the clock, but will also be able to analyse10 dimensions of thinking to determine the best next step to try based on the results of the experiment. After 8 days and 688 trials, the smart mobile AI chemists have discovered the first photocatalyst mixture that is six times more active than the original formulation. This marks the beginning of an important role for 'AI + computers' to drive automation and intelligence in synthetic chemistry, which is set to drive a new paradigm shift in chemical research.

Astronomical telescopes, as the main tool for capturing celestial information in astronomy research centers, have evolved from the initial optical and reflecting telescopes to today's radio telescopes that can capture the full range of wavelengths. With the development of astronomical observing devices, data from astronomical observations have exploded, information technology such as network technology, data storage and computational processing has provided the basis for today's astronomical research, and information technology in scientific research has provided astronomers with a wealth of resources and powerful services, and has driven the interconnection and collaboration of large scientific devices in astronomy, promoting the globalization and integration of astronomical research 1970s The existence of gravitational waves was indirectly confirmed by the observation of the phenomenon of binary neutron stars entwined with each other by Professor Taylor and his student

Hirsch at the Massachusetts Institute of Technology (MIT), who used Einstein's theory to accurately calculate the orbital variations of the two neutron stars. It was only on 11 February 2016 that three American scientists detected the existence of gravitational waves, a discovery that the scientific community believes completes the last missing piece of the "puzzle" of Einstein's prediction of general relativity. The initial gravitational wave discovery was the result of five months of data analysis of nearly two billion nuclear hours of observational data, and high performance calculations that allowed the precise timing and location of gravitational wave observations to be released to the world in a matter of seconds. In 2018, scientists at the National Center for Supercomputing (NCSA) at the University of Illinois at Urbana-Champaign combined deep learning algorithms with a numerical black hole merger relativity model and data items obtained through LIGO to develop an end-to-end temporal data model. They have developed an end-to-end time-series signal processing method, Deep Filtering, which is more computationally efficient and has smaller computational errors, enabling faster processing of gravitational waves using LIGO data to detect novel gravitational wave sources. This shows that the development of information technology and the application of information technology in research has profoundly influenced the way and methods of research in the field of astronomy, with supercomputing facilitating the discovery of gravitational waves, and the combination of big data and AI technology accelerating discoveries and promoting a profound change in the paradigm of astronomical research.

In the past 20 years, life science research has evolved from the experimental, observational and investigative studies of the past to the stage of genetic analysis research, which is typical of the scientific paradigm revolution in the life sciences and has given rise to the discipline of bioinformatics. The bioinformatics research paradigm basically relies on modern computational techniques coupled with relevant models and algorithms to analyze and process biological DNA data to achieve the research objectives. In the field of bioinformatics, with the accumulation of DNA sequencing data, there has been a small paradigm shift towards the Scientific Research Model (SRM),i.e. pattern recognition using models trained on known patterns of gene structure, which has led to the discovery of a number of new genes. well-known bioinformatics tools and databases such as CLUSTAL W, MEGA, PDB, etc. have been used to identify new genes. Bioinformatics tools and databases have all been developed within SRM. The traditional research paradigm can only discover new genes one by one through time-consuming and laborious methods. However, complex biological systems often function through interactions between many genes, proteins or other components, as well as pathways, modules or networks. Bioinformatics has rapidly and efficiently facilitated the accelerated development of the First Posture Strategy life sciences through high-throughput computational methods, making it possible to study biological and medical problems on applied information systems. The Microarray technology, yeast two-hybrid experiments, and evolutionary modeling techniques have facilitated a paradigm shift towards systems biology, aiming to reconstruct interaction or synergistic networks to explain emergent properties in systems. Systems biology data informatics tools such as Gene Ontology, KEGG and Cytoscape have been developed and widely used [7]. 2020 AlphaFold2 algorithm researched by Google DeepMind in the US has reached the level of human observation and prediction using sophisticated equipment such as cryoelectron microscopy in the field of protein structure prediction scientific research, which is unprecedented and tremendous progress. This is an unprecedented advancement and a major breakthrough in the application of "artificial intelligence + computers" in science.

The rapid development of information technology, the rapid development of research information technology, and the continuous upgrading and optimization of the network infrastructure environment have promoted the globalization of research collaboration, the integration of widely distributed computer resources in the Grid, the interconnection and sharing of large scientific devices, the explosive growth of research data, the establishment of large scientific data centers in various disciplines, the increasing complexity of scientific research, and the acceleration of the This has accelerated the shift from computational science (the third paradigm) to data-intensive science (the fourth paradigm) and AI for Science (the fifth paradigm) in the scientific disciplines. Modern scientific research has entered the deep waters of scientific research on complex systems, especially in the fields of astronomical observation, physical systems, earth systems, biological systems and medical science, where scientific and technological elements such as data, computing power and algorithms have gradually become important drivers of scientific innovation and breakthroughs, and emerging technologies such as big data and artificial intelligence have become important tools for scientific research. Therefore, in the field of natural science, a series of the major achievements in this area are, on the one hand, the enormous contribution of the tools of research informatics [computers, software (including algorithms) and data] to the paradigm shift in research in the subject areas and, on the other hand, the paradigm shift that has effectively promoted the intersection and integration of research in these areas, which is no longer driven by a single first or second research paradigm.

3.3 Research Informatization Promotes a Paradigm Shift in Research in Engineering Science and Technology

Engineering technology links scientific discovery more directly to industrial development and is a major driver of economic and social development. Every industrial revolution is inextricably linked to a technological revolution. eighteenth century, the steam engine triggered the first industrial revolution, bringing mankind into the age of mechanization; from the end of the nineteenth century to the first half of the twentieth century, electric motors and chemicals triggered the second industrial revolution; in the second half of the twentieth century, information technology triggered the third industrial revolution, transforming social production and consumption from industrialization to automation and intelligence, with social productivity The second half of the twentieth century saw the third industrial revolution, which led to a shift in social production and consumption from industrialization to automation and intelligence.

In the field of engineering science and technology research, the overall role of information technology is mainly reflected in the combination of advanced information technology with industry and production, and more so in the construction of major projects. In recent years, emerging information technologies such as the Internet of Things, blockchain, 5G, artificial intelligence, big data and digital twin have been widely applied and combined in different industries and industrial fields such as aerospace, transportation, equipment manufacturing, biomedicine and new energy. The Internet of Things (IoT) has enabled the interaction and seamless connection of information between people and things, and is widely used in areas such as smart transportation, smart homes, and food traceability; 5G has made smart driving and smart transportation possible; artificial intelligence has greatly improved the efficiency of drug development and testing; digital twins have enabled digital modeling and smart manufacturing in large manufacturing and construction engineering fields such as aerospace; blockchain technology has been widely used in digital currency and finance. blockchain technology in digital currency and finance.

In 1996, the United Nations General Assembly adopted a comprehensive nuclear test ban, but this does not mean that countries have stopped the development of nuclear weapons, in fact, countries in the previous decades of nuclear testing have accumulated enough data, and according to these data can be derived from a set of nuclear weapons experimental model, the use of modern supercomputing simulation technology, you can simulate the nuclear explosion in the laboratory, information technology The development of information technology not only avoids the negative environmental impact of nuclear explosions, reduces the cost of research, but also completely transforms the original approach to nuclear explosion research and experimentation.

In the construction of water conservancy projects, the construction period is generally long, the construction content is also more, the construction costs are high, the construction techniques involved are complex and diverse, and the problems that need to be dealt with during the construction process are also relatively numerous. If the traditional manual organization and management are followed, it is basically impossible to keep to the schedule. Without the use of information technology for scientific design, there is also no way to ensure the scientific nature and quality of construction and to achieve cost control. In order to improve the science and quality of construction and cost control of water conservancy projects, new technologies must be introduced. Information technology can be integrated into the construction management, design and cost control of water conservancy projects in order to effectively improve the quality of water conservancy construction and better develop the water conservancy engineering business. In addition, the current big data technology based on information technology in water conservancy projects can better promote the construction and development of information technology in water conservancy projects. Under the premise of maintaining the status quo, most of China's water conservancy management units have established relevant information management systems for water conservancy projects according to the actual situation, and use water conservancy engineering information management systems to effectively develop engineering construction plans, optimize and effectively manage construction conditions, construction quality, construction progress, construction costs, construction safety, etc. [8]

Introduced by the US Air Force Research Laboratory in 2011, the "digital twin" is a composite technology that combines 3D modelling, simulation and optimization, IoT and sensors, artificial intelligence and virtual reality, among other emerging technologies, initially applied to solve problems such as health maintenance and assurance of aerospace vehicles. In recent years, digital twin technology has rapidly become a craze and is being applied to smart manufacturing. The digital twin can virtually build a digital model of a product, and simulate it for testing and verification. During manufacturing, the operation of the equipment can be simulated, along with the changes brought about by parameter adjustments. Through the simulation of previous production and maintenance data are collected to create a realistic digital model, which engineers can debug and experiment within the virtual space, enabling the best possible results for the products to be produced. By applying digital twin technology, the quality of the final product and its application can be greatly improved while increasing production efficiency. The application of digital twin technology is a typical example of the use of information technology to promote industrial change. It integrates the digital world with the physical world, providing complete lifecycle data for industrial equipment and the like, and is gradually becoming an important application trend in industries such as smart manufacturing.

As can be seen from several of the above examples, various information technologies are used in an integrated manner in various aspects of engineering construction and product innovation. Research informatization has effectively facilitated the effective organization and management of large-scale research activities in the field of engineering science and technology, and a variety of research paradigms are widely available in various engineering science and technology fields. The application of research informatization has effectively changed the means of organizing and managing engineering science and technology and research in the past, and has effectively integrated the knowledge of various disciplines, thus improving the efficiency of engineering science and technology research output and labor production.

3.4 *Research Informatization Promotes Paradigm Shift in Research in Humanities and Social Sciences*

The objects of research in the humanities and social sciences are fundamentally different from those of the natural sciences and engineering sciences: the former is oriented toward people, the latter towards things; the former are spiritual worlds, the latter are material worlds, so there are certain deviations in research methods between the social sciences and the natural sciences. The traditional social science research approach is still "theoretically hypothesis-driven", but with the spread of the Internet and the development of information technology, big data and scientific computing technologies are also pushing for a paradigm shift in the field of social science research, thus giving rise to "computational economics" and "computational social science".

In the early twentieth century, e-Social Science (e-SS) was gradually introduced and developed as an analogy to information technology in research, and refers to two main aspects: firstly, it describes the collaboration between computer experts and humanities and social scientists; secondly, it encompasses the development of social science methods and tools. The University of Reading in the UK established the Big Data Research Centre in 2014 and has used big data technology to successfully complete many social science research tasks such as the analysis of the UK referendum, immigration research, house sales and data analysis. The Chinese Academy of Social Sciences has also completed various database and management software systems in an attempt to fully apply electronic informatics to scientific research and management. As a tool and environment to support social science research, research informatization has changed the methods and means of social science research, and has also increased the research field and broadened the scope of research [9].

The development of big data technologies has had a significant impact on social science research. In addition to the direct impact of "full-sample" data, big data technologies and data-driven knowledge discovery, big data has further contributed to changes in the paradigm of social science research at three levels: firstly, changes in research paths. First, a change in research paths, where the "bigdata-driven" paradigm is combined with the current "theoretical hypothesis-driven" paradigm in the social sciences to form a new research paradigm; second, a change in research tools, where big data and related technologies will become a powerful weapon for causal discovery; third, a change in functions, where predictive and causal questions will receive equal attention and will be organically integrated into [10].

Compared to the close integration of research with informatization in the natural sciences and engineering science and technology, informatization in the social sciences still has a certain distance to go. The informatization of research in the humanities and social sciences has contributed to a paradigm shift in research, more so from a single field survey, literature analysis and opinion formation to a paradigm shift in research based on data and computational models to present results. Based on the current development of big data, the field of humanities and social sciences needs to make full use of the massive amount of data for in-depth mining, so as to more objectively and deeply understand and study the cause-effect relationship and development trend of social phenomena, and thus better support leaders' decision-making and grasp the direction of social-related fields. Further integration of humanities and social sciences with information technology to promote the paradigm shift in social research remains a challenge and a trend in the current informatization of social sciences.

4 The New Requirements of Research Informatics for Paradigm Shift in Research

Research informatics provides the tools (software, hardware, networks, platforms, data, etc.) for paradigm shifts in various scientific fields and facilitates them. This is mainly due to the fact that research informatization can change the original organizational and management models of research and can provide the needs of all types of researchers for their research activities, from data, algorithms, software, and hardware to various platforms. It can be said that research informatics is the revolutionary force that drives the paradigm shift in research, and in the process of the paradigm shift, it only creates new requirements for the construction of research informatics.

4.1 Information Technology Construction Projects to Guarantee Openness, Sharing and Cooperation of Data Elements in Various Fields

The opening, sharing and cooperation of scientific data elements are on the one hand to maximize the value density of scientific data, and on the other hand to promote greater cooperation among scientists in the open sharing of scientific data to jointly promote scientific and technological progress and accelerate scientific discovery in the field of research. The information technology construction project should guarantee that these scientific data elements can be opened and shared. At the same time, a good cross-platform service mechanism should be established to accelerate the flow of scientific data and guarantee a larger scope of scientific research collaboration. Big data in scientific research is the basis for the academic community to carry out disciplinary integration research. The application platform and service system of scientific research informatization does not only serve a certain unit and institution, a certain discipline or a certain geographical area, it can provide data resource sharing and mining and analysis services for researchers in different places and disciplines, so that researchers from different disciplines can set research goals and carry out collaborative research around major scientific and technological problems. This enables researchers from different disciplines to work together to set research goals and conduct collaborative research on major scientific and technological issues. Theories, knowledge, methods, tools and data from different scientific fields intertwine and influence each other, resulting in a common understanding of research methods, research frameworks, scientific language and shared scientific data, as well as a common research paradigm, which further enhances the speed, quality and effectiveness of research and innovation activities.

4.2 Research Information Technology Construction Projects Should Be Selected to Facilitate the Formation of a New Research Paradigm

Research informatization is a revolutionary force for the transformation of the research paradigm. Therefore, it is an important mission of the "platform system + services" of research informatization projects to promote the transformation of the research paradigm, as well as an important guarantee for their operation. In the selection of research informatization projects, it is important to choose those projects that can transform from limited resources based on local space to unlimited resources based on digital space, so that the informatization projects can cover as wide a range of research activities as possible, achieve effective large team cooperation, and fully realize the gathering of research elements (such as research data), thus promoting the paradigm shift from small to large teams through informatization projects. The project will facilitate the transformation from a small team to a large team. In terms of the operational mechanism of research informatization projects, the selected projects should be able to transfer the offline physical space to the online digital space, realize the operation and guarantee of the digital space, and promote the transformation of research data elements and spatial organization elements through informatization construction projects. In terms of research decision- making, research informatization construction projects should transform vague and subjective decision-making into accurate decision-making based on big data and human–machine collaboration, prevent subjective single decision-making, and promote the transformation of research decision-making mechanisms. In terms of evaluating the effectiveness of research informatization project construction, it is important to shift from a single project construction as the main result-oriented indicator to a whole-domain collaborative indicator of whether the service covers the whole innovation chain, and promote the transformation of the research evaluation mechanism. Therefore, the selection of research informatization construction projects should be prevented from becoming a single IT application project, and the selection of research informatization projects should consider whether they have contributed to the change from relying on experience and simple experimental simulations to computational simulations that rely on data, knowledge and arithmetic power, and whether they have contributed to the scientific field of the discipline Paradigm change, etc.

4.3 Information Construction Should Be Systematized to Ensure a Modern "Platform System + Services" Model of Research Activities

First of all, the planning of information technology should be carried out with rigorous strategic research and analysis and foresight, forming a rigorous logical system, so that the various platforms of information technology construction can form a pattern of mutual use of technology, data interoperability, reuse of infrastructure and mutual use of support teams. We should not only focus on the construction of a certain platform, but also ignore the elements of the service mechanism, and form a comprehensive "platform system + service" construction mode for the characteristics of scientific research activities in different fields. Secondly, we should realize data services and workflow intelligence for all kinds of research objects based on the demand for intelligent research activities. In the various research information platforms, intelligent services for research data management, storage and processing can be supported so that researchers can use various methods and models of artificial intelligence to extract useful information from large-scale research data in a timely and convenient manner and improve research efficiency.

5 Conclusion

Scientific paradigm change is revolutionary, an intrinsic driver of disciplinary change, fundamental in its impact on a disciplinary field and immense in its contribution to the advancement of human science and technology. Scientific paradigms are the concretization of scientific paradigms that are instrumental and paradigmatic. The transformation of the scientific paradigm is an external driving force for the change of the scientific paradigm. Research informatization is an important driver for the transformation of the scientific paradigm. With the advancement of modern technology, it is becoming increasingly evident that research informatics promotes paradigm change in research. In the contemporary scientific paradigm, revolution is still relatively slow, and the core theory of scientific dominance has not changed fundamentally, but because the complexity of scientific problems is no longer the same as in the past, the scientific and academic community should strive to promote scientific paradigm change through new tools, new environments, and new methods; on the basis of the rapid development of information technology, through the various means derived from scientific information technology, to Promote a paradigm shift in scientific research. A new era of paradigm change in scientific research is approaching, and we need to embrace it, actively plan for it and adapt to it.

References

1. Xi J (2022) Speech at the 20th congress of the Chinese Academy of Sciences, the 15th congress of the Chinese Academy of Engineering and the 10th National Congress of the Chinese Association for Science and Technology [EB/OL]. [2022-11-08]. https://www.ccps.gov.cn/xxsxk/zyls/202105/t20210529_148977.shtml
2. Tieniu T (2016) Analysis of the development trend and development mode of scientific research informatization. Science Press, Beijing
3. Kuhn TS (1970) The structure of scientific revolutions. The University of Chicago Press, Chicago

4. Xueqi C, Hong M, Wei Z et al (2020) Data science and computational intelligence: connotations, paradigms and opportunities. Proc Chin Acad Sci 35(12):1470–1481
5. Ritzer G (1975) Sociology: a multiple paradigm science. Allyn and Bacon, Boston
6. Wu Jia Rui (2021) Characteristics of the "elite-centric" research paradigm and the challenges it faces. Sci Bull 66(27):3509–3514
7. Yin H, Sun Z, Wang Z et al (2021) The data-intensive scientific revolution occurring where two-dimensional materials meet machine learning. Cell Rep Phys Sci 2(7):100482
8. Chen R, Zhang W, Zhao L, et al (2019) Research on Crack Propagation Law of Assembly Concrete Structure in Hydraulic Engineering Based on Large Data Analysis. J Phys: Conf Series IOP Publ 1314(1):012141
9. Xiaoxia W (2017) The application of electronic information technology in social science research - e-Social Science research model and its development trend. Digit Technol Appl 2:100–102
10. Liu T-X, Yin D-C (2017) Big data era and paradigm change in social science research. Theor Inq (6):27–32

Jiu-Lin Sun, Member of the Chinese Academy of Engineering, Researcher, Institute of Geographical Sciences and Resources, Chinese Academy of Sciences D. supervisor, national outstanding contribution expert, famous resource scientist, one of the leading experts in the discipline of agricultural and resource environment information engineering, expert in many fields such as resource information science and technology, environmental information science and tourism information science, data mining and key technology of earth system science, etc. He is the pioneer and founder of scientific data sharing and data rendezvous in China.

Practice of Big Data Knowledge Service in Engineering Sciences and Technology—Construction and Application of CKCEST

Yunhe Pan, Zuoning Chen, and The Team Participating in the Construction of CKCEST

Abstract China Knowledge Center for Engineering Sciences and Technology (CKCEST) aims to give full play to the advantages of the Chinese Academy of Engineering (CAE) in discipline distribution and its demonstration and leading role in engineering sciences and technology, establish a collaborative innovation system for knowledge services, implement big data strategies, and serve the development of engineering sciences and technology. CKCEST is mainly designed for two types of users, i.e., personnel of think tanks, engineering and technical personnel. Focusing on data resources of different types from multiple sources and different disciplines in the field of engineering sciences and technology, it has formed a collaborative innovation system of knowledge services together with leading information technology enterprises and authoritative data service agencies, created an interdisciplinary knowledge service cooperation mechanism through technological innovation, knowledge innovation and service innovation, and formulated standards and specifications for knowledge services in the field of engineering sciences and technology. Besides, it has launched national big data applications and knowledge engineering projects, established a high-level think tank support and knowledge service system, cultivated large international and domestic user groups, and made remarkable achievements. Based on CKCEST's construction process in the past nine years, this paper systematically analyzes **the construction background and significance, implementation framework and achievements, and future prospects of operation and direction**.

Keywords CKCEST · Engineering sciences and technology · Big data · Knowledge service

Y. Pan (✉) · Z. Chen · The Team Participating in the Construction of CKCEST
Chinese Academy of Engineering, Mianyang City, China
e-mail: panyh@cae.cn

China's e-Science Blue Book 2023, https://doi.org/10.1007/978-981-99-8270-7_2

1 Construction Background, Significance and Goals

1.1 Construction Background

The knowledge revolution driven by rapid scientific and technological development has exerted far-reaching impacts on economic and social development worldwide. The possession, allocation, creation and utilization of knowledge resources represented by scientific and technological information resources have increasingly become an important factor affecting the development of engineering sciences and technology. Big data service platforms that integrate data from multiple sources, such as papers on journals, patent data, scientific research projects, scientific and technological reports, and news information, have become highly competitive highlands of knowledge services. Currently, knowledge service platforms can be divided into three categories according to different service modes, i.e., data discovery platform focusing on general needs, thematic knowledge service platform focusing on specific topics, and knowledge analysis platform focusing on individual needs.

Data discovery platform focusing on general demands refers to an intelligent knowledge search engine with universality in specific fields, which is underpinned by key technologies such as knowledge warehouse and semantic knowledge grid, semantic search and context awareness of scientific and technological big data. It provides services such as one-stop search for professional knowledge, multi-dimensional clustering, accurate positioning, intelligent recommendation, knowledge association, visualized analysis, and reliable knowledge acquisition, helping users discover valuable knowledge from massive resources quickly. Knowledge service platforms in this category include AI-driven knowledge discovery platform—Yewno, academic search platform—Semantic Scholar, AI-driven search platform—Magi and data platform—ELIXIR, etc.

Thematic knowledge service platform focusing on specific topics is designed for differentiated and refined demands of different disciplines, industry directions, major strategies, professional focuses, and different service groups, which uses convenient topic building tools to create special professional knowledge sessions, and effectively connect and aggregate various knowledge resources for special online topics. Examples of this kind include Arabidopsis Information Resource (TAIR) under UNESCO ELIXIR interoperability platform (EIP), etc.

Knowledge analysis platform focusing on individual needs refers to individualized knowledge application platform based on "data, people and tool", which provides knowledge services to address demands related to emergency, complexity, interactivity and openness, and covers expert and academic circles, knowledge context analysis, dynamic monitoring of scientific and technological development, etc. One example is Medscape—Tools.

Facing the rapid development of engineering sciences and technology at home and abroad, considering the development trend of big data, the CAE decided to build "China Knowledge Center for Engineering Sciences and Technology" [1], which would reorganize and refine digital information resources, and encourage users to create knowledge more actively, and shift from obtaining information to sharing and creating knowledge. In March 2012, the construction project of China Knowledge Center for Engineering Sciences and Technology (hereinafter referred to as "CKCEST") kicked off.

1.2 Significance

CKCEST is jointly constructed by the CAE, national research institutes, intelligence institutes of various ministries, universities, industry information centers and large enterprises, aiming to coordinate resource construction in specific fields and professional knowledge services. Its significance is reflected in the following aspects:

(1) It can meet the demands of strategic consulting research and construction of national think tanks in engineering sciences and technology. The basic work for strategic consulting research is analyzing, studying data and materials, and mining the potential value of data. Building a knowledge center that can provide in-depth services covering data, information, knowledge and value is the basic work for strategic consulting and construction of think tanks in engineering sciences and technology.

(2) It can meet the demands of innovation-driven development strategies. Since innovation needs talents, platforms, technologies and information, CKCEST will integrate domestic and foreign resources in engineering sciences and technology, spread scientific and technological knowledge and ideas, promote academic exchanges in this field, achieve innovation and development and cultivate innovation talent teams.

(3) It can meet the long-term development demands of engineering sciences and technology strategies. The innovation and development of engineering sciences and technology are impossible without support and guarantee of scientific and technological information resources. Construction and services of CKCEST can improve national infrastructure of scientific and technological information resources and make strategic layout for China's scientific and technological information resources. At the same time, it can integrate international data resources in engineering sciences and technology, promote the construction of an international knowledge base for engineering sciences and technology, strive to seize the opportunity in the new wave of big data, and play an important role in enhancing international competitiveness of national economy, science and technology.

1.3 Goals

CKCEST is a non-profit, open resource integration and knowledge service platform designed for interdisciplinary professional data fusion and in-depth knowledge mining. With the goal of providing information support and knowledge services for major decision-making, major engineering sciences and technology activities, corporate innovation and personnel training in the field of engineering sciences and technology, it will eventually build itself into an information clustering center, data mining center and knowledge service center in the field.

The construction of CKCEST focuses on the following goals. First, promoting data aggregation and resource sharing to achieve scientific allocation and efficient connection; second, promoting technological innovation and capability openness to achieve knowledge fusion and ecological prosperity; third, improving knowledge value and promoting featured services to achieve intelligent leadership and value creation; fourth, promoting coordinated operation and innovative practices to achieve win–win cooperation and sustainable development.

2 Construction Philosophy and Implementation Framework

2.1 Construction Philosophy

Taking serving users as the purpose, CKCEST serves the construction and systematic application of think tanks in engineering sciences and technology, masses of scientific and technological workers, and the development of engineering sciences and technology.

CKCEST takes innovation as the soul. Innovation is the primary driving force for development. The construction of CKCEST shall pay great attention to innovation, and seek continuous innovation in philosophy, technology, management and services in terms of platform alliance construction, R&D and application of key and core technologies, and interdisciplinary resource integration.

Taking openness as the characteristic, CKCEST regards openness as the main feature and advantage in the construction process. It will strengthen domestic and international opening-up and cooperation according to the open thinking of "mutual benefits, drawing on each other's strength, joint contribution, shared results, and common development", and especially promote the construction of International Knowledge Center for Engineering Sciences and Technology (IKCEST) [2] under the auspices of UNESCO with a global perspective.

2.2 *Implementation Framework*

(1) Organization structure

In terms of organization and management, CKCEST mainly relies on the project steering team, expert committee, project management office, technical expert team, sub-centers and technology research and development center of the CAE, and gradually forms a complete organization structure (as shown in Fig. 1). Under the overall coordination of the project management office of CKCEST, the main center, technology research and development center and sub-centers work together to promote project construction.

The main center is the core organization that implements various specifications and standards, gathers and integrates professional data resources, develops and promotes application technologies, aggregates various professional knowledge service applications, and maintains daily operation of CKCEST. As the infrastructure for CKCEST's operation and management, it consists of the general platform that gathers resources of each sub-center and provides portal services, interdisciplinary services and ubiquitous services, the application system platform that provides system and facility support, and the sharing platform that offers data resource support.

The professional sub-center is a subsystem for a specific and professional field of engineering sciences and technology, and an important part of CKCEST that covers all professional fields of engineering sciences and technology. It's responsible for making overall plans for resources in the professional field, sorting out data resources, promoting data processing and integration, facilitating research and development of

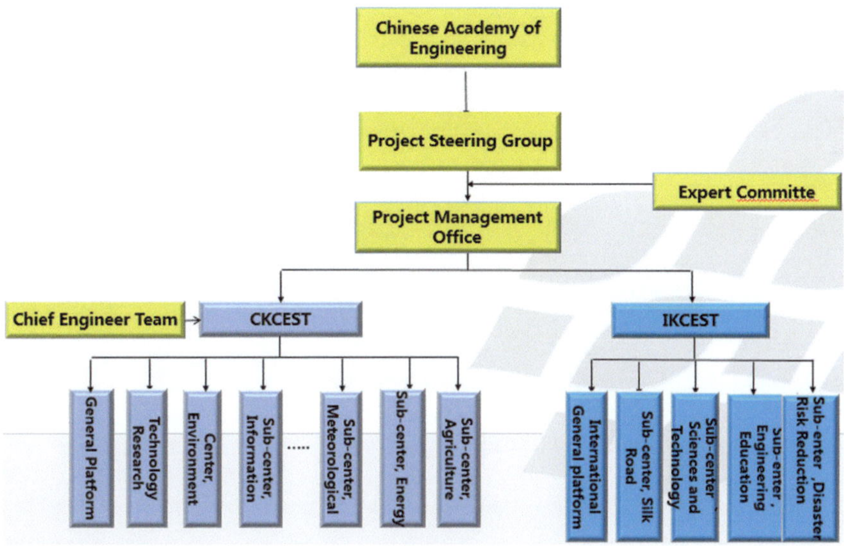

Fig. 1 Organization structure of CKCEST

professional knowledge service systems, establishing sub-platforms in professional fields, providing knowledge services for user groups, and supporting operation and maintenance of sub-center systems.

The technology research and development center is an R&D institution for key generic technologies and cutting-edge new technologies.

(2) Construction framework

The construction contents of CKCEST mainly include cloud infrastructure environment, resource system, technology system, product system, operation system, management system and portal system (as shown in Fig. 2). Specifically, cloud infrastructure environment is the basic storage, computing and network environment of the entire center; the construction of resource system focuses on gathering knowledge resources in the field of engineering sciences and technology, and establishing a data governance service system that integrates resource aggregation, resource organization, and resource management and control; the construction of technology system mainly involves capability openness, data circulation, data processing, knowledge organization and calculation, and application service support; the construction of product system mainly involves interdisciplinary knowledge services, knowledge services in professional fields, and knowledge services related to international engineering sciences and technology; the construction of operation system mainly includes operation mechanism, operation mode, evaluation system, publicity and promotion, operation maintenance guarantee; the construction of management system mainly includes standards and specifications, cyber security, intellectual property rights and talent teams.

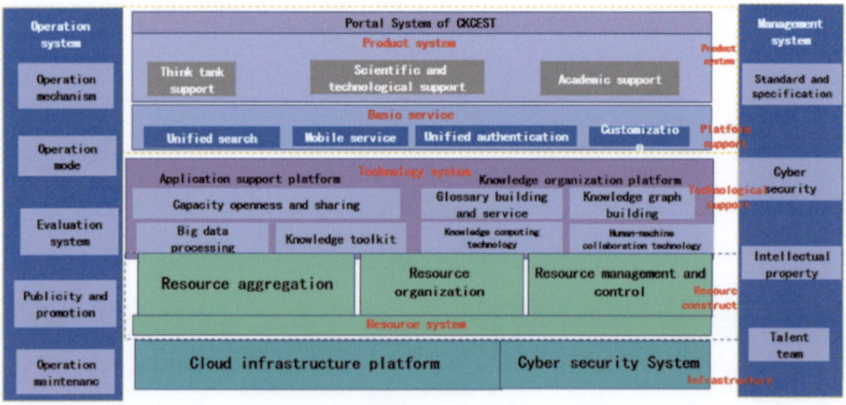

Fig. 2 Construction framework of CKCEST

2.3 Technical Solutions

Creating a cloud infrastructure environment. The focus is placed on improving cloud storage, computing facilities and network operation environment. CKCEST builds a hybrid cloud that combines public cloud and private cloud, establishes a disaster recovery center, provides unified cloud resource services for the general platform and knowledge service platforms in various fields, and realizes classified storage of data of different types and different importance.

Creating a big data processing environment. CKCEST establishes a metadata standard and specification system for different types of resources to support resource convergence, association, integration and services, and a complete big data governance system to realize whole-process management of heterogeneous data from multiple sources. It classifies engineering sciences and technology to sort out and label resources, support deep intelligent search, and builds a ledger system for database resources to monitor data resources, gather data quality statistics, make statistical analysis, and conduct full-process refined management of heterogeneous data resources from multiple sources.

Establishing a resource system. CKCEST gathers knowledge resources in the field of engineering sciences and technology through methods such as construction by itself, alliance, procurement and open access, and establishes a data governance service system that integrates resource aggregation, resource organization, and resource management and control. Under unified management of the main center, sub-centers organize various data resources in the field and send them to the main center to increase basic public data resources in the field of engineering sciences and technology, promote the development of special resources and data convergence in various fields, boost the development of data integration and expansion from multiple resources, and realize integration, optimization and continuous updating of metadata in engineering sciences and technology.

Establishing a knowledge organization and computing technology system. CKCEST promotes integrated innovation and independent research and development of key generic knowledge computing technologies, facilitates research and development of data aggregation and indexing technologies, knowledge processing and reorganization technologies, deep search and multi-dimensional navigation technologies, collaborative crowd-sourcing technology, big data visualization analysis technology, knowledge network technology, and intelligent question answering technology. It carries out key technology integration testing, establishes a reference system, and builds a reference model and a supporting software KS-Studio for professional knowledge service system. It promotes projects and application of knowledge computing technology, and realizes named-entity recognition and linking, hierarchical concept indexing, extraction of entity-attribute relations, integration of data from multiple sources, efficient collaboration and deep integration of expert group's wisdom and knowledge computing tools [3]. It builds a one-stop human–machine collaborative crowd-sourcing service system for knowledge processing, conducts research on knowledge organization methods and tools, builds a management system for establishment of knowledge organization system and a glossary for engineering sciences and technology, and develops construction and supporting technology tools of knowledge graph, new word discovery tools and relation recommendation tools to effectively support the processing of engineering sciences and technology resources and development of knowledge applications [4].

Promoting research and development of intelligent search services. CKCEST creates an open environment for technology integration, cooperates with Baidu on search engine technology, integrates artificial intelligence (AI) technology and knowledge graph technology of key universities and institutions, and builds a leading knowledge search engine in China. It cooperates with Baidu Scholar on resources to establish the largest resource database in engineering sciences and technology in China, coordinates processing of original text resources and front-end product development, jointly provides more professional intelligent search services. It keeps optimizing the user management system, including sensitive word management, log management, original text service and authority management. It continues to explore the application of knowledge graphs in search, optimize and improve search breadth, search depth, search efficiency and search quality based on knowledge graphs, so as to enhance the accuracy and comprehensiveness of search results. It improves knowledge disclosure of data fusion in various fields of engineering sciences and technology based on semantic understanding and knowledge correlation. It enables users to surf different online platforms freely with single sign-on, look for interdisciplinary resources efficiently, and obtain original texts and entity data conveniently, and provides one-stop discovery service of "finding and obtaining what you need easily" for them.

Establishing a knowledge service system. CKCEST establishes a knowledge service system that coordinates comprehensive services of the main center and professional services of sub-centers. It develops professional knowledge applications to find knowledge from big data, gather collective wisdom, and provide in-depth knowledge services for users. It provides a series of services for high-end think tanks and strategic consulting agencies, such as one-stop discovery of high-value data, informatization support in the full life cycle, intelligent tool support, automatic report writing and push of high-value intelligence. It continuously focuses on scientific research demands of the engineering sciences and technology community, and improves a series of high-quality services. It focuses on national strategies, responds to emergencies, and keeps enriching knowledge applications in various professional fields.

Establishing a product system. By developing products such as think tank support platform, think tank viewpoint and intelligent strategic consulting support system, CKCEST provides high-quality knowledge and application services for strategic consulting and research, initially realizes full-process support from data collection to data analysis and collaborative report writing, and forms a strategic consulting support service system for engineering sciences and technology with experts as the core, procedures as the norm, data as the support, and interaction as the means.

Establishing standard and specification system. Considering different types of data resources and the full life cycle of resource creation, description, organization, search, service and long-term preservation, CKCEST puts forward the overall standard and specification framework, including data resource specification, technical specification, application service specification, security specification and operation management specification, etc.

Establishing an operation management system. CKCEST has developed corresponding management methods and systems from various aspects. For example, in terms of comprehensive management system, the *Intellectual Property Management Measures* and *Network Information Security Management Measures* have been formulated; in terms of resource construction system, *Resource Construction Measures*, *Resource Sharing Management Measures*, and *Resource Security Management and Long-term Storage Management Measures* have been proposed; in terms of service management system, *Knowledge Service Management Measures* and *Platform Operation Service Management Measures* have been introduced. In terms of publicity and promotion, CKCEST has pooled strength of all parties, innovated operation mechanisms, created a unified knowledge service brand, and promoted ecological cooperation, forming a service operation and promotion system in the field of engineering sciences and technology.

3 Achievements

3.1 Developing a Complete Interdisciplinary Knowledge Service Alliance Mechanism

The CAE and its partners jointly build CKCEST through an alliance mechanism, which makes it possible to combine the opinions of field researchers and literature and information experts in decision-making. Through provisions of rights and obligations, CKCEST has straightened the relations between alliance units and overall goals of the project and between alliance units themselves, and developed a new management model of joint contribution and shared benefits related to data and applications. By the end of 2021, the CAE had built alliances with more than 50 information service units, including intelligence departments of ministries and commissions, industry information centers, leading domestic universities, national scientific research institutes, leading domestic information technology enterprises, and authoritative data service agencies. It had pooled the strength of industries, universities and research institutions to jointly build big data industry ecology in the field of engineering sciences and technology.

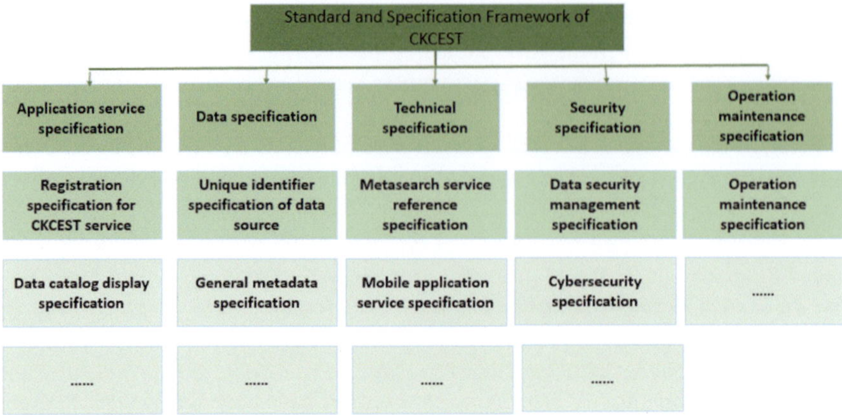

Fig. 3 Standard and specification framework of CKCEST

3.2 *Establishing a Standard and Specification System and Management System of Knowledge Services in Engineering Sciences and Technology*

According to business and management needs, CKCEST has formulated and improved the corresponding standards and specifications, necessary rules and regulations, which has not only promoted effective integration of interdisciplinary data, but also ensured efficient operation relying on the alliance mechanism and participation of multiple institutions. It has made the overall standard and specification framework (as shown in Fig. 3), including application service specification, data specification, technical specification, security specification and operation management specification. It has formulated 24 specifications for general metadata and 95 specifications for special resource metadata.

3.3 *Establishing a Massive Resource System for Engineering Sciences and Technology*

By the end of 2021, the CKCEST's resources reached 7.3 billion items or 100 TB in total, covering 100% of first-level disciplines and 90% of second-level disciplines in the field of engineering sciences and technology. CKCEST coordinated efforts of the main center and sub-centers to build 34 big data knowledge service platforms in professional fields, and integrated nearly 3,000 high-quality, special, professional datasets in various fields. The main center and sub-centers jointly built a knowledge base and a professional glossary system in the field of engineering sciences and technology. The main table had 196,000 core entries and 3.2 million expanding entries. There were more than 600,000 entries in sub-fields. CKCEST formulated a classification system with 973 main categories and 620 subcategories, and produced knowledge graphs in the fields of traditional Chinese medicine, water conservancy and tea-oil, and knowledge graphs of experts, scholars, projects and achievements in engineering sciences and technology.

3.4 *Creating a Technological Path for Big Data Application and Knowledge Engineering*

The CKCEST's technological support and achievements in the construction of integrated infrastructure and creation of technological environment, knowledge organization and computing technology, big data mining and knowledge discovery provide a pioneering example and exert demonstration effects on how to promote interdisciplinary collaboration among organizations, resource aggregation, integration of connections, big data mining and knowledge services, and on how to turn data into information and knowledge. The integrated technological environment, standard and specification formulation and big data processing environment provide guarantee for the basic environment and supporting facilities of knowledge service platforms in various fields, break through organizational barriers between units, thus ensuring the horizontal flow of interdisciplinary data resources, and enabling users to surf on different platforms and have the same experience. Construction of knowledge organization system and semantic annotation enhances knowledge disclosure of data fusion in various fields of engineering sciences and technology. Big data mining, knowledge computing, cross-media computing and deep search technologies provide guarantee for the construction of intelligent, personalized and special data service platform for engineering sciences and technology.

3.5 *Forming a Gridded Knowledge Service System for Engineering Sciences and Technology*

CKCEST has provided various types of knowledge services at different levels, including basic service, in-depth service, special service and customized service for users, basically forming a three-dimensional and gridded knowledge service system. The main center mainly provides basic knowledge services, integration of all knowledge services in the field of engineering sciences and technology, and one-stop knowledge discovery services in the field. It integrates access to professional sub-centers, and develops and integrates comprehensive knowledge service products, such as "Hall of Academicians", "Expert Database", "Belt and Road", "Zhiling Livestreaming" and "Zhiling Video". Professional sub-centers delve into vertical fields and provide in-depth knowledge services, search of professional resources, gathering and display of characteristics of each field, display of professional knowledge correlation, in-depth knowledge analysis and customized services in respective fields.

By the end of 2021, CKCEST had launched more than 260 knowledge applications in various fields online. For example, the intelligent map search tool launched by the geographic information sub-center provides reference for scientific and technological workers to write technological reports or check the figures of academic papers by themselves. The application of special knowledge in science, education, culture and health in the geographical resources sub-center shows temporal and spatial distribution and changes in these fields in the past 40 years since China's reform and opening up. The dynamic environmental products of the marine sub-center provide environmental support for the production in shipping and coastal fishery industries. The application services of environmental protection products from the environmental sub-center provide current quantity analysis of multi-dimensional environmental protection products. The oil-tea knowledge map application system has been established for oil-tea related knowledge application products of the forestry sub-center. The water-related knowledge map "water-related question" from the water conservancy sub-center has straightened relations between the main research objects of water conservancy and provided visualized display. The intelligent analysis of infectious diseases of the medical sub-center has compared and made visualized analysis of the data of 28 common statutory notifiable Class A and B infectious diseases in China in recent years.

3.6 Establishing a Science and Technology Service Support System for Think Tanks

The strategic consulting support service system in engineering sciences and technology has been established by developing various think tank knowledge products, such as think tank support platform, think tank viewpoint, and intelligent strategic consulting support system. CKCEST has embedded offline services into the whole process of consulting and research, and initially built a think tank information service model that combines field tracking, information reference and data analysis. It has provided intelligent support tools to meet different needs, use data to drive the establishment of new think tanks, and comprehensively facilitate high-quality development of high-end think tanks.

3.7 *Continuously Expanding International Influence of the IKCEST*

Under the framework of UNESCO, the IKCEST has actively participated in open services, cooperation and exchanges in international engineering sciences and technology, formed a "1 + N" layout, i.e., one general platform and four sub-platforms (i.e., Disaster Risk Reduction, Silk Road Sciences and Technology, Engineering Education and Intelligent City), and initially developed an integrated management and operation system of the general platform and sub-platforms. Under the unified standard and specification system, it continuously gathers international resources in engineering sciences and technology and special international resources in advantageous fields. The IKCEST platform has 130 million pieces of data, and 51 online knowledge applications such as disaster risk reduction, public health security, AI ethics, cultural heritage protection, global engineering, scientific and technological trends, and the "Belt and Road" index. It continuously provides talent training services for developing countries and higher education institutions. From 2015 to 2021, IKCEST held a total of 93 international training sessions, involving 28 topics, 115 countries and regions, and more than 13,000 students (with women accounting for 35%). The satisfaction degree of the training exceeded 90%. IKCEST is committed to building an international high-end academic exchange platform, and strives to hold high-end international seminars in fields such as big data, AI and online engineering education and provide an academic exchange platform for well-known experts and scholars at home and abroad.

4 Future Prospects

The "14th Five-Year Plan" period will be the first five years following China's achievement of the First Centenary Goal to build a moderately prosperous society in all respects, and we shall ride that momentum to embark on a new journey of building a modern socialist China and to advance toward the Second Centenary Goal. *The Recommendations of the Central Committee of the Communist Party of China for Formulating the 14th Five-Year Plan for National Economic and Social Development and Long-Range Objectives Through the Year 2035* proposes that China will build a high-end exchange platform for national scientific research papers and scientific and technological information. The *14th Five-Year Plan for National Informatization* proposes to give full play to the key role of data as new production factors, focus on the development, utilization, sharing, circulation, full life cycle governance and security assurance of data resources, establish and improve data element resource system, explore the value of data elements, enhance their facilitating role, lead and create new demands via innovation and high-quality supply, form a strong domestic market, and promote the establishment of a new development pattern.

In order to fully implement the "14th Five-Year Plan" and vigorously implement the innovation-driven development strategy, CKCEST will build a knowledge discovery, gathering, processing, management and service system, continuously promote the gathering and integration of resources in the field of engineering sciences and technology, deeply tap the value of data, promote the transformation from data into knowledge, establish a knowledge resource system in the field, and develop its own brands. It will make breakthroughs in key technologies such as big data, artificial intelligence and knowledge services, improve independent research and development of knowledge products, build a high-end national information exchange platform for engineering sciences and technology, provide full-process knowledge services such as knowledge search, intelligent Q&A, in-depth analysis, online communication and group collaboration, and become a major scientific and technological infrastructure supporting national strategic decision-making and scientific and technological innovation. It will actively integrate into the global innovation network, connect to global innovation resources, and form an open and inclusive ecological environment for information exchange in engineering science and technology to enhance the ability and influence of international exchanges.

References

1. Pan Y (2012) Research on the construction of Digital Library of Chinese Academy of engineering and demonstration of implementation scheme design. Chinese Academy of Engineering, Beijing
2. Wang J, Bu K, Yang F et al (2020) Disaster risk reduction knowledge service: a paradigm shift from disaster data towards knowledge services. Pure Appl Geophys 177:135–148
3. Pan Y (3 Jun 2020) Multiple knowledge representation of AI. Engineering, 216–217
4. Gang P, Yunliang Z, Qinghong Z (5 Apr 2018) Thinking and practice of knowledge service in engineering sciences and technology. Inf Eng, 4–12

Yunhe Pan member of the Chinese Academy of Engineering, served as President of Zhejiang University from 1995 to 2006, Executive Vice President of the Chinese Academy of Engineering from 2006 to 2014. He has been engaged in the research of computer graphics, AI, CAD and industrial design for a long time.

Big Earth Data for Sustainable Development

Huadong Guo, Dong Liang, Zhongchang Sun, and Fang Chen

Abstract In 2015, 193 member states of the United Nations adopted the 2030 Agenda for Sustainable Development, in which, 17 Sustainable Development Goals (SDGs), 169 targets, and more than 230 indicators were proposed to guide the economic, social, and environmental aspects of development. However, several factors have constrained the implementation of the SDGs, including lack of data, technical obstacles, and shortage of funds. With the development of science and technology, the global data volume is growing exponentially. Advances in computing and data technologies have made real-time processing and analysis of big data a reality, while new types of data combined with traditional data, such as statistical and survey data, can create more detailed, timely, high-quality information. Big Earth Data technology, through its extensive use and further innovation, can be an effective way to address the data divide and the lack of information and tools for sustainable development. The Big Earth Data Science Engineering Program of the Chinese Academy of Sciences carried out research on improving the monitoring and evaluation of SDGs and achieved a series of goals. To this end, the International Research Center of Big Data for Sustainable Development Goals (CBAS) was established to strengthen national and international efforts through improved scientific support driven by innovative big data solutions for SDGs.

Keywords Big Earth Data · Digital Earth · Scientific big data · Sustainable Development Goals

1 Development Trends of Big Earth Data

Big Earth Data is a core element of science, technology, and innovation—a strategic highland in the digital economy era and a new type of strategic resource, nationally and globally. Developed from Digital Earth and scientific big data, Big Earth Data is

H. Guo (✉) · D. Liang · Z. Sun · F. Chen
International Research Center of Big Data for Sustainable Development Goals, Beijing 100094, China
e-mail: hdguo@radi.ac.cn

China's e-Science Blue Book 2023, https://doi.org/10.1007/978-981-99-8270-7_3

now becoming a new key for understanding Earth, a new engine for discovery, and a new paradigm for scientific cognition [1].

1.1 Digital Earth

Digital Earth refers to the virtual Earth that integrates massive, multi-resolution, three-dimensional, dynamic data about Earth in accordance with geographical location. The concept calls for a high integration of Earth science, spatial science, and information science [2]. The development of Digital Earth represents a profound scientific and technological revolution and a far-reaching change in Earth science research.

The concept of Digital Earth was proposed in 1998 and it soon began its global development when the "Beijing Declaration on Digital Earth" [3] was passed at the 1st International Symposium on Digital Earth in 1999. Digital Earth has been linked to China since those early years: China was one of the first countries to carry out Digital Earth research, and the International Society for Digital Earth (ISDE), initiated by Chinese scientists and headquartered in China, has become the world's foremost academic organization in this field. The *International Journal of Digital Earth*, founded by ISDE, is included in the Science Citation Index (SCI) and is the only international academic journal in this field [4].

During its 20-year evolution, Digital Earth has faced severe challenges in the management of Big Earth Data, construction of Digital Earth platforms and ecosystems, expansion of fundamental research, response to social complexity, and education on Digital Earth, among others. However, Digital Earth research is encountering both new development opportunities and challenges in services, applications, and scientific research paradigms with the rise and sustainable development of new technologies such as big data, cloud computing, artificial intelligence (AI), blockchain, and the Internet of Things (IoT).

The *Manual of Digital Earth*, the first international academic book on Digital Earth and Big Earth Data, was completed and published online by Springer-Verlag with open access. The book analyzes Digital Earth technology, combs through multi-field applications of Digital Earth, introduces the development of Digital Earth regions and countries, and discusses the education and ethics of Digital Earth [5]. It is a forward-looking and highly professional academic work that boosts the positive development of Digital Earth around the world and makes academic contributions to the Digital China strategy.

1.2 Scientific Big Data

The generation of data is no longer limited by time and space with the in-depth development of AI and network technology as well as the emergence of semi-structured and unstructured data, which has led to the explosive growth of data. Numerous, complex data types have exceeded the capabilities of traditional data management systems and processing modes. For this reason, humans have started their journey into the era of big data [6].

In September 2013, the concept of scientific big data was put forward in an article titled "Scientific Big Data and Digital Earth" published in the *Chinese Science Bulletin*. It was pointed out that scientific big data has unique scientific connotations and characteristics, making it different from Internet big data and commercial big data in essential attributes and characteristics [7].

Scientific big data, a branch of big data, is becoming a new driving force of scientific discovery that brings a new paradigm of scientific discovery, namely, the data-intensive scientific research paradigm. First, the scientific research objects have changed. Large-scale, complex instruments instead of telescopes are used to "observe" objects by transmitting data to data centers. Second, the tools for scientific discovery have changed. "Data rush" has become scientists' most important work in the big data era, and data mining is the main tool of scientific discovery. Third, scientific data and knowledge products have changed. Data rules have emerged along with the full data model, where scientific big data no longer focuses on high precision but high volume and comprehensiveness. Last, the division of labor and the process of scientific discovery have changed. A part of scientific research has been socialized or automated, and many people have unconsciously engaged in the production of scientific data.

Scientific big data is a new model that partially relies on causality and mainly on correlations to discover new knowledge. It has become a typical example of the data-intensive scientific paradigm after the empirical, theoretical, and computing models [8].

1.3 Big Earth Data

Big Earth Data, an important part of scientific big data, is an emerging research frontier combining Earth science and information science. It is a novel, data-intensive research method for Earth science [2].

Big Earth Data refers to big data in the field of Earth science with spatial attributes, especially the massive Earth observation data based on spatial technology. It is mainly generated from large-scale scientific experimental devices, detection equipment, sensors, socio-economic observations, and computer simulation processes. On the one hand, Big Earth Data, like big data, generally is massive, multi-source, heterogeneous, multi-temporal, multi-scale, and non-stationary. On the other hand, it boasts strong spatiotemporal correlation, physical interdependency, and the controllability of data generation methods and data sources. Big Earth Data science is the product of the cross-integration of natural science, social science, and engineering. The correlation and coupling of Earth systems are systematically studied based on Big Earth Data analysis; that is, Earth is observed and studied as a whole to understand the complex interaction as well as the development and evolution between Earth's natural system and human social system through the comprehensive application of big data, AI, and cloud computing. In this way, Big Earth Data makes important contributions to a community with a shared future for humankind, global sustainable development, and China's "Belt and Road" initiative [9].

Fully recognizing the significance of Big Earth Data, the Chinese Academy of Sciences (CAS) established the Strategic Priority Research Program (Class A) "Big Earth Data Science Engineering Program" (CASEarth) at the beginning of 2018. Systematic research on Big Earth Data has been carried out through the program, aiming to promote and accelerate the transformation from simple Earth data systems and data sharing to a Digital Earth data integration system. The research will facilitate the sharing of data, knowledge, and experience around the world, thus contributing to scientific research, decision support, and knowledge dissemination [10].

2 Big Earth Data Science

2.1 Overview of Big Earth Data Science

The progress of scientific research is mainly driven by observed or measured data, and its prosperity arises from these data and powerful computational science. The same is true for Earth science research. Big Earth Data science involves establishing a Digital Earth science platform for managing, processing, and analyzing Earth observation data, and comprehensively using advanced scientific and technological means such as big data, AI, and cloud computing to observe and test the Earth system as a whole including the natural world, physical world, and data world [9].

As a branch of data-driven science, Big Earth Data science can also be regarded as a subfield of data science. Big Earth Data science is crucial for studying the design and architecture of the Big Earth Data ecosystem and its role in the digital transformation of society and global sustainable development. Big Earth Data requires the adoption of integrated methods based on engineering sciences such as natural science, social science, big data, and AI. It is necessary to investigate the ecosystem of the Big Earth Data platform and establish a programmed "organism" to understand the macroscopic phenomena of Earth and obtain operable, reliable knowledge (see Fig. 1).

2.2 Research Objectives of Big Earth Data Science

Big Earth Data Science involves methods and technical activities for studying the Big Earth Data analysis ecosystem. It is an "organism" that systematically discovers information from data related to Earth processing, e.g., developing and deploying methods and technologies that enable the collection, storage, retrieval, and access

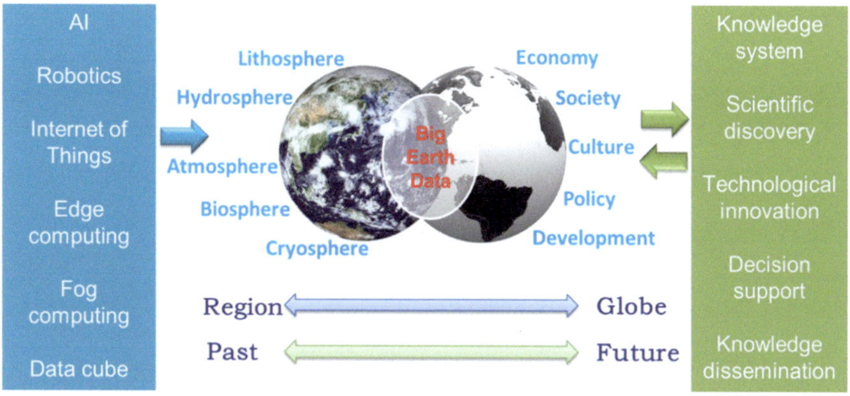

Fig. 1 Big Earth Data Science ecosystem [11]

of data in different natural and social domains in an efficient analysis environment. Different inputs must be integrated by the Big Earth Data ecosystem in a geographical environment so that different communities can have access to the ecosystem and the democratization of data and information can be ensured (Fig. 2).

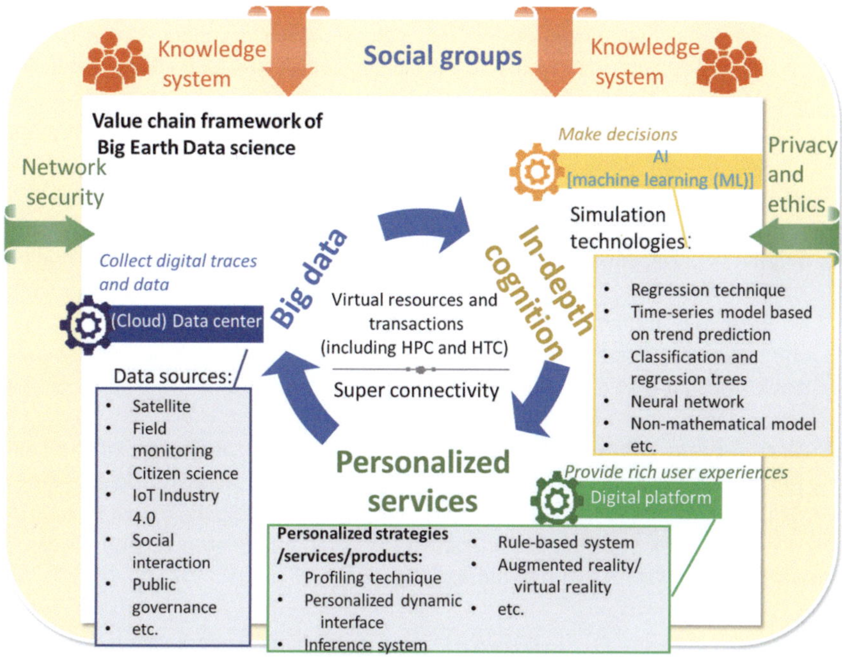

Fig. 2 Value chain framework of Big Earth Data Science [9]

An important objective of Big Earth Data Science is to scientifically understand, model, and apply the process of transforming data into information, and to provide knowledge for global sustainable development. Research on Big Earth Data Science is of great importance to solve major social problems [12].

2.3 Technical Systems of Big Earth Data Science

Big Earth Data science aims to acquire knowledge from multi-source, massive, and complex Big Earth Data, and develop relevant theories to explain the operation and evolution mechanism of the socio-physical system through various tools and algorithms, thereby building a sustainable human society to protect Earth.

The main technical systems of Big Earth Data science include: (1) ubiquitous data perception, (2) data reliability sharing, (3) multivariate data fusion, (4) digital twin and complex simulation, and (5) spatial Earth intelligence [9].

(1) Ubiquitous data perception means the efficient observation and integration of comprehensive data by making the most of data collection facilities in the full-space system, aiming to provide instant and available data sources for data fusion, association analysis, and spatial statistics.
(2) Data reliability sharing refers to accurately recording all the processing flows of geospatial data in the entire life cycle and their precision levels through distributed accounting books to ensure the traceability of data, the credibility of decisions, and the protective use of private data.
(3) Multivariate data fusion aims to enhance value through multi-level, multi-angle, and multi-scale association, transformation, filtering, and integration of data to fully mine the correlations and values of multivariate data, thus providing knowledge for decision-making.
(4) Digital twin and complex simulation mean simulating the evolution and development mechanisms of the Earth system under the constraints of many factors such as geography, human activity, society, and the economy through nonlinear, high-dimensional, complex systems. They perform self-learning based on the feedback from multisource data, and present the real situations of physical entities in the digital world in a near real-time fashion.
(5) Spatial Earth intelligence cognition is based on the completion of the basic functions of machine vision such as element extraction, recognition, and classification, and supplemented by AI, ML, and software analysis, so that simulation systems can recognize and understand the complex phenomena and processes of the Earth system just like humans.

3 Applied Research on Big Earth Data in Support of the Sustainable Development Goals (SDGs)

In 2015, the United Nations (UN) passed 17 SDGs covering three major fields, economics, society, and the environment, indicating the direction of comprehensive transformation into sustainable development in various countries [13]. However, the realization of SDGs has been restricted due to missing data, development disequilibrium, and association and mutual restriction between SDGs [14]. Even worse, the outbreak of COVID-19 throughout the globe in 2020 aggravated the challenges faced by various countries in realizing the SDGs.

It has been an international consensus that science and technology play an important role in promoting the realization of SDGs. As further highlighted in the *Global Sustainable Development Report 2019*, science and technology are a significant force driving sustainable transformation and global development and reform [15]. Big Earth Data, as an important aspect of scientific and technological innovation, exerts a significant effect on supporting the realization of SDGs. Hence, the "Big Earth Data Science Engineering Program (CASEarth)" has deeply investigated Big Earth Data in serving SDG 2 (Zero Hunger), SDG 6 (Clean Water and Sanitation), SDG 11 (Sustainable Cities and Communities), SDG 13 (Climate Action), SDG 14 (Life below Water), SDG 15 (Life on Land) and multiple indicators of SDGs as well by combining the advantages and features of Big Earth Data. Providing a solution to global cross-domain and interdisciplinary collaboration, this program is an innovative practice aiming to support the realization of SDGs via the UN Technology Facilitation Mechanism [16].

For four years, CASEarth has conducted 64 case studies for 24 targets and displayed the research, monitoring, and evaluation results of relevant SDGs and their indicators through 53 data products, 33 methods and models, and 42 decision support analyses from four scales: local, national, regional, and global. China's sustainable development progress (Table 1) [1] during 2010–2020 was evaluated based on existing research results by aiming at 20 indicators in six SDGs.

Among the 20 evaluated indicators, four have already approached or reached requirements in China, including SDG 2.2.1 (prevalence of stunting among children under five years of age), SDG 11.2.1 (proportion of population that has convenient access to public transport), SDG 15.3.1 (proportion of land that is degraded over total land area) and SDG 15.4.2 (mountain green cover index). Though improved somewhat in recent years, SDG 6.4.1 (change in water-use efficiency over time) and SDG 15.5.1 (red list index) still face challenges. In the future, stress should be laid on and more efforts should be put into water conservation and the protection of wild animals and plants. Results show that from 2010 to 2015, two indicators deteriorated while 11 improved. During 2015–2020, no indicators became worse, and 16 were improving. In short, China is forging ahead to realize sustainable development by 2030, and the magnitude of improvement has been large, especially since 2015. Four cases already approaching or achieving the target requirements will be introduced.

Table 1 Evaluation of china's progress on SDGs based on Big Earth Data (2010–2020) [17]

	Indicators	Evaluated items	2010-2015	2015-2020
SDG 2	2.2.1 Prevalence of stunting among children under 5 years of age	Under-five stunting	Challenging, Improving	Close or achieved, Improving
	2.4.1 Proportion of agricultural area under productive and sustainable agriculture	Land productivity, water resources use, and excess fertilizer application	Challenging, Improving	
SDG 6	6.3.2 Proportion of bodies of water with good ambient water quality	Water clarity in lakes and reservoirs	Challenging, Improving	Challenging, Stable
	6.4.1 Change in water-use efficiency over time	Water-use efficiency		Very challenging, Improving
	6.5.1 Degree of integrated water resources management	Integrated water resources management		Challenging, Improving
	6.6.1 Change in the extent of water-related ecosystems over time	Areal extent of wetlands (riverine and lacustrine systems)	Challenging, Improving	Challenging, Improving
SDG 11	11.2.1 Proportion of population that has convenient access to public transport, by sex, age and persons with disabilities	Proportion of population that has convenient access to public transport		Close or achieved, Improving
	11.3.1 Ratio of land consumption rate to population growth rate	Ratio of land consumption rate to population growth rate	Very challenging, Worsening	Challenging, Improving
	11.4.1 Total per capita expenditure on the preservation, protection and conservation of all cultural and natural heritage	Unit Input in heritage sites	Challenging, Improving	Challenging, Stable
	11.5.1 Number of deaths, missing persons and directly affected persons attributed to disasters per 100,000 population	Interannual change in total loss from natural disasters at prefectural level	Challenging, Improving	Challenging, Improving
	11.6.2 Annual mean levels of fine particulate matter (e.g., $PM_{2.5}$ and PM_{10}) in cities (population weighted)	Fine particulate matter in cities	Very challenging, Worsening	Challenging, Improving
	11.7.1 Average share of the built-up area of cities that is open space for public use for all, by sex, age and persons with disabilities	Area of open space in cities		Challenging, Improving
SDG 13	13.1.1 Number of deaths, missing persons and directly affected persons attributed to disasters per 100,000 population	Number of deaths, missing persons and directly affected persons attributed to disasters per 100,000 population	Challenging, Improving	Challenging, Improving
	13.2.2 Total greenhouse gas emissions per year	Concentration of greenhouse gas		Challenging, Improving
SDG 14	14.1.1 (a) Index of coastal eutrophication; and (b) plastic debris density	Marine debris and microplastics in coastal waters	Challenging, Improving	Challenging, Improving
	14.2.1 Number of countries using ecosystem-based approaches to managing marine areas	Area of mangrove	Challenging, Improving	Challenging, Improving
	14.7.1 Sustainable fisheries as a proportion of GDP	Area of raft culture in coastal waters		Challenging, Improving
SDG 15	15.3.1 Proportion of land that is degraded over total land area	Proportion of degraded land	Close or achieved, Improving	Close or achieved, Improving
	15.4.2 Mountain Green Cover Index	Mountain Green Cover Index		Close or achieved, Stable
	15.5.1 Red List Index	Red List Index	Very challenging, Improving	Very challenging, Improving

Colored circles indicate state of implementation: Very challenging; Challenging; Close or achieved

Arrows indicate trend: Improving; Stable; Worsening

Bright green bars indicate the years covered by the evaluation during the period of 2010-2020. The example below indicates 2010-2018.

(1) Prevalence of stunting among children under five years of age

Great attention has always been given to children's health in China. The improvement of children's nutritional status is not only a focal point of SDGs but also a significant constituent in boosting the construction of a healthy China. In recent years, the nutritional status of Chinese children below five years old has been substantially improved thanks to Chinese childhood nutrition improvement policies and the promotion and implementation of sustainable projects. From 2002 to 2017, the undernourishment rate of Chinese children below five years old declined from 18.8 to 4.8% (already reaching the corresponding goals in SDG 2.2), where that in cities declined from 7.8 to 3.4% and that in rural areas from 25.6 to 5.8%, indicating that the urban–rural gap was considerably narrowed [1]. The childhood nutrition improvement policies carried out by the Chinese government have achieved prominent results, manifested by the significant improvement of the nutritional status among Chinese children below five years and the narrowing of urban–rural gaps (see Fig. 3).

(2) Proportion of Chinese population that has convenient access to public transport

Urban public transport is an indispensable part of urban transport, the artery guaranteeing normal urban production and life, and a key factor in realizing sustainable development related to education, food security, and the environment. China has energetically prioritized and implemented a public transport development strategy and promulgated and carried out multiple plans to perfect the top-level design of

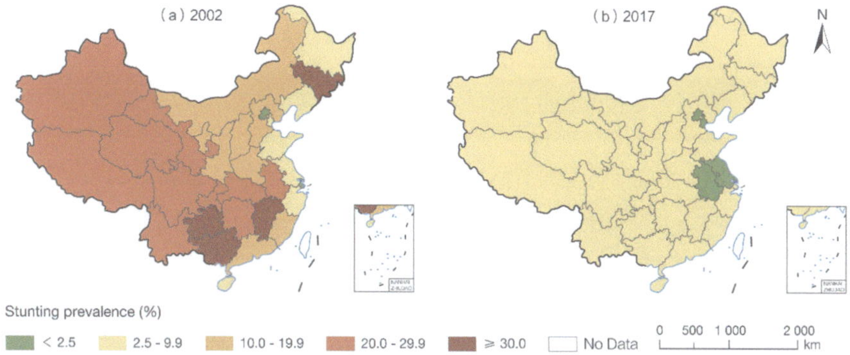

Fig. 3 Stunting under-five in 2002 (**a**) and 2017 (**b**) by province in China (no data available for Hong Kong, Macao, and Taiwan, China) [18]

urban public transport. In 2020, the overall proportion of Chinese that had convenient access to public transport was 90.15%, which was 9.59% higher than that in 2018. This indicator grew, to different extents, in about 96.90% of cities across China and declined by a small margin in about 3% of Chinese cities [17] (see Fig. 4).

(3) **Follow-up assessment of zero growth in land degradation and its global contributions**

China has made remarkable achievements in the governance of land degradation. Based on the framework of the United Nations Convention to Combat Desertification (UNCCD) and Big Earth Data, China has monitored and evaluated the national-scale zero-growth criteria for land degradation, and the progress achieved, with global consistency and spatial comparability. The results reveal that land degradation continued a favorable zero-growth trend in China. In 2018, the net area of land recovered reached a year-on-year growth rate of 60.30% compared with that in

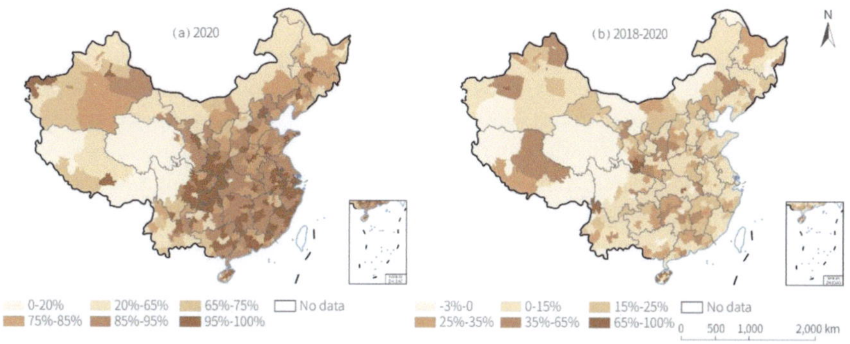

Fig. 4 **a** Proportion of the population with convenient access to public transportation in China, 2020; **b** variations in distribution of the proportion of the population with convenient access to public transportation in China in 2018–2020 [17]

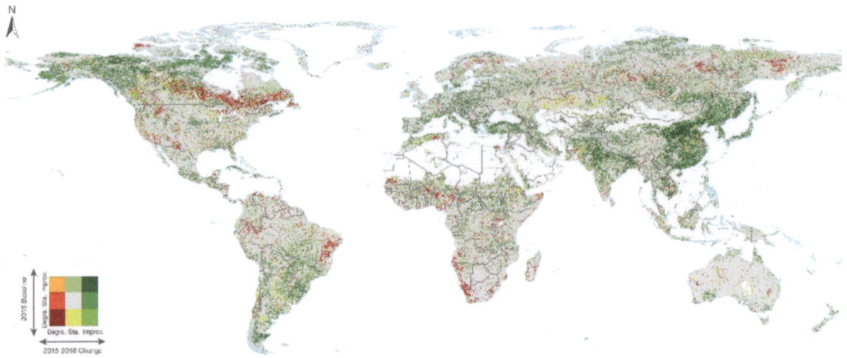

Fig. 5 Criteria for global land degradation and dynamic spatial distribution during 2015–2018 [19]

2015, and accounted for about 1/5 of the total net area of land recovered in the globe, making the greatest contribution to zero growth in global land degradation [19] (see Fig. 5).

(4) **High-resolution monitoring of the global mountain green cover index**

China is a big country, with the most mountains in the world and a proportion of mountain area as high as 64.59%. The idea of the sustainable development of mountain green cover—"clear waters and lush mountains are invaluable assets"—proposed and practiced by China has been deeply rooted in the heart of Chinese people. Moreover, China has dynamically monitored the global mountain green cover index through high-resolution raster imagery based on the national-scale mountain green cover index estimation scheme of the Food and Agriculture Organization of the United Nations (FAO) and Big Earth Data. The results showed that the global average mountain green cover index was 80.56% in 2020, while that in China reached 82.05%, which was nearly even with that in 2015 [17, 20]. Given the environmental constraints of high-altitude areas, China has basically realized its goal for the mountain green cover index, and future attention should be paid to interregional differences (see Fig. 6).

The annual report on *Big Earth Data in Support of the Sustainable Development Goals*, compiled under the organization of CASEarth, has been released by the state leadership of China for three consecutive years. *Big Earth Data in Support of the Sustainable Development Goals (2019)* [21] is listed as one of the four formal documents and one of the two documents with which the Chinese government participated in the 74th United Nations General Assembly and the UN Summit on the SDGs, respectively. This document fills the void of data and methods for the international community and provides a new perspective and new support for accelerating the implementation of the *UN 2030 Agenda for Sustainable Development*. On the occasion of the 75th anniversary of the founding of the UN and the 5th anniversary of the *UN 2030 Agenda for Sustainable Development* was passed, *Big Earth Data in Support of the Sustainable Development Goals (2020)* [18, 22] was released by

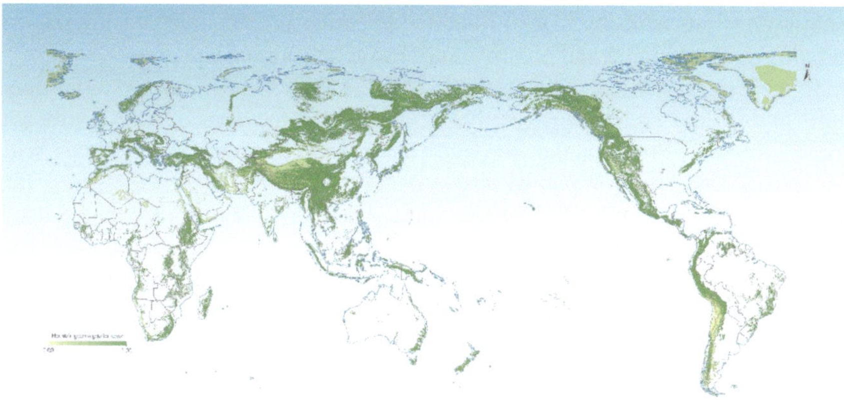

Fig. 6 Global mountain green cover index distribution in 2020 [17, 20]

the state leadership of China during the High-level Video Conference on Poverty Eradication and South-South Cooperation in September 2020, which provided a reference for various countries to monitor and evaluate the implementation of the *UN 2030 Agenda for Sustainable Development*. Then, *Big Earth Data in Support of the Sustainable Development Goals (2021)* was released by the state leadership of China at the 2021 Forum on Sustainable Development, serving as a beneficial reference for implementing the *UN 2030 Agenda for Sustainable Development* in the international community.

4 International Research Center of Big Data for Sustainable Development Goals (CBAS)

On September 6, 2021, the inaugural event of CBAS and the 2021 International Forum on Big Data for Sustainable Development Goals was held, announcing the formal establishment of CBAS, which is the first international scientific research organization serving the *UN 2030 Agenda for Sustainable Development* with big data [23].

Adhering to the TFM, CBAS has exploited a new Big Earth Data-driven research paradigm for SDGs and established a global SDG monitoring and evaluation system to provide data sharing, scientific and technological support, and decision support for related UN agencies and member states, expecting to construct a first-class international scientific research organization.

The five major tasks of CBAS include: developing and constructing a big data platform for SDGs, carrying out scientific research on the monitoring and evaluation of SDG indicators, developing and operating scientific SDG satellites (SDGSATs), establishing think tanks for SDGs promoting scientific and technological innovation, and providing education and training services for developing countries.

To meet the major needs for realizing SDGs, CBAS has constructed a multidisciplinary big data cloud service platform for SDGs on the basis of a CAS Strategic Priority Research Program, the "Big Earth Data Science Engineering Project". With a novel self-designed hyper-converged system architecture, this platform integrates four major subsystems: supercomputing, big data cloud, data storage, and a high-speed network. It has a double-precision floating point supercomputing power of 1,000 trillion operations per second, 50 PB data storage capacity, and 10,000 CPU cores. In addition, this platform is equipped with self-developed core software for big data management, computational analysis, and visualization, where the converged data size reaches 11 PB.

The platform has systematically integrated a variety of geographic, remote sensing, ground monitoring, and social statistical data; coordinated the whole process of "big data storage-management-computational analysis-visualization"; and fused over one hundred dedicated data analysis and AI algorithm tools. Through the unified Chinese-English bilingual portal system, the platform has provided SDG-related research and decisions with diversified core functions, such as the production of required data products, online SDG indicator computing, interactive analysis and decision support, and SDG-dedicated data repositories. Just by gaining access to the internet with a personal computer, scientific researchers can realize terabyte-scale interactive data analysis, produce the data products they require, and perform online computing, visualization, and presentation for SDG indicators.

Meanwhile, the platform has created an SDG application-oriented integrated service environment, operational control and data sharing service management of SDGSAT-1, collaborative analysis functions for SDG indicators, and a world-class visual simulation platform for SDG decision support and comprehensive analysis. By September 2021, the number of users had already exceeded 370,000 (unique IP addresses), covering 174 countries and regions around the globe. Moreover, the number of visits in the sharing system exceeded 60 million, providing over 13 TB of data support and a guarantee for SDG research.

5 Suggestions and Conclusions

Research remains to be done on continuously deepening the use of Big Earth Data to promote the progress of Earth system science and to accurately understand and predict the complex scientific problems in sustainable development.

(1) **Strengthen Big Earth Data infrastructure**

Given the features of Big Earth Data and the typical needs for SDG-related applications, future Big Earth Data infrastructure needs to further condense and abstract typical loads and application models. Moreover, pertinent system design and construction should be done based on benchmark results, and efforts should be made to integrate and schedule services and resources (e.g., high-performance computing, high-throughput computing, intelligent computing and cloud computing). This should be done by focusing on transparent data access and efficient data circulation, thus achieving an integrated software and hardware architecture in full consideration of performance, capacity, and flexibility.

The following should also be done: collect computing and storage resources into a high-speed network infrastructure, provide safe and reliable data storage management services for SDG-related mass data, support a unified standard data interface and a complete authority management mechanism, and precisely satisfy computational analysis needs with different scales and patterns based on dynamic resources and adaptable application features. The foundation platform is required to have a single system mirror image and a single service entry point and provide global research organizations and scholars with a one-stop integrated system for data retrieval, online analysis, remote visualization, and decision support services for SDGs.

(2) **Improve Big Earth Data analysis and application abilities**

Analysis paradigms, algorithms, and models constitute the foundation and core for the application of Big Earth Data. Therefore, follow-up efforts should continuously perfect Big Earth Data analysis paradigms; construct an analytical framework for Big Earth Data; deeply integrate cloud computing, ML algorithms, deep learning algorithms, mathematical methods, and spatial analysis methods with Big Earth Data; and deepen the research on fundamental analysis of Big Earth Data, application analysis modeling, and model optimization. This would facilitate the transformation of Big Earth Data into information and knowledge.

An interdisciplinary, collaborative analysis ecosystem is required to establish the methodology of Big Earth Data and apply it. In the future, intelligent analysis algorithms should be developed and policies and sharing mechanisms should be perfected based on a cloud computing infrastructure. Scientific applications of Big Earth Data should be guided by the evaluation of global SDG indicators, thus forging an ecosystem of open-source exchange of methods and algorithms for Big Earth Data acquisition, processing, analysis, and applications.

(3) **Promote Big Earth Data sharing and knowledge services**

The understanding of Big Earth Data should be further deepened from an engineering angle. Following the principal line of "data-knowledge-service", data engineering should be carried out during the entire life cycle of Big Earth Data to elevate its governance level. Especially in recent decades, the data products of developed countries have played a dominant role in supporting significant international reports and even major decisions. Given this, China should accelerate the R&D of high-quality global public Big Earth Data products under the support of engineering approaches, and provide data solutions with Chinese characteristics for evaluating the progress of the *UN 2030 Agenda for Sustainable Development*, carbon neutrality, and peak carbon emissions.

In the meantime, scientific and technological innovation should be further strengthened, and a shared service pattern should be developed to promote interdisciplinary fusion applications and knowledge discovery. For instance, advanced technologies like AI should be used to enable Big Earth Data by combining data, computing, and services. What is shared under this pattern is not just data but also algorithms, models, and services, thus realizing interdisciplinary data association analysis and fusion applications, and driving significant scientific discovery and decision support.

References

1. Huadong G, Dong L, Fang C et al (2021) Big Earth Data facilitates sustainable development goals. Bull Chin Acad Sci 36(8):874–884. https://doi.org/10.16418/j.issn.1000-3045.20210707006
2. Guo H (2017) Big Earth data: a new frontier in Earth and information sciences. Big Earth Data 1(1–2):4–20. https://doi.org/10.1080/20964471.2017.1403062
3. International Society for Digital Earth (03 Nov 2021) Beijing declaration on Digital Earth (1999). https://www.cas.cn/zt/hyzt/gjszdqhy/ljhy/dyj/200909/t20090903_2463154.html
4. Guo H, Wang C (2013) Digital Earth: fifteen years experience and perspective. Bull Chin Acad Sci z1:59–66. https://doi.org/10.3969/j.issn.1000-3045.2013.Z01.007
5. Guo H, Goodchild MF, Annoni A (2020) Manual of Digital Earth. Springer, Singapore. https://doi.org/10.1007/978-981-32-9915-3
6. Huadong G (2018) Scientific big data—a footstone of national strategy for big data. Bull Chin Acad Sci 33(8):768–773. https://doi.org/10.16418/j.issn.1000-3045.2018.08.001
7. Guo H, Wang L, Chen F et al (2014) Scientific big data and Digital Earth. Chin Sci Bull 59(35):5066–5073. https://doi.org/10.1007/s11434-014-0645-3
8. Huadong G (2014) Big data, big science, big discovery—review of CODATA workshop on big data for international scientific programmes. Bull Chin Acad Sci 29(4):500–506. https://doi.org/10.3969/j.issn.1000-3045.2014.04.014
9. Guo H, Nativi S, Liang D et al (2020) Big earth data science: an information framework for a sustainable planet. Int J Digit Earth 13(7):743–767. https://doi.org/10.1080/17538947.2020.1743785
10. Huadong G (2018) A project on big earth data science engineering. Bull Chin Acad Sci 33(8):818–824. https://doi.org/10.16418/j.issn.1000-3045.2018.08.008
11. Guo H, Liang D, Sun Z et al (2022) Measuring and evaluating SDG indicators with Big Earth Data. Sci Bull 67(17):1792–1801. https://doi.org/10.1016/j.scib.2022.07.015
12. Guo H (2018) Steps to the digital Silk Road. Nature 554:25–27. https://doi.org/10.1038/d41586-018-01303-y
13. United Nations (2015) Transforming our world: the 2030 agenda for sustainable development. United Nations, New York. https://sustainabledevelopment.un.org/content/documents/21252030%20Agenda%20for%20Sustainable%20Development%20web.pdf
14. United Nations (2017) Global indicator framework for the sustainable development goals and targets of the 2030 agenda for sustainable development. United Nations, New York. https://unstats.un.org/sdgs/indicators/Global%20Indicator%20Framework%20after%202020%20review_Eng.pdf
15. United Nations (2019) Global sustainable development report 2019: the future is now—science for achieving sustainable development. United Nations, New York. https://sustainabledevelopment.un.org/content/documents/24797GSDR_report_2019.pdf

16. Guo H, Chen F, Sun Z et al (2021) Big Earth Data: a practice of sustainability science to achieve the sustainable development goals. Sci Bull. https://www.sciencedirect.com/science/article/pii/S2095927321000487. https://doi.org/10.1016/j.scib.2021.01.012
17. Big Earth Data Science Engineering Program (2021) Big Earth Data in support of the sustainable development goals (2021). Chinese Academy of Sciences, Beijing. http://www.cbas.ac.cn/en/publications/reports/202109/U020220401630362820943.pdf
18. Guo H (2020) Big earth data in support of the sustainable development goals (2020): China. Science Press and EDP Sciences, Beijing
19. Li X, Lu Q, Jia X (2021) Harnessing Big Earth Data to facilitate land degradation neutrality goals—practices and prospects. Bull Chin Acad Sci 36(8):896–903https://doi.org/10.16418/j.issn.1000-3045.20210705002
20. Bian J, Li A, Lei G et al (2020) Global high-resolution mountain green cover index mapping based on Landsat images and Google Earth Engine. ISPRS J Photogramm Remote Sens 162:63–76. https://doi.org/10.1016/j.isprsjprs.2020.02.011
21. Guo H (2019) Big earth data in support of the sustainable development goals (2019). Science Press and EDP Sciences, Beijing
22. Guo H (2020) Big earth data in support of the sustainable development goals (2020): the belt and road. Science Press and EDP Sciences, Beijing
23. Guo H, Liang D, Chen F et al (2021) Innovative approaches to the sustainable development goals using Big Earth Data. Big Earth Data 5(3):263–276. https://doi.org/10.1080/20964471.2021.1939989

Guo Huadong is a professor of the Chinese Academy of Sciences (CAS) Aerospace Information Research Institute, and Director-General of the International Research Center of Big Data for Sustainable Development Goals (CBAS). He is an Academician of CAS, a Foreign Member of the Russian Academy of Sciences, a Foreign Member of the Finnish Society of Sciences and Letters, and a Fellow of the World Academy of Sciences (TWAS). He presently serves as Honorary President of the International Society for Digital Earth (ISDE), Director of the International Center on Space Technologies for Natural and Cultural Heritage under the Auspices of UNESCO, Member of the International Science Council (ISC) Global Commission on Science Missions for Sustainability, Chair of the Digital Belt and Road Program (DBAR), and Editor-in-Chief of two scientific journals, *International Journal of Digital Earth and Big Earth* Data. He served on the UN 10-Member Group to support the Technology Facilitation Mechanism for SDGs (2018–2021), was Chairman of the International Committee of Remote Sensing of Environment (2017–2020), and was President of ISDE (2015–2019) and the ICSU Committee on Data for Science and Technology (2010–2014). Prof. Guo specializes in remote sensing, radar for Earth observation, and Digital Earth science. He has published more than 500 papers and 24 books, and is the awardee of 19 international and domestic prizes.

Practical Applications

Information System of China Spallation Neutron Source and High Energy Photon Source and Its Applications

Hesheng Chen, Gang Chen, Fazhi Qi, Hao Hu, Yakang Li, and Fengyao Hou

Abstract The China Spallation Neutron Source (CSNS), the fourth pulse spallation neutron source in the world built during China's 11th Five-Year Plan (2006–2010), has begun its operation after passing the national acceptance test in August 2018. There are three spectrometers built and put into operation and another eight to be built in the Phase I of CSNS (CSNS I), and there are more spectrometers will be built before 2027 in the Phase II of CSNS (CSNS II). The High Energy Photon Source (HEPS), a large-scale scientific facility which is given priority to build during China's 13th Five-Year Plan (2016–2020), will be used for high-throughput synchrotron radiation experiments with ultra-high spatial resolution, ultra-high time resolution and ultra-high energy resolution as a research platform for frontier fundamental research and national strategic research. CSNS and HEPS, as powerful advanced interdisciplinary research platforms for materials science, physics, chemistry, chemical industry, life science, environment and resources, and new energy, will serve experimental users from different institutions domestically and abroad to carry out their interdisciplinary

H. Chen (✉) · G. Chen · F. Qi · H. Hu · Y. Li · F. Hou
Institute of High Energy Physics, Chinese Academy of Sciences (CAS), Beijing 100049, China
e-mail: chenhs@ihep.ac.cn

G. Chen
e-mail: gang.chen@ihep.ac.cn

F. Qi
e-mail: qfz@ihep.ac.cn

H. Hu
e-mail: huhao@ihep.ac.cn

Y. Li
e-mail: liyk@ihep.ac.cn

F. Hou
e-mail: houfy@ihep.ac.cn

H. Chen · G. Chen · F. Qi · Y. Li · F. Hou
Spallation Neutron Source Science Center (SNSSC), Dongguan 523803, China

H. Hu
University of Science and Technology of China, Hefei 230026, China

China's e-Science Blue Book 2023, https://doi.org/10.1007/978-981-99-8270-7_4

experiments. The experimental data of the CSNS and HEPS produced every day will be rather huge and difficult to process promptly. To save and share raw data, analyze and process experimental data, normalize and manage scientific data, it is crucial to set up an information system to operate the facilities, process scientific data efficiently, and support experimental users with good experiences. The information system will provide not only basic services like network, computing, and storage but also data services (transmission, storage, analysis, sharing, etc.), research collaboration, software support (general software and scientific software), information management and information security for researchers, engineers, and users in CSNS and HEPS. The explorations in the development and application of the information system in CSNS and HEPS will help to establish an open and independent scientific software ecology evolving with hardware for fundamental researches, and a solid informatization foundation for the successful operation of large-scale scientific facilities in China.

Keywords China Spallation Neutron Source (CSNS) · High Energy Photon Source (HEPS) · Informatization · Scientific Software · Scientific Data Processing · Data Sharing

1 Introduction

The China Spallation Neutron Source (CSNS) [1] is one of the first major large-scale scientific facilities built during China's 11th Five-Year Plan (2006–2010). The Phase I of CSNS (CSNS I) passed the national acceptance test in August 2018. There are three spectrometers built and put into operation and another eight being built to be put into operation in CSNS I before 2023. There are more spectrometers to be built up in Phase II of CSNS (CSNS II) by 2027 when the power of the accelerator will be upgraded from 100 kW to 500 kW. As the fourth pulsed spallation neutron source built in the world after the U.K., U.S., and Japan, CSNS provides a powerful advanced interdisciplinary research platform for the research of material science, physics, chemistry, chemical industry, life science, resources, environment, and new energy in China.

The High Energy Photon Source (HEPS) [2], as a large-scale scientific facility, had a priority to build for national strategic needs and frontier fundamental research during China's 13th Five-Year Plan (2016–2020). HEPS is the fourth-generation synchrotron radiation light source and located at the northern core area of Huairou Science City in Beijing. It will be the first high-energy synchrotron radiation light source in China and have the highest brightness in the world when it is built up.

As open large-scale interdisciplinary research platforms, CSNS and HEPS can provide researchers domestically and abroad with an interdisciplinary experimental research environment, with various experimental methods, such as neutron scattering, X-ray diffraction, scattering, imaging, and spectroscopy.

Benefited from the latest technological achievements in accelerators, target stations, and spectrometers in the world, the flux of effective pulsed neutron in CSNS I reaches 2.0×10^{16} cm^{-2} s^{-1}. After the upgrade of the beam power from 100 kW to 500 kW in CSNS II, the flux of spallation neutron will reach megawatt which is comparable to the flux of spallation neutron sources in the United States and in Japan. The pulse repetition frequency of CSNS chosen as 25 Hz can greatly improve the flux of effective longwave neutron and the utilization efficiency of each pulsed neutron, which will benefit the researches in biological, chemical macromolecules, and molecular clusters. The spectrometers built in CSNS, such as high throughput powder diffractometer, high resolution powder Diffractometer, small angle neutron scatterer, multi-purpose reflectometer, and high-energy direct-geometric inelastic spectrometer, will meet the different requirements in frontier research fields including life science, material science, nanoscience, physics, and chemistry. As a fourth-generation synchrotron radiation facility, HEPS has a great improvement compared to the existing light sources, and can provide an X-ray with millielectron volt energy resolution, nano spatial resolution, and femtosecond time resolution [2].

With the development of optics technology and electronics technology, advanced optical instruments and advanced detectors are adopted, and the volume of experimental data is increasing explosively in the process of experiments [3]. It is necessary for these huge raw data and meta data to be collected, transmitted, stored, analyzed, and shared effectively and safely, and all kinds of needs of the facility, beamline stations, and experimental users should be met with to promote the output of scientific experiments. The high energy physics in the international has formed effective working modes to develop the scientific research information system in large-scale scientific facilities. For knowledge sharing, the research teams established corresponding software cooperation organizations, such as the particle physics Software Development Foundation (HSF), a framework for future high energy physics experiments (Key4hep), in the process of collaborative development of experimental software framework and software systems around the specific experimental needs of large-scale scientific facilities in high energy physics.

The Mantid project, a scientific data processing software framework for spallation neutron sources, is supported by the Science and Technology Facilities Council (STFC) of the UK and the Department of Energy (DOE) of the US. The project was initiated by ISIS, the spallation neutron sources of UK in 2007, and joined in by the Spallation Neutron Source (SNS) of the Oak Ridge National Laboratory in the U.S. and the High Flux Isotope Reactor (HFIR) in 2020. Mantid, composed of a highly modularized architecture with C++/Python, provides an extensible and high-performance software framework of data operation, data analysis, and visualization for the measurement of neutron scattering and muon spin. As a large-scale international cooperative software development project, the Institute Laue-Langevin (ILL), the Paul Scherrer Institute (PSI), the Australian Nuclear Science and Technology Organization (ANSTO), the European Spallation Source (ESS), and CSNS have also joined in the project of Mantid recently to develop the software framework and made significant progress in data analysis and visualization until now. ESS, the

Swedish Synchrotron Radiation Laboratory MAX IV, and PSI are also developing the scientific data management system, SciCat Data Catalog System, together.

However, a comprehensive but large-scale software framework for advanced light source facilities in Europe and the United States has not still been established. The Photon and Neutron data infrastructure (PaNdata) project, funded by the EU seventh framework program, brings 14 European research facilities together to solve the general problems such as the standards and corresponding tools of data interaction. The research teams of the German synchrotron light facility Petra III at the Deutsches Elektronen-Synchrotron (DESY), the European XFEL (EU-XFEL), and the Linac Coherent Light Source (LCLS) of the United States have developed the online analysis software OnDA together.

The Center for Advanced Mathematics for Energy Research Application (CAMERA) of Lawrence Berkeley National Laboratory focuses on the research requirements of coherent diffraction imaging to develop basic mathematics, new algorithms, codes, and software and applies the achievements obtained to the Advanced Light Source (ALS) and other facilities. In China, a joint research teams from Shanghai Synchrotron Radiation Facility (SSRF), Shanghai Free Electron Laser (SFEL), Hefei Light Source (HLS), Beijing Synchrotron Radiation Facility (BSRF), HEPS and other related light source facilities have established a cooperation mode to develop a software framework for life cycle management and processing of scientific data in order to promote an open, shared, and fully utilized scientific data, so as to mostly adapt and utilize the hardware of large-scale scientific facilities in China,and fully leverage the important role of large-scale scientific facilities in technological innovation.

2 Informatization Challenges of CSNS and HEPS

CSNS and HEPS built with high-energy physics technology, are interdisciplinary research platforms for experimental users who come from different units domestically and abroad. It is necessary to provide experimental users information services to carry out their research conveniently at these facilities while ensuring the stable operation of the neutron source and the light source, and its spectrometers, and beamline stations.

2.1 Storage Challenges of Huge Data

Among of 20 spectrometers planned to build in CSNS, the spectrometers of the Small-Angle Neutron Scattering, Multi-Purpose Reflectometer, and General-Purpose Powder Diffractometer have been built up and put into operation in CSNS I, and will produce over 30 TB experimental raw data every year now and up to 120 TB every year after the power is upgraded to 500 kW. The spectrometers, which are building and will be put into operation successively before the end of 2023, are estimated to produce about 3 TB of raw data every day and up to 12 TB every day after CSNS is upgraded to 500 kW (Table 1). The Energy-Resolved Neutron Imaging Spectrometer will produce 2 TB raw data every day, significantly exceeding the data produced by others.

After the accelerator power is upgraded from 100 to 500 kW, the spectrometers will produce 3–4 times raw data than before, which should be transferred to the central storage system for offline data analysis and long-term storage by the data processing systems and transmission systems.

There are one test beamline and 14 beamlines (B1-BE) in HEPS, which will produce 700 TB of raw data per day on average (the peak volume of the data produced per day can reach 1200 TB) (Table 2). However, the data production rates of different line stations vary greatly. For beamline station B7, the raw data produced is about 10 GB per second, which will increase to 23.6 GB/s in future, and the amount of data produced within the sampling time per day will be 680 TB-1100 TB. The raw data production rates of beamline station B2, BA and BE are about 500 MB/s to 1.5 GB/s, and the amount of data produced within sampling time per day will be between 15 TB and 40 TB. The data production rates of other beamlines are less than 200 MB/s, and the amount of data produced in each beamline within sampling time per day will be less than 4 TB. There is an order of magnitude difference in the data throughput among the three kinds of beamline stations, and the requirements for storage space and data read–write of the beamline station B7 exceed the total of other beamline

Table 1 Raw data rates of spectrometers at CSNS

Instrument name	Average amount of data produced per day (byte) (100 kW)	Average amount of data produced per day (byte) (500 kW)
Energy-resolved neutron imaging instrument (BL13)	2048 G	8192 G
High-pressure neutron diffractometer (BL15)	150 G	600 G
Engineering material neutron diffractometer (BL08A)	150 G	600 G
High-resolution neutron diffractometer (BL09)	50 G	200 G
Very small-angle neutron scattering (BL14)	60 G	240 G
Multi-physical instrument (BL16)	512 G	2048 G
Atmospheric neutron irradiation spectrometer (BL11)	21 G	84 G
High-energy direct-geometric inelastic spectrometer (BL05)	150 G	600 G
Total	3.14 T	12.56 T

Table 2 Raw data rates of beamline stations at HEPS

Line station	Average amount of data produced per day (byte)	Peak amount of data produced per day (byte)
B1	820 G	4 T
B2	14 T	20 T
B3	112.5 G	1.4 T
B4	2 T	2 T
B5	6.8 G	10 G
B6	20 G	50 G
B7	680 T	1100 T
B8	10.32 G	30 G
B9	5 G	10 G
BA	35 T	35 T
BB	91.08 G	165.6 G
BC	1 T	1 T
BD	275 M	500 M
BE	11.2 T	25 T
Total	744.27 T	1188.67 T

stations. To solve these problems, these raw data will be collected by pixel blocks of detectors and written into the storage system by multiple streams.

2.2 *Challenges of Real-Time Data Processing*

The scientific data produced on CSNS and HEPS need to be processed, analyzed, stored, and shared in a timely and rapid way, in order to provide experimental users rapid feedback on the results of real-time data analysis to make their decision or guide the experimental process or guide the experimental parameters.

From the perspective of facility scientists, an online analysis environment that requires sufficient data storage resources and different computing sources (CPU, GPU, FPGA, etc.) are necessary to operate and maintain the facilities. For some beamlines with relatively high data production rates like the beamline B7 having high-resolution area detectors at HEPS, it requires a high-performance scientific data processing platform, a fusion environment of data, computing, network, and software, to realize the real-time data processing, feedback, and utilization of experimental data in research.

From the perspective of experimental users, a fast, convenient, easy-to-use, and friendly environment is necessary for them to complete the analysis of experimental data, obtain ideal research results in a short time, track, access, and download experimental data and relative results, and ensure data security, data integrity, and long-term storage. For some experiments, due to the huge amount of data (for example, diffraction tomography experiments, each experiment will produce up to 54 TB of data), a high-performance network is required to support the rapid data transmission, meanwhile, a powerful computing platform and software environment are required to process and analyze the data rapidly.

2.3 Challenges of Experimental Data Management

To deal with user experiments in CSNS and HEPS, it is necessary to process a large amount of experimental data from different experimental groups domestically and abroad, and these data with different sizes (KB~GB), different formats (ASCII, image, HDF5), and different generating sources (different experimental technologies, different beamlines, different front ends, etc.) should be unified and managed. The management of scientific databases (experimental database, metadata database, database of instrument and equipment, etc.), data management policy, data standardization and data interfaces will powerfully support and guarantee the utilization and sharing of scientific data.

The beam power will be increased significantly in CSNS II, and more than 90 beamlines will be built up and put into operation in HEPS. Then the experimental data produced per day will increase rapidly, as will bring more pressure on the storage, analysis, management, and sharing of these scientific data. Therefore, the scalability of the architecture and technology of scientific data processing system should be considered when planning.

HEPS is a synchrotron radiation facility with ultrahigh brightness, multi-beamline stations, and a large volume of experimental data. However, the corresponding software framework domestically and abroad is still immature and developing rapidly. Therefore, it is more urgent to vigorously develop the corresponding essential software framework for the management and analysis of scientific data.

2.4 Challenges of User Service

As open interdisciplinary research platforms for users all over the world, CSNS and HEPS will provide more than 5000 hours beamtime every year on average. Many users from around the world submit proposals, apply for beamtime, do their experiments, obtain and analyze experimental data, and upload their research results in CSNS and HEPS every year.

With more spectrometers built and put into operation in CSNS II, the number of scientific experiments, the number of users, and the scale of data will grow explosively and bring great challenges to the information system for comprehensiveness service, such as service scope, service process, data relevance, user convenience, service internationalization, and other aspects. A user service system, with which users can carry out their experiments after their proposals are reviewed by experts, is required to deal with the processes of user registration, proposal submission, beamtime application, etc., and provide users with a clear and procedural experience. Meanwhile, the system will also carry out a series of management processes of user security training, schedule arrangement, experiment data, publication collection, and user feedback. So, it can improve the efficiency of application and utilization of beamlines, enhance the operational efficiency of facilities, and accelerate the output of scientific achievements.

3 Information System for User

As comprehensive interdisciplinary research platforms and major scientific and technological infrastructures in China, the operation status of experimental devices, the processing efficiency of scientific data, and the users' experience of scientific experiments in CSNS and HEPS are very important. The information system will provide CSNS and HEPS facilities, line station scientists, engineering technicians, and users with network, computing, storage, and other information infrastructure and information services including data transmission, data storage, data analysis, data sharing, and scientific research cooperation, as well as scientific software, general software, general information system, network information security, data normalization and interface standardization between different systems.

For the information system, it is necessary to focus on the users of scientific experiments, meet the requirements of research and management of scientists, build the information infrastructure and information services system, and provide corresponding integrated information comprehensive services for scientific research throughout the whole stage of the scientific experiment including the early preparation stage, the experimental operation stage and the post-experimental stage, to realize the management of the whole life cycle of scientific experiment data (Fig. 1), provide multi-dimensional and process-oriented information support and technical services for CSNS and HEPS, and automate the scientific experiment processes including experimental data analysis and sharing, scientific achievements release, scientific research project management, and other scientific research activities.

From the perspective of scientific experimental users, the system can realize the informatization, automation, and conveniences of the whole process of user experiments. From the perspective of facility operation, it can realize the digitization and intellectualization of the experimental operation process, and ensure the efficient, reliable, and safe operation of facilities. From the perspective of information technology, it can also provide the hardware architecture heterogeneity, software function modularization, functional interface, and data management standardization of the information systems and its applications. From a comprehensive perspective, it can realize the whole life cycle management of scientific experiment process

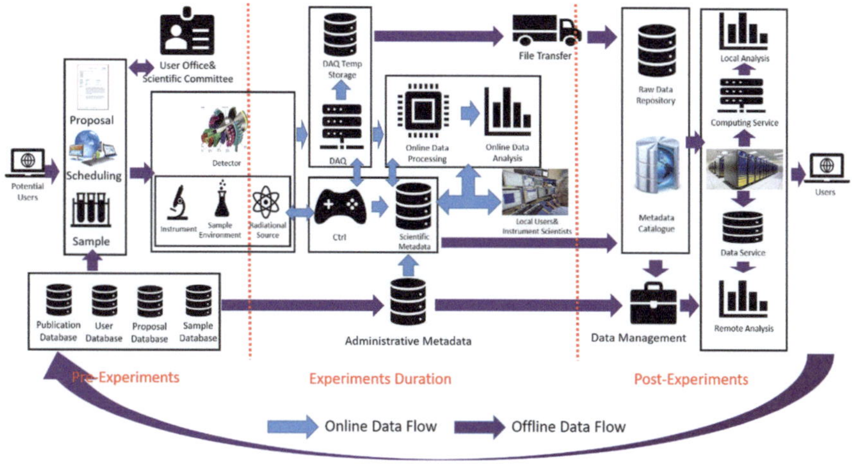

Fig. 1 Experiment and science data lifecycle management

and scientific data in CSNS and HEPS, promote the effective output of research achievements, deepen interdisciplinary and research cooperation, and promote the development of physics, chemistry, life science, material science and technology, resources and environment, etc.

3.1 *Technological Architecture of Information System and Its Implementation*

According to the functions of the informatization system, the system includes scientific data storage system, scientific data processing software system, scientific data management system, user comprehensive service system, the systems supporting the operation and maintenance of information infrastructure, such as network system, storage system, computing system, public information system, network security system, intelligent operation system, etc., as well as open source software, such as OpenStack, Lustre, SLURM, MySQL, Open LDAP, CASTOR, CernVM File System. The informatization system provides users with secure and reliable "one-stop" and "full-process" information services, a series of high-reliability and high-availability database clusters, which provides an external database-as-a-service with a unified database access service and a series of comprehensive informatization services for scientific experiments and daily office routines to carry out experiments safely and conveniently.

3.2 Scientific Data Processing Software System

The scientific data processing software system is an important component of the scientific data processing platform of CSNS and HEPS. Based on the services of data storage, data access, computing resources, network environment provided by the platform, the software system is used to generate scientific data with clear physical significance from the measurement data collected from the experiments, and obtain the scientific understanding of the experimental objects by analyzing these scientific data. The scientific data processing software system of CSNS and HEPS is designed to support research models in a different domain, and ensure that the models, scientific data processing software framework, and the scientific data processing platform of CSNS and HEPS can be upgraded iteratively without impacting on each other.

Due to the diversity of experimental users, the differences in equipment and instruments, and the richness in methodologies, it is decided that the development of scientific data processing software should be driven by solving and supporting specific scientific applications, that is the Domain Driven Design (DDD). DDD requires that the development forms the domain model first by sort out specific domain knowledge and then design and realize the scientific data processing software.

There are two typical operating environments for data processing software, (1) Online real-time data processing environment, which will process the collected data in real-time using limited online computing sources and send the results to line station scientists and experimental users to judge the experimental progress and make a dynamic adjustment for the subsequent experiment while the experiment is being executed, and (2) Offline data processing environment, which will make detailed, comprehensive, and in-depth processing and analysis of the collected data to draw some conclusion as the evidence of the experimental users' scientific discoveries by using users' own computing resources or the resources of computing center in IHEP (IHEPCC). Due to the development of computing technology and experimental methodology, many data that used to require offline processing can now be processed in real-time. Meanwhile, some technologies and methods in modern experiments, such as the high-throughput experimental methods in materials science, genomics, and other fields, and the computed X-ray tomography method in the imaging beamline stations, will produce a large amount of measurement data in short time, which leads a result that offline data processing also needs to learn from of online data processing systems in automation and integrated computing. Therefore, the scientific data processing software system in HEPS and CSNS is designed to share computing components to the utmost extent between online and offline data processing systems. The integration design of online and offline software systems can not only reduce the workload of development and maintenance but also provide users with a unified operation experience and lower learning costs.

To achieve the goal above, the scientific data processing software system is divided into four components, (1) Software Framework, (2) Visualization and Data Analysis Software, (3) Distributed Middleware based on Microservices, and (4) The Third-party Software Library and External Program Integration. Since the operation of

CSNS and HEPS will last for decades, during which the instrument technologies, experimental methods, and computing hardware will change greatly, the overall design of the software system must consider the lowest coupling between computing components and the scalability of the computing platform. Therefore, the design of the system should follow the following principles, (1) interface-oriented system design, the interface is the protocol of interaction between computing components, and the flexibility of the computing system can be ensured by defining interfaces rather than binding implementations, and (2) plug-in development mode, the software platform expands the functions of the system through plug-in which completes specific functions according to the interface definition as a computing component.

3.2.1 Software Framework

DAISY, a scientific data processing software framework [4] (Fig. 2), is developed based on SNiPER [5], and will provide general functional components required for scientific computing in CSNS and HEPS. Due to a data-centric interface-oriented software architecture, C++ development environment and a Python interface, DAISY, providing the efficiency of computing and the flexibility of software framework, has been applied to the data processing system of Jiangmen Neutrino Experiment, CSNS [6] and Large High Altitude Air Shower Observatory (LHAASO). The two most important components of the framework are the transient data object and algorithm object. The transient data object is a data object stored in memory during software operation in a tree structure, and can be directly accessed by other components of the framework through the path name.

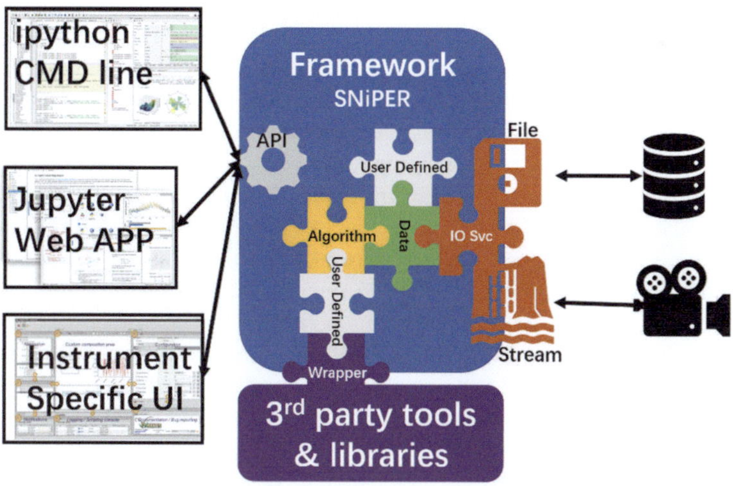

Fig. 2 Software framework

Since a large number of scientific computing programs are developed based on the Python scientific computing library NumPy now, and ndarray, the basic data object of NumPy, can be easily called by C++. The basic transient data object of HEPS also adopts the ndarray, and users can expand custom data objects on this basis. The algorithm object is a specific computing component that performs data processing, and the domain model is packaged in a specific algorithm implemented by Python or C++. The input and output of algorithm object are generally specific transient data objects, and the interface of algorithm objects supports framework implementation, so that the framework can call algorithm through the interface to realize the decoupling between the algorithm and the framework.

Scientific data processing software in HEPS adopts two kinds of data for input and output. one is File type. Through the specified file path, the software can read or write the file which has a definite size and boundary (the label of file header and file end). The other is Data Stream type. The software obtains data through the specified port, and the data stream has no clear boundary. The software has to support reading and writing in HDF5 format and other formats required by beamline stations. Therefore, the input and output module of the software should have a plug-in architecture, by which developers flexibly replace the data source and convert it into the specified transient data object, and realize the decoupling of external data source and transient data object. The software framework provides the Application Programming Interface (API) of C++ and Python. Other applications can access all functions of the software through the API.

3.2.2 Visualization and Data Analysis Software

Visualization and data analysis software is an important application built in the software framework (Fig. 3). Due to the wide application of Python in the field of data processing, it is implemented by Python, and its functions are accessed through the Python API provided by the framework. It provides three kinds of user interfaces, Script Interface based on ipython, Graphical User Interface (GUI) based on matplotlib and PyQt, and Web Interface based on Jupiter. For the general GUI interface, a variety of common graphical widgets will be developed, including the project explorer, file browsing, data browsing, data viewing, dataset plotting, algorithm browsing, log and console. The combination and arrangement of these widgets are defined through Qt Modeling Language (QML) by Qt, to provide their unique user interface for each line station and application. The traditional interactive data processing software follows the Read-Evaluate-Print Loop (REPL) programming model. All operations are executed circularly in the same single process or thread in this model, so that the data analysis algorithm usually runs too long time due to the large amount of light source data to cause a running failure with no response temporarily. Therefore, the front-end user interface and the execution module belong to different processes, and communicate through ZeroMQ and Jupiter Message. Also, this programing model has good expansibility and adapts to the migration from a stand-alone program to distributed parallel program.

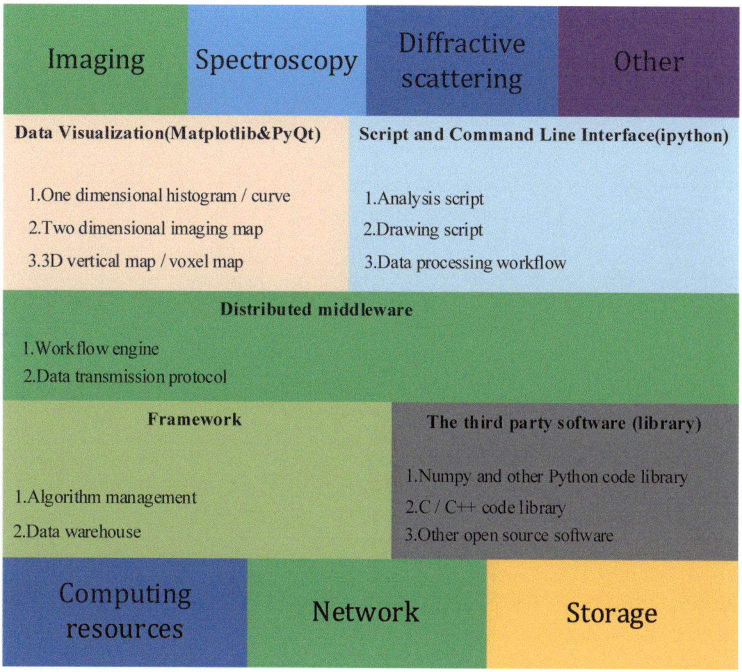

Fig. 3 Software for visualization and analysis

3.2.3 Distributed Middleware Based on Microservices

Data processing software in CSNS and HEPS needs to run in different computing environments, including online real-time processing environments, offline high-performance computing environments, and personal computer environments. However, it is not expected that beamline station scientists focus on configuring and managing computing resources so that the computing resources required by experiments can be automatically scaled according to needs. Microservice architecture (Fig. 4) is a better solution to this problem. Microservice is an accessible application with independent functions, which is usually deployed in the form of the container in the heterogeneous distributed computing environment provided by a scientific data analysis system. Microservice gateway is the entrance and reverse proxy of all microservices. The call of application request first reaches the gateway, then is redirected to a specific microservice according to the service route by the gateway. According to the business application and initial parameters submitted by the users, the workflow engine generates a microservice call sequence, i.e., workflow or pipeline, into which the data are input from the data source, and produces a result output finally after a series of micro-service calls. A distributed middleware based on microservice can shield the implementation details of the distributed environment through the microservice gateway, and meet the adaptation of different

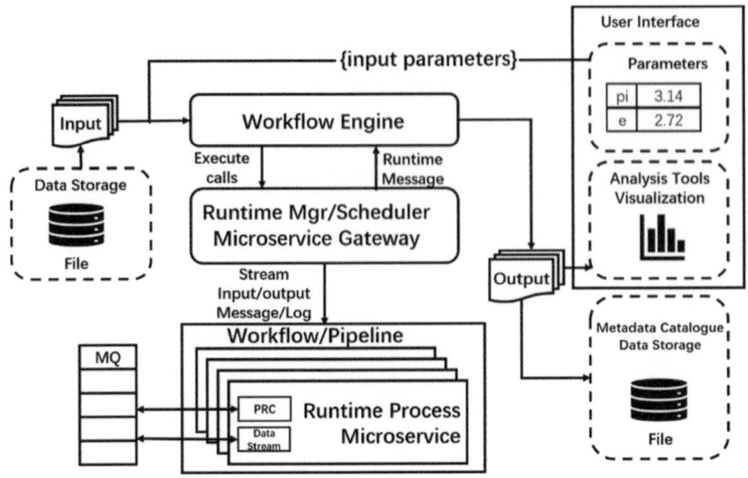

Fig. 4 Workflow based on microservice

computing environments from a personal computer to a computing cluster without modifying the program code. At the same time, the domain model is controlled inside the boundary of microservices. Different microservices interact through lightweight communication protocols, which can better realize the high-cohesion and low-coupling computing components. Message Queue (MQ) is used to realize the communication between microservices. Because scientific data processing software needs to transmit a large amount of data, two types of messages are distinguished, one is the Remote Procedure Call (RPC) message with control instructions, and the other is the data flow generated after data object serialization. The former has a small amount of data but requires a timely response, JSON data object expression conforming to Jupiter Message standard is adopted. The latter focuses on throughput. The data format is defined by the data sender and receiver, and the serialization and deserialization of transient data objects are realized by themselves.

3.2.4 The Third-Party Software Library and External Program Integration

It is still important work to inherit the legacy code accumulated by predecessors while developing advanced software architecture on the latest computing system. The third-party software and external programs can be integrated into the data processing software system in two ways. One is to package the external program through the algorithm, provide the algorithm interface defined by the framework, and complete the conversion of data objects between frame standard and external program inside the algorithm. The other is to package external programs through microservices. For the former, the external program can directly call all the functions of the framework through the algorithm, with full integration. The data object is directly accessed and managed through the transient data object mechanism of the framework, with high efficiency. However, this integration way requires an in-depth understanding of the internal implementation of the framework and external program, which has a large workload, so it is a better choice for the integration of scientific computing components. For the latter, external programs are packaged into microservices, deployed, and run independently, and they interact with other components through MQ. Although the integration is not as good as the first way and requires the support of the distributed environment, it is relatively easy to implement, which is a better choice for other systems to access scientific data processing software systems.

Based on the requirement of providing convenient and easy-to-use services for scientific users, the software framework will be used to run the functions of the equipment control system, the data acquisition system and the data processing software system, and user interaction functions in the online data processing environment, so that the data processing software can access the real-time collected measurement data through the data acquisition system, obtain the quick-view results after rapid data processing, and provide feedback to the control system for next experimental plan. All of those constitute a closed-loop process of data acquisition, data analysis, and control feedback. Therefore, it is necessary to develop an integrated environment as a unified interface for experimental users to access the data acquisition system, the online data processing system, and the control system.

The interaction between an integrated environment and a software system can be divided into two levels. The first level is the visualization and data analysis software, as a subsystem of the integrated environment, which provides general analysis interfaces and functions, such as file browsing, data browsing, data viewing, dataset plotting, algorithm browsing, and ipython terminal. Since the visualization and data analysis software adopts the front-end user interface and the execution module, which belong to different computer processes, the integrated environment can configure the analysis interface components by using the Qt running environment and loading the required Widget through QML, to realize the online quick-view analysis interface customized by each line station. The execution module runs the data analysis algorithm in the background and interacts through MQ. The second level is the integrated environment, as a client, sends an access request to the online data processing system by calling the API opened by data processing software system in order to obtain a specific microservice response, or sends a business application to the workflow engine in order to start the workflow of required data processing.

3.3 Scientific Data Management System

The huge scientific data of CSNS and HEPS needs standardized management, and the standards and specifications of scientific data management need to be formulated.

Through the scientific data management system, the scientific data produced by the experiment can be managed and tracked in the whole life cycle at all stages of data acquisition, transmission, storage, analysis, and achievement release. At different stages of the experiment, data and metadata are obtained from the control system, user service system, and data analysis system, the scientific metadata catalog management architecture is designed to manage the whole life cycle of scientific data, the integrity and traceability of scientific data are ensured, the standard interfaces are provided, the collaboration and communication between other systems are met, and the efficient and convenient user data services are provided, therefore, the data can be viewed, downloaded, shared and utilized under the data standards and specifications.

The diagram of data flow for the scientific data management system (Fig. 5) describes all the processes of data flow in the whole experimental process. On the one hand, the raw data and scientific metadata received from the spectrometers or beamline stations are stored in the data center in the format of files. On the other hand, the analysis results obtained from the raw data analyzed through the online analysis system are also saved in the data center. If the raw data file obtained from the spectrometer or beamline station are not in the standard HDF5 format, it is necessary to convert the data format and encapsulate the data (including the raw data and all metadata), form the standard HDF5 file, register the metadata, and save it for long time. At the same time, the metadata extractor gets the relevant information of proposals, users, and samples from the user service system, obtains some key experimental metadata from the experimental parameter files, and stores them in the metadata directory database for the query, search, and sharing of experimental data. Through the data service portal, users can search, view, download, and analyze the data offline, meanwhile, the result data obtained by offline computation and analysis will also be stored in the central storage file system for long time.

Opening and sharing of scientific data have become an inevitable trend in international scientific research, relevant policies have been released successively in European and American countries. At present, there is still a lack of relevant national

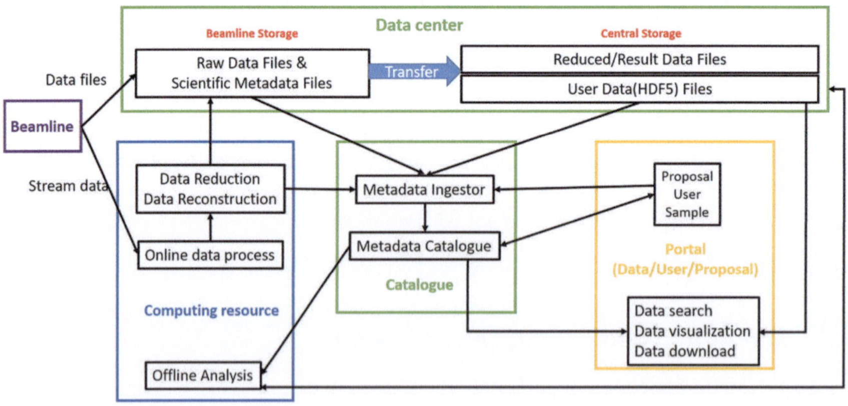

Fig. 5 The data flow for data management system

laws and regulations to clearly explain the ownership and use of the right of scientific data of user facilities. Therefore, the formulation of scientific data policy [7] is of great significance to the implementation of CSNS and HEPS scientific data management and the opening and sharing of scientific data in future. To further strengthen and standardize scientific data management, ensure the safety of scientific data, and improve the level of opening and sharing, the General Office of the State Council of the People's Republic of China has issued the Scientific Data Management Measures [8]. The scientific data policy is the formulation and specification of the whole data management process. The scientific data management of CSNS and HEPS must comply with the provisions of the data policy.

The scientific data policy of CSNS and HEPS includes the constraints and instructions on the ownership, management, storage, and access of all data and metadata obtained by the facility. For the storage of the data produced in CSNS and HEPS, disk storage for at least three months and tape storage for permanent will be required for long-term storage of raw data, result data, calibration data and user data, and short-term storage of reduced data and processed data. Each data set is allocated a unique Permanent Identifier (PID), and any published article related to the data set must reference the PID. The embargo period of data is established, such as three years, and only authorized users have data ownership during the period. After the period, the data can be provided to other legitimate users for access and download.

3.4 Scientific Data Analysis System

The beamtime of CSNS and HEPS are so valuable that it is expected to complete multiple experiments as soon as possible within the given beamtime and obtain accurate and satisfactory experimental results. This requires the information system to provide fast and high-performance online data processing so that users can judge whether the completed experiments meet expectations, and provide a judgment for the parameter adjustment of the next experiment. Therefore, it is very important to establish an efficient scientific data analysis system to improve the efficiency of experiments and accelerate data processing and data sharing.

The scientific data analysis system is built to meet the scientific research requirements of CSNS and HEPS experiments, and provides two main functions, online data processing service and offline data processing service. Figure 6 shows the data flow process in HEPS experimental including data production, data processing, storage, and transmission. The system will provide a large-scale heterogeneous computing system integrating CPU, GPU, FPGA, and other types of computing resources, which will provide corresponding computing service solutions to ensure the efficient execution of scientific computing tasks according to different experiments and usage patterns of different users.

The real-time and fast online data analysis system can allocate computing resources according to the amount of experimental data and computational complexity, to meet the requirements of different experimental stations, provide

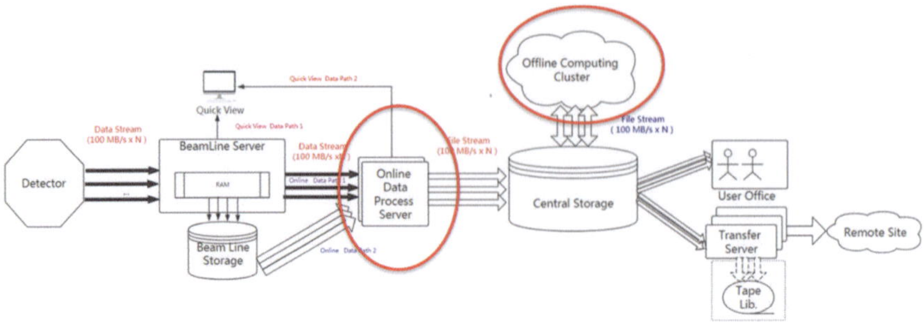

Fig. 6 Data flow for experiments

customized high-performance data analysis services for users, and give feedback on the data processing results in real-time so that users can easily and quickly obtain the preliminary experimental results. After the experiment, some users also need large-scale computing resources for offline data analysis and further analysis of the experimental data to obtain better results. The system architecture, shown in Fig. 7, provides four computing service modes, including stream computing based on CPU, data analysis based on WEB, cloud hosting analysis, and batch job analysis, to meet the requirements of online and offline computing services.

(1) Stream Computing based on CPU

The data stream from the detector directly enters the stream computing cluster based on the CPU provided by the online computing system (Fig. 8) and is quickly displayed after data processing. The key technologies involved are Spark streaming computing based on memory computing and the reception and distribution of MQ.

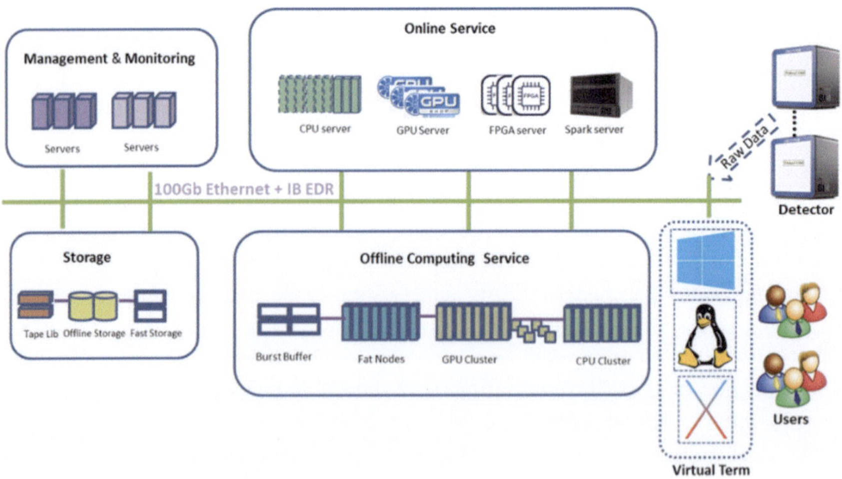

Fig. 7 The struct of data analysis system

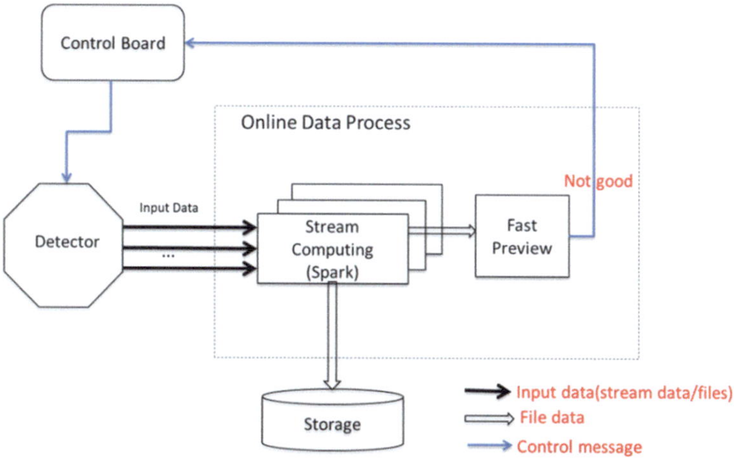

Fig. 8 Streaming computing mode

This computing mode is suitable for beamline stations with a large amount of data and high real-time requirements of data processing.

(2) Data Analysis based on WEB

The computing mode is composed of the user front-end layer, the middleware software layer, and the underlying resource management layer. Based on Jupyter Notebook, the user front-end layer provides a data analysis interface for multi-user access, development Jupyter Notebook of data analysis for different beamline stations, and fixed data analysis process. Based on Jupyter Hub, the middleware software layer provides user authentication and authorization, application of computing resources, and management of Jupyter Notebook. Based on Kubernetes, the underlying resource management layer provides dynamic creation, management, and scheduling of computing resource according to the requirements of middleware software. The system, which is suitable for lightweight data processing of small-and-medium beamline stations, supports one-click data analysis and visualization, and flexible customization of analysis steps by users.

(3) Cloud Hosting Analysis

The underlying computing resources of cloud hosting analysis system, which are composed of heterogeneous resources such as CPU and GPU and transparent to user, are managed and scheduled by cloud software (Fig. 9). According to the data analysis requirements of different beamline stations, different data processing environments, including integrated user identity information, experimental data storage systems, and software depository, are created, and corresponding experimental data access is integrated. CVMFS depository is used to manage the update and maintenance of the analysis environment and the cloud analysis software. The system supports

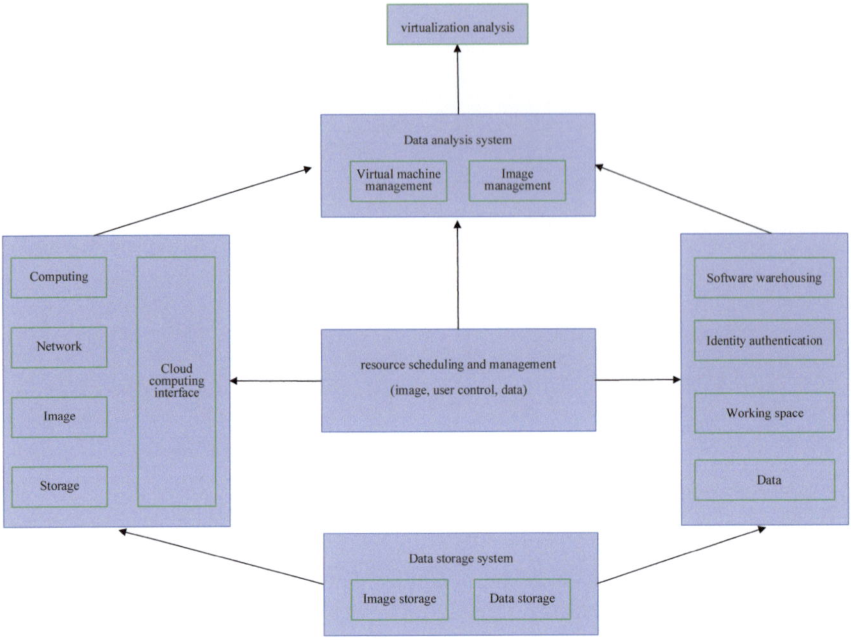

Fig. 9 Cloud computing mode

on-demand requests for computing resources, high-security isolation, computing-intensive data analysis, and more complex data visualization services.

(4) Batch Job Processing

Batch job processing system are composed of a high-performance cluster with CPU and GPU nodes, which are managed and scheduled by SLURM, and provides an interface to submit batch job processing for users to conduct large-scale offline data processing and analysis.

All available computing resources will be integrated, uniformly managed, and allocated by the scientific data analysis system (Fig. 10), and the system can quickly schedule and allocate online and offline computing resources. According to the proposal information of users, the computing resources required for the experiment and the data processing environment (including operating system, software, data, etc.) will be figured out and the corresponding computing cluster/cloud hosting can be generated to match the hardware resources such as CPU/GPU/FPGA. The scheduling and allocation of the whole resources and the details of the underlying hardware are transparent to users. Users do not need to spend time deploying the data processing environment, which greatly improves the work efficiency of scientists.

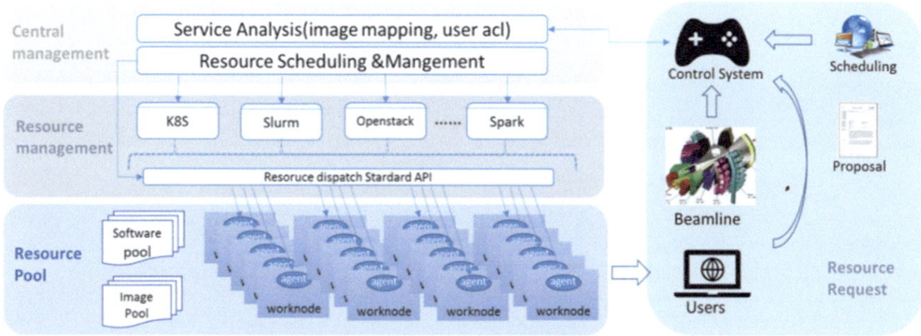

Fig. 10 Computing resources scheduling and management

The system will develop the upper scheduling management software based on virtualization technology, reduce the coupling degree between application and infrastructure through the technology of virtual machine and container, provide flexible computing services, meet different experimental and application needs, and maximize resource utilization.

3.5 User Integrated Service System

User integrated service system in CSNS and HPES is an open integrated information service platform, which provides users all over the world with concise and fast services including user registration, proposal submission, beamtime application, and more. At the same time, the system will carry out a series of management processes for user security training, schedule arrangement, user experimental data and experimental achievements, provide users with a clear and procedural experience, accept user feedback, improve the efficiency of experimental application and usage, enhance the operational efficiency of the facility, and accelerate the output of scientific achievements.

The public service system mainly realizes the following four management procedures, the user service procedure, the experimental process management procedure, the facility operation status management procedure, and the data sharing management procedure, to realize the business procedure centered on "people", the scientific research procedure centered on "data", the management procedure centered on "project", and the operation and maintenance procedure centered on "facility". The user service procedure includes a life cycle service process for facility users from proposal application, beamtime allocation, experimental process, and data service to result in feedback (Fig. 11). The experimental process management procedure provides user management, sample management, experimental process management and experimental result analysis for spectrometer, experiment users, and administrators of beamline station. The facility operation status management procedure provides facility operation status and operation analysis for experimental users and operation

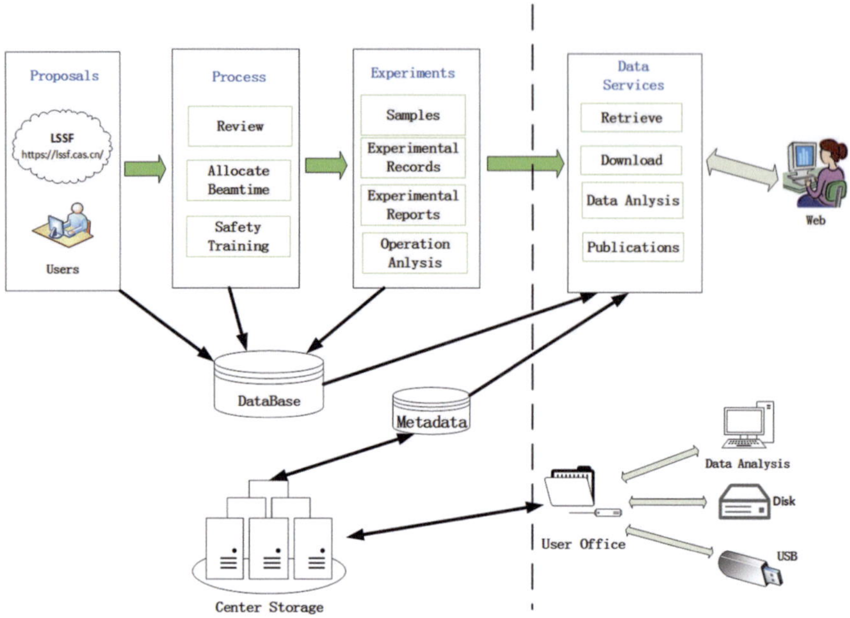

Fig. 11 Data flow for public services

managers of spectrometer or beamline station. The data sharing management procedure provides a data management process for data administrators to allocate data sharing according to the data sharing strategy and analyze the data usage.

User integrated service system is a Web-based service system, which takes unified authentication to authenticates and authorizes users to access the system by OAuth and LDAP. The implementation of the system adopts the technical architecture of SpringBoot + JPA + Thymeleaf, which is divided into six layers. The bottom layer is data layer which places relational databases and files. The entity layer uses the JPA framework to map the relationships between tables in the database into entity classes and configuration files, and all operations on the database are carried out to these entity classes. The next layer is the data access layer, and all operations on entity classes are carried out in the files of this layer. The service layer calls the data access layer to complete the business logic of the system. In this layer, the transaction management of the SpringBoot framework is used to ensure the integrity and consistency of data. The control layer processes the page logic and judges the jump relationship requested by the user, and records the user's operation log in this layer. The view layer is oriented for users and is the input and output interface of the control layer. The technologies of JQuery, Thymeleaf, and Ajax are used to dynamically form pages to provide services to users.

4 The Future of Information Systems in Large-Scale Scientific Facilities

As an important support and service system for carrying out scientific research and scientific experiments relying on national major scientific and technological infrastructures, the information system faces great challenges in terms of functionality, performance, ease of use, and other aspects.

With the rapid progress of accelerator technology and detector technology, the scientific data produced on CSNS and HEPS and the requirements for scientific data and computing systems will constantly increase. To meet the functional requirements, a modular and scalable technology architecture is adopted and the performance and function of the system should be continuously improved according to the increase in data and the change in scientific research activity mode.

The development of new computing architectures including GPU, ARM, and FPGA, as well as advanced computing technologies such as cloud computing and artificial intelligence, provides new opportunities for the application of data acquisition, data trigger selection, data real-time online processing, and offline processing on CSNS and HEPS. Through the exploration and research of advanced technologies such as computable storage and artificial intelligence, cooperation and communications of scientific research between institutes and enterprises in key and frontier technologies are strengthened, and the efficiency of scientific data processing is further improved.

As a fundamental system framework software of CSNS and HEPS connecting large-scale scientific facilities, scientific data computing platforms, and scientific experimental users, the information system software provides an important guarantee for the stable operation of large-scale scientific facilities and the timely processing and analysis of scientific data, and promotes scientific research achievement and technological innovation effectively.

At present, there are several major scientific and technological infrastructures of light sources (including spallation neutron source and free electron laser) in China, which are operating, building, or planning to build, such as Beijing Synchrotron Radiation Facility (BSRF), Hefei Light Source (HLS), Dalian Coherent Light Source (DCLS), Shanghai Synchrotron Radiation Facility (SSRF), CSNS, HEPS, Shanghai Free Electron Laser (SFEL), Hefei Advanced Light Facility (HALF), Southern Advanced Photon Source (SAPS, in pre-research),, and so on. There are similar requirements and challenges in scientific data processing and certain commonalities in technology. By now, a consensus on cooperative development the framework of basic scientific software among the research teams from the domestic major scientific infrastructure of light source has been spontaneously formed, a cooperative research and development project team has been established, and extensive cooperation in this field is expected to be built.

Through the exploration and practice of information system software on CSNS and HEPS, the cooperative development mode of basic scientific software on large-scale scientific facilities in China has been gradually formed, a national cooperative development organization of basic scientific software has been established, and corresponding software standards and technical specifications will be also set up, the advanced computing architecture and computing technology will be fully utilized to form an open and independent basic scientific software ecology with coordinated development of software and hardware, and a solid information foundation for the successful operation of large-scale scientific facilities in China will be strengthened then.

References

1. Chen HS (2016) China spallation neutron source. Modern Phys 28(01):3–10
2. Jiang XM, Wang JQ, Qin Q et al (2014) Chinese high energy photon source and the test facility. SCIENTIA SINICA Physica, Mechanica & Astronomica 44(10):1075–1094
3. Vesel S, Schwarz N, Schmitz C (2018) APS data management system. J Synchrotron Radiat 25:1574–1580
4. Hu H, Qi FZ, Zhang HM et al (2021) The design of a data management system at HEPS. J Synchrotron Radiat 28:169–175
5. Zou JH, Huang XT, Li WD et al (2015) SNiPER: an offline software framework for non-collider physics experiments. J Phys: Conf Ser 664(7):072053
6. Tian HL, Zhang JR, Yan LL et al (2016) Distributed data processing and analysis environment for neutron scattering experiments at CSNS. Nucl Instrum Methods Phys Res, Sect A 834(21):24–29
7. Wang C, Steiner U, Sepe A (2018) Synchrotron big data science. Small 14(46):e1802291
8. General Office of State Council of the People's Republic of China (2018) Notice on printing the measures for the management of scientific data [EB/OL]. 17 Mar 2018. http://www.gov.cn/zhengce/content/2018-04/02/content_5279272.htm

Hesheng Chen born in 1946, particle physicist, member of the Chinese Academy of Sciences, professor of the Institute of High Energy Physics, director of National Laboratory of the Beijing Positive-positron Collider (BEPC), the chief director of the project of China Spallation Neutron Source (CSNS).

Gang Chen born in 1961, professor of the Institute of High Energy Physics, and director of the National High Energy Physics Science Data Center (NHEPSDC).

Fazhi Qi born in 1978, male, Ph.D., Head and researcher of Computing Center, Institute of High Energy Physics, the leader of the HEPS and CSNS computing and networking group.

Hao Hu born in 1981, female, M.D., engineer of Computing Center, Institute of High Energy Physics, responsible for the HEPS scientific data management system.

Yakang Li born in 1987, M.D., senior engineer of Dongguan Campus, Institute of High Energy Physics, works for the CSNS Computing and Communication system.

Fengyao Hou born in 1978, male, Ph.D., senior engineer of Dongguan Campus, Institute of High Energy Physics, responsible for the CSNS high-performance computing environment development and performance optimization.

The Construction and Application of Data System of LHAASO Project

Zhen Cao, Zhiguo Yao, Minhao Gu, and Yaodong Cheng

Abstract The Large High Altitude Air Shower Observatory (LHAASO) is a major national science and technology infrastructure project. The project collects trillions of cosmic ray events every year, generating about 10 PB of data annually, providing valuable scientific data resources for physicists all over the world to explore the origin of high-energy cosmic rays, the related evolution of high-energy celestial bodies, and search for dark matter. In this paper, we firstly give a brief introduction to the LHAASO experiment including its detectors, and then explain the LHAASO data system in detail, including data acquisition system, data processing platform, data processing software, and scientific applications. Finally, we summarize the design and construction of LHAASO data system. It is expected that the experience could be useful to similar projects in future.

Keywords LHAASO · Cosmic ray · Data acquisition · Data processing · Data storage and management

1 Overview

1.1 Introduction

The Large High Altitude Air Shower Observatory (LHAASO) [1] is a major national science and technology infrastructure project, which is listed as one of the 16 high-priority projects in the National Medium-and-Long-Term Construction Plan for Major Science and Technology Infrastructure (2012–2030) issued by the State

Z. Cao · Z. Yao · M. Gu · Y. Cheng (✉)
Institute of High Energy Physics, Chinese Academy of Sciences, Beijing 100049, China
e-mail: chyd@ihep.ac.cn

Z. Cao
e-mail: caozh@ihep.ac.cn

Z. Cao · Z. Yao · Y. Cheng
University of Chinese Academy of Sciences, Beijing 101408, China

China's e-Science Blue Book 2023, https://doi.org/10.1007/978-981-99-8270-7_5

Council of China. The project was approved by the government in December, 2015, and started to construct in November, 2017. It launched its first quarter of detectors on April 26, 2019, and finished construction of all the detectors in July, 2021. LHAASO was put into operation while it was under construction, and it discovered a dozen pevatrons and photons exceeding 1 PeV, including one at 1.4 PeV, which is the highest energy photon ever observed [2, 2].

The main scientific objectives of LHAASO are to search for the origin of high energy cosmic rays, to study the evolution of the universe and high energy celestial bodies, and to push forward the frontier of new physics. The particular scientific objectives of LHAASO are given as follows:

(1) To explore the origin of high-energy cosmic ray. Through the accurate measurement of the broad-range energy spectra of gamma-ray sources, especially, by searching the cosmic Pevatrons in the energy range above 100 TeV, to study the particle features of high-energy radiation sources, to explore the evidence for the existence of hadron accelerator in the Galaxy, and to make a breakthrough in the aspect of cosmic ray source detection. To measure accurately the energy spectrum and composition of cosmic ray, and to study the acceleration and propagation mechanisms.
(2) To make the full sky gamma-ray source scanning and searching, to detect a large number of new gamma-ray sources, especially the extragalactic sources, to accumulate the statistical sample of various sources, to explore their high-energy radiation mechanisms, including the mechanism to produce the violent time-variation phenomena, to study the evolutionary rules of active galactic nuclei with the central super-massive black holes, to catch out the high-energy GRB events in the universe, and to explore their eruption mechanisms.
(3) To search the new physical phenomena, such as the dark matter, quantum gravity, or Lorenz invariance defect etc., and to find new rules.

1.2 Experimental Equipment of LHAASO

The site of LHAASO is selected to be the Haizi Mountain in Daocheng of Sichuan Province with an altitude up to 4400 m above the sea level, which is close to the provincial road 217, and only 8 km from the Daocheng Yading airport, hence the transportation is very convenient. The LHAASO experiment mainly includes three detector arrays [4] (see Fig. 1): the ground-based particle detector array KM2A of one square kilometer has the largest distributed area, where the ground-based electromagnetic particle detectors (ED) and underground muon detectors (MD) are uniformly distributed in an area of 1.3 km^2; the central part is the Water Cherenkov light Detector Array (WCDA) with a full coverage, low energy threshold, and the total area of 78,000 m^2; there is also a telescope array (WFCTA), which is composed of 20 Wide Field Cherenkov light Telescopes (WFCTs) that may be distributed flexibly to fit the different physical needs.

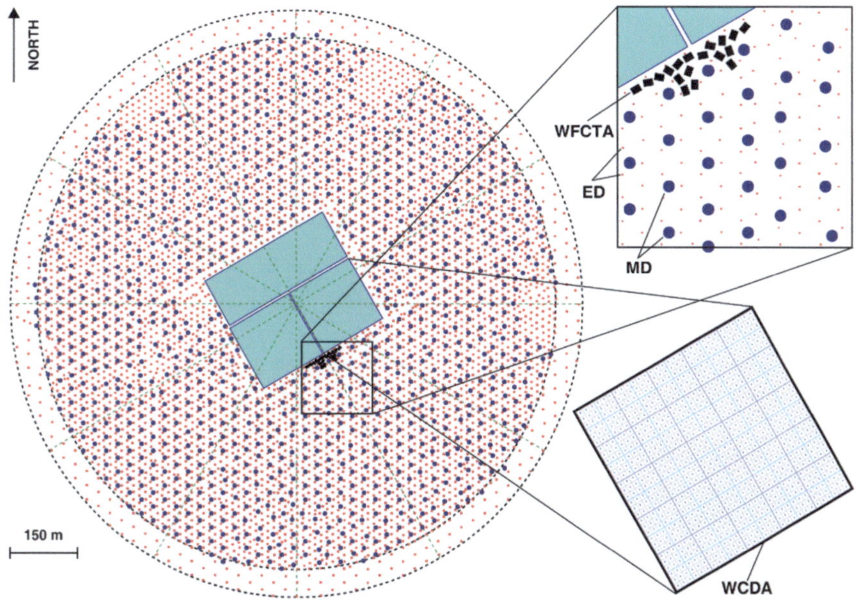

Fig. 1 The layout of LHAASO detectors

1.3 Requirements and Challenges of Data System

The event rate of the joint operation of the full array of WCDA is about 160 kHz, with an effective operation time of approximately 93%. It can acquire 14 billion cosmic ray events per day. After noise filtering, the raw data storage capacity is 12 TB per day, which is equivalent to 4380 TB per year. In addition to the normal trigger mode, WCDA records all the data of each gamma-ray burst (GRB) event in the field of view for 2.5 h, during which 140 million detector signals (hits) are generated every second, for a total of 1.26 trillion hits per GRB event, with a data storage capacity of about 20 TB. It is expected that one GRB event will be recorded every 3 days in the field of view of WCDA, and a total of 153.3 trillion hits will be obtained in one year, with a raw data storage capacity of 2.4 PB annually. The KM2A array contains 5216 EDs and 1188 MDs, and the trigger event rate is about 2500 Hz with the duty cycle of more than 95%, acquiring 170 million cosmic ray events one day, and the raw data storage capacity is about 800 TB per year. WFCTA acquires a total of about 800 million cosmic ray events, laser events and LED events per year, with a raw data storage capacity of 150 TB annually.

After the acquisition of the raw data from detectors, the reconstructed data are generated by decoding and reconstruction processing tasks. 540 TB reconstructed data for WCDA normal mode and 240 TB for WCDA GRB mode are generated every year with 510 TB for KM2A and 120 TB for WFCTA at the same time. The volume of reconstructed data of the three detector arrays is totally 1410 TB per year,

Table 1 the annual data volume of the LHAASO

Dataset	Event (Billon/year)	Volume (TB/year)
Trigger data of WCDA	4300	4380
GRB data of WCDA	153,300 (hit)	2400
Trigger data of KM2A	62.05	800
Trigger data of WFCTA	0.8	150
Reconstruction and simulation	–	2820
Total	–	10,550

and an equal amount of simulation data also needs to be generated for subsequent data analysis. The total data volume produced by the LHAASO annually shown in Table 1.

Such a tremendous amount of data brings unprecedented challenges to the data system of LHAASO including data acquisition, data transmission, data processing, data storage and management.

2 The LHAASO Data System

2.1 Objectives

The LHAASO data system is an important part of the LHAASO experiment, responsible for the whole life cycle of data management including acquisition, storage, transmission, processing, sharing, long-term preservation of massive scientific data, supporting the realization of scientific goals. The main objectives of the LHAASO data system are as follows:

(1) High-speed data acquisition system. The main objective of the data acquisition system is to complete the data readout of the detector electronics channels, event triggering, online data processing and storage while controlling the detector operation to meet the requirement of 4.5 GB/s raw data acquisition from detector arrays.
(2) High-efficient data processing platform. The main objective of the data processing platform is to achieve timely data transmission, storage, processing, long-term preservation of massive scientific data, consisting of computing system, storage system, network system and metadata management system, in order to meet the efficient processing needs of annual multi-petabyte experiment data.
(3) Easy-to-use data processing software. The LHAASO data processing software provides users with a simple, easy-to-use, secure and reliable data processing environment with full services. It allows offline data processing and analysis

software modules to be flexibly integrated and organically combined to complete simulation, reconstruction, physical analysis and other data processing tasks.

2.2 The Overall Architecture

The LHAASO data system mainly consists of a timing system, a data acquisition system, a data processing platform, and an offline software system which are located in a small on-site data center, a large data center and a control room respectively. The small on-site data center is built at LHAASO observation site in Haizi Mountain, Daocheng, and is equipped with a timing system, a data acquisition system (DAQ), and an online data processing platform. The control room is located at the measurement and control site in Daocheng county. The large data center in Beijing is located at the Institute of High Energy Physics, Chinese Academy of Sciences, and is equipped with an offline data processing platform and an offline software system. These subsystems are connected through network and work together based on the flows of data, clock, control and status, as shown in Fig. 2.

The LHAASO timing system [5] is equipped with an internal atomic clock and receives GPS signals in real time to generate an absolute clock. The clock signal can be distributed to the electronics of the detector array through a data transmission network connecting the detectors and the central computer, enabling clock synchronization between detectors with an accuracy of 0.2 ns.

The electronics system of each detector array obtains the high-precision absolute time through the timing system and transmits the data to the data acquisition system through the network simultaneously; after the data acquisition system obtains the data, it executes the software trigger tasks on the computing cluster, then assembles signals into events after the trigger, records it into the data file, and transmits it to the online data processing platform; the online data processing platform receives the data and processes it in real time on a computing cluster. The raw data of WCDA needs to be preliminarily reconstructed to filter out most of the noise signals and then stored into the buffer disk, while the raw data of KM2A and WFTCA are directly stored into the disk without online reconstruction. The data in the on-site data center

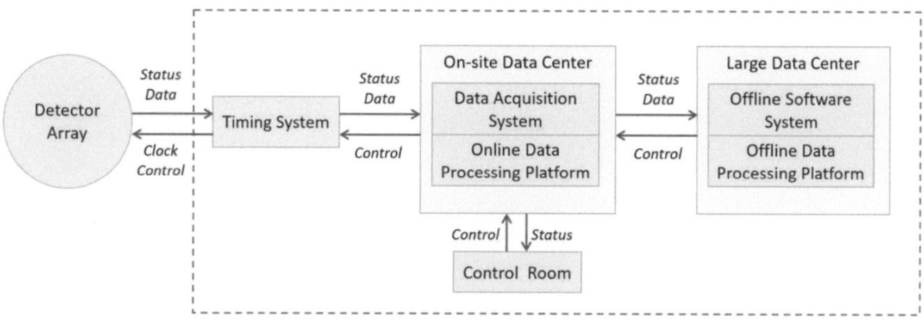

Fig. 2 The overall architecture of LHAASO data system

at Daocheng are transferred in real time via the dedicated network link to the offline data processing platform in the large data center at Beijing. After receiving the data, the offline data processing platform stores it on disk, and writes it to tapes as a permanent backup.

Computers located in the control room, data center or other remote locations can send control commands to the online data processing platform and the data acquisition system through the network to control these two systems. Control commands can also be transmitted to the timing system and even further to the detector array through the data acquisition system, thus achieve network control and remote management of the entire observatory.

3 Data Acquisition System

The LHAASO Data Acquisition System (DAQ) is responsible for data readout, software triggering, online data processing, storage and control of the experiment for four types of detector electronics channels: ED, MD, WCDA, and WFCTA. The DAQ system consists of a hardware system composed of a computer cluster and readout network and a set of data acquisition software. WCDA and KM2A have no global hardware triggers, and the front-end electronics system digitalizes the detector signals into raw data fragments, then the DAQ software packs these read-out data fragments into time slices according to the time stamp given by the electronics system and then distributes them to the online computer cluster for trigger calculation; Each telescope of WFCTA has an independent hardware trigger, and the data is packed in events, and the DAQ software provides flexible software triggers and online computing interfaces, and users can extend online data processing algorithms. The performance requirements of the LHAASO data acquisition system are shown in Table 2.

Table 2 Performance requirements of the DAQ system

Parameter	KM2A	WCDA	WFCTA
Number of channels	e: 5252	3120	18
	μ: 1188		
Hit rate of single channel (kHz)	e: 2	50	–
	μ: 10		
Input data of DAQ[a] (MB/s)	546	2400	240
Output data after trigger[a] (MB/s)	≈10	430	15

[a] MB/s stands for Megabytes per Second

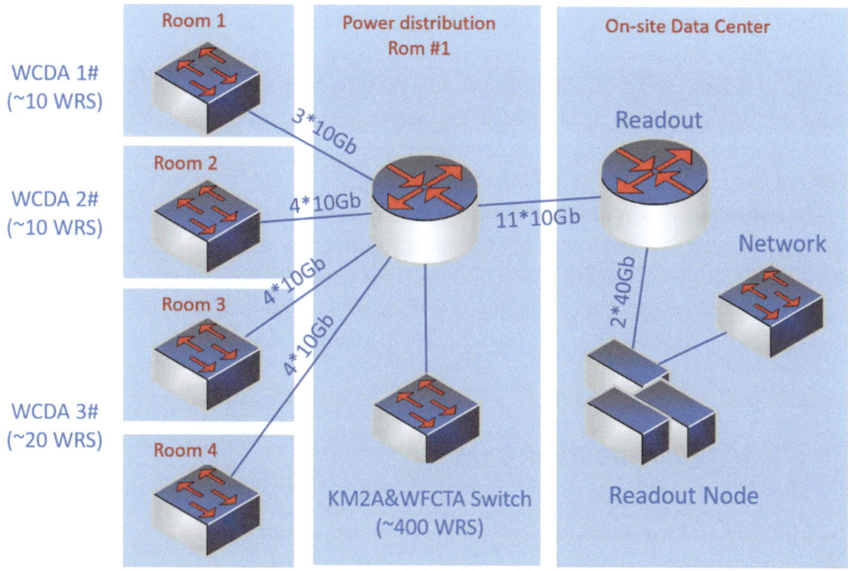

Fig. 3 The structure of readout network

3.1 Readout Network

The readout network is a layer-2 network composed of commercial and self-developed switches. The self-developed switch, called WRS (White Rabbit Switch), is based on the White Rabbit clock synchronization protocol [6] which was proposed by the European Organization for Nuclear Research (CERN) and can achieve sub-nanosecond synchronization accuracy for a large-scale network based on Ethernet. The LHAASO project has improved the remote synchronization accuracy to reach 0.2 ns. The readout network access nodes consist of 5252 KM2A-ED front-end electronics, 1188 KM2A-MD front-end electronics, 3120 WCDA front-end electronics, 18 WFCTA readout electronics, and about 60 computing nodes. The readout network provides a channel for downstream configuration information and upstream raw data channels for the electronics and data acquisition system, as shown in Fig. 3.

3.2 Online Computing Cluster and Storage

The online computing cluster is used for readout, time slice assembly, software triggering and other applications in data acquisition software. The online storage system is built to save triggered event data, which is transmitted to the Institute of High Energy Physics of the Chinese Academy of Sciences and then deleted.

The DAQ system contains a set of IO-intensive hardware and software systems. After multiple stages of software pipeline processing, the data is saved. It realizes the readout and processing of 36Gbps raw data and the online cluster computing nodes consist of blade servers and high-density servers. The network between computing nodes adopts 10Gbps network interconnection and 40Gbps convergence to meet the needs of experiment readout and data processing.

The online storage hardware is composed of rack-mounted servers and disk arrays, and the GlusterFS distributed file system is deployed to manage storage resources. The KM2A and WFCTA use 3 rack-mounted servers with built-in storage and GlusterFS to store physical data. The WCDA detector is equipped with one set of disk array and two rack-mounted servers with 40Gbps network link to save physical data.

3.3 *Data Acquisition Software*

The LHAASO data acquisition software [7] consists of four parts as shown in Fig. 4:

(1) Online service software. It provides functions such as deployment configuration, process status monitoring, information sharing, and control command distribution for the data flow module.
(2) Data flow software. It consists of five modules: readout, assembly, data flow management, online processing, and storage. Each data flow module runs one or more copies (i.e. data flow process) in the computing cluster, and the data flow

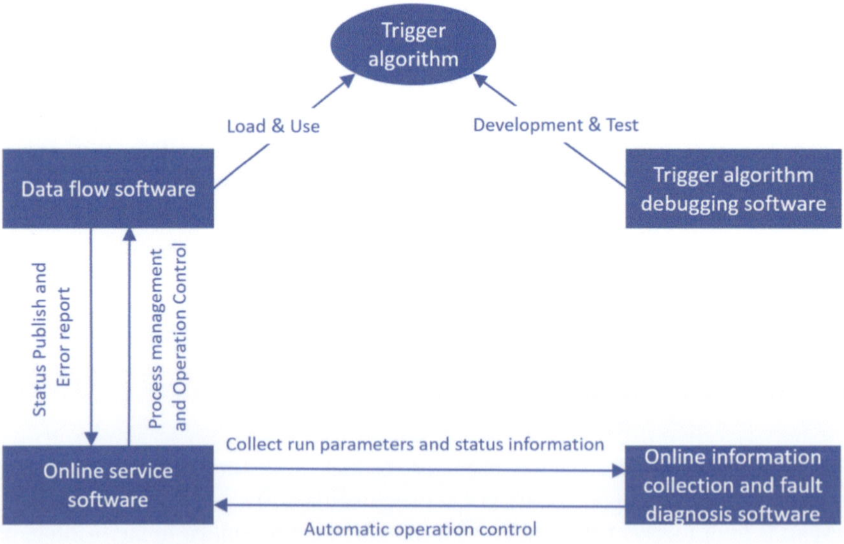

Fig. 4 The components of data acquisition software

processes collaborate with each other to complete data acquisition and online data processing tasks.

(3) Online information collection and fault diagnosis software. It collects the running status parameters of the data acquisition software and saves some parameters (e.g., single channel hit rate, alarm information) to the backend MongoDB database. Based on information such as the DAQ data rate of writing disk, the single channel hit rate, whether the data flow process has exited, the memory usage of the data flow process, the ping status of the computing node, the system determines whether the operation is normal. If it's abnormal, the fault diagnosis system automatically restarts DAQ software, and switches the program to the backup node to run.

(4) Trigger algorithm debugging software. The software reads pre-trigger files, calls the trigger algorithm, and generates the event data file. The output file format is consistent with the online system, and is used to develop and debug the trigger algorithm.

The data flow software is the core module of the entire data acquisition software, which consists of readout (Readout), time slice assembly (Builder), trigger and online processing (PS), and event storage (ES). These components are coordinated by Data Flow Manager (DFM) to complete online data processing tasks step by step. The data flow software adopts a three-level data processing model and a two-level data cache (see Fig. 5), with the subsystem readout function implemented in the Readout module, the data flow merging function implemented in the Builder module, and the online processing and screening module implemented in the PS module.

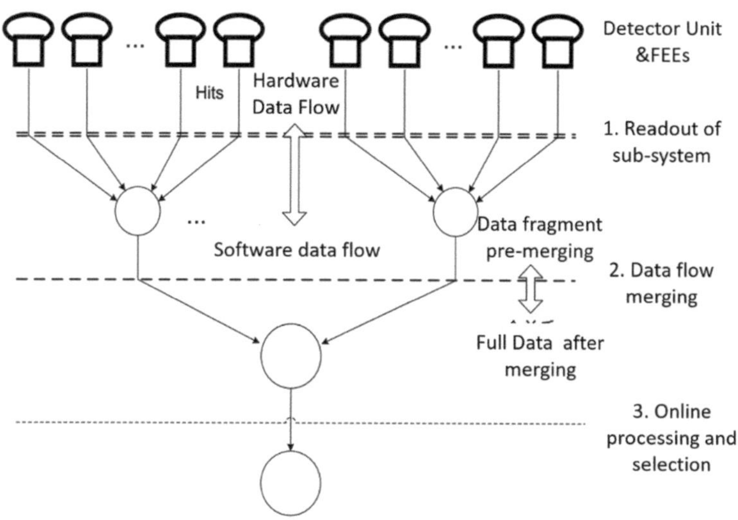

Fig. 5 Three-level processing model of the DAQ data flow software

4 Data Processing Platform

The data processing platform provides fundamental computing and network services including operation monitoring, data transmission, storage, decoding, calibration, reconstruction, simulation, and data analysis. It is composed of a high-speed data transfer system, a massive data storage system, a high throughput computing (HTC) system, a metadata management system, distributed data sharing and other components. which are the foundation for physicists to carry out daily work.

The data processing platform is distributed in different location, including a small on-site data center on Haizi Mountain, a large data center in Beijing and a control room in Daocheng county, as shown in Fig. 6.

The small on-site data center on Haizi Mountain, mainly performs rapid pre-processing and compression of raw data acquired by the DAQ system, and save the output data into the disk buffer for a certain period of time. Then the output data is transferred to the large data center at Beijing. A small-scale HTC cluster and disk storage system are installed in the on-site data center.

The large data center at Beijing is located in the Institute of High Energy Physics, and takes the responsibilities of data storage and processing, supporting simulation, reconstruction, physics analysis and long-term data preservation etc. The data center is equipped with large-scale CPU cluster, disk arrays and tape library, and also deploys corresponding software systems such as job scheduling system, distributed file system, tape library management system, data transfer system, data backup system, operation management system, metadata management system, etc.

As of October 2021, the LHAASO data processing platform was equipped with a high-performance disk file system with 25 PB of storage capacity, one large tape library with 25 PB online tapes and 20 LTO7 drives, a high throughput computing

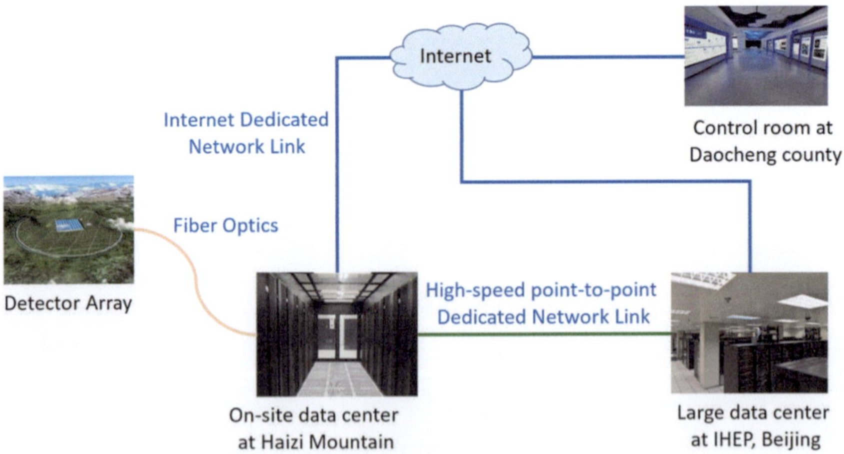

Fig. 6 The topology of the data processing platform

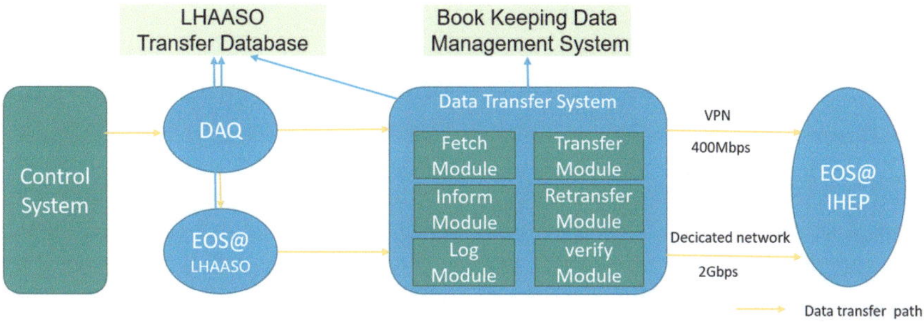

Fig. 7 The architecture of LHAASO data transfer system

system with 12,250 CPU cores and a dedicated WAN link with the bandwidth of 2.4Gbps between Daocheng and Beijing.

4.1 Data Transfer System

The data transfer system needs to transmit the massive scientific data generated at the Haizi Mountain observatory to the large data center at Beijing in a timely manner. The real-time transfer of massive data is the basis of data processing. For the reason, we developed the LHAASO data transfer system, which supports some advanced features such as real-time data discovery, multi-target, multi-link, multi-stream, verification and retransmission, etc. Once DAQ generates data and registers it in the database, the data transfer system immediately transmits the data to Beijing via a dedicated network link and performs verification to ensure correctness and integrity. The bandwidth utilization can reach near 100% due to these advanced technologies. The architecture of the data transfer system is shown in Fig. 7.

At present, there are two network links between observatory at Haizi Mountain and data center at Beijing, one is a 2Gbps CSTNet (China Science and Technology Network) point-to-point dedicated link, and another is a 400Mbps China Telecom Internet link, and the two links are completely independent. According to the network link conditions, the "Best-effort delivery" strategy is adopted for data transmission, and the two models of "load balancing" and "fault tolerance" are implemented. The two links are fully utilized and data can still be transmitted to Beijing when one link fails.

4.2 Data Storage System

The data storage system manages the storage resources including disks and tapes distributed in the on-site data center and Beijing data center. The Lustre parallel file system [8] is used as a home directory to store user programs, job scripts and analysis results, etc. The EOS distributed storage system [9] is used as a storage space for massive experiment data. The Amanda software [10] is used as a user data backup. The CASTOR [11] software is used to manage tape library to backup experiment data, and it will be upgraded to the CTA software [12] in the near future. The LHAASO data storage system is entirely built based on commodity hardware, open source software and self-development software providing high-performance, high-reliability and easy-to-use data storage services. The overall deployment of the system is shown in Fig. 8.

The LHAASO data storage system adopts several cutting-edge IT technologies, including memory database cluster, hardware-assisted data redundancy technology, and distributed hierarchical storage technology, etc. Billions of files will be stored in the LHAASO data storage system, and thousands of concurrent jobs will access these files, which puts tremendous pressure on the metadata server of the storage system. So we deployed QuarkDB [13] memory database cluster to save metadata information and support large-scale file system operations, and synchronize the in-memory database through RAFT protocol [14] to achieve very high scalability and reliability.

Traditional RAID (Redundant Array of Independent Disks) uses controllers to implement data redundancy within the array, but as more and more hard disks are added to the array, the controller becomes a performance bottleneck. Therefore, the LHAASO data storage system abandons the traditional RAID approach and adopts the model of hard disk direct connection to the server (i.e. JBOD), and the

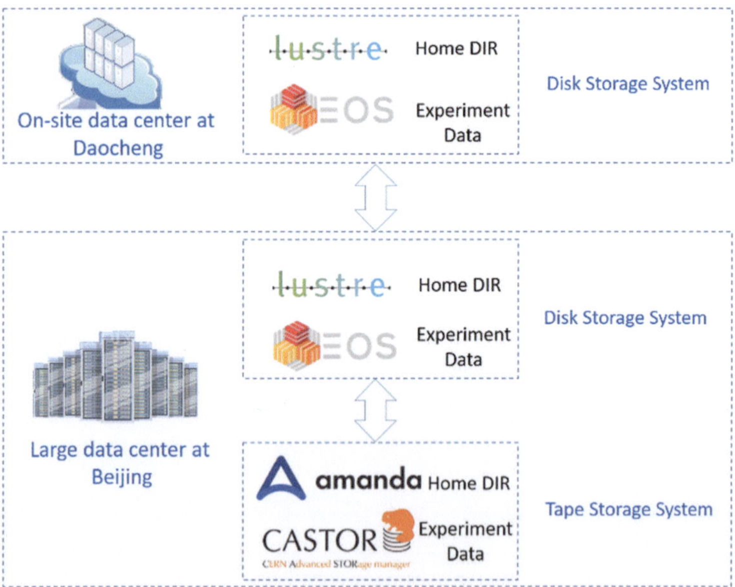

Fig. 8 The deployment of data storage system

data redundancy functions are completely implemented by the upper-level storage software, such as replication, erasure code, and checksum. But the calculation of the erasure code consumes a lot of CPU power of storage server. So, we develop a method to call the CPU internal acceleration engine to accelerate the calculation. For example, ISA-l is an Intelligent Storage Acceleration Library [15] developed by Intel, which implements a hardware-assisted data redundancy technique. In the case of using JBOD with 84 SATA hard disks, the performance of this method is more than four times higher than traditional RAID, and it also achieves data redundancy between servers, thereby improving data reliability and availability.

With the rapid development of SSD, we have built a distributed hierarchical storage system based on SSD-SATA-TAPE to meet the needs of high-performance access and long-term storage of massive data. Firstly, three storage pools are built inside the storage system, namely SSD pool, SATA disk pool and TAPE pool. These storage pools are managed by the distributed storage system EOS. Then, when a user writes a file, an artificial intelligence method is used to determine which storage pool this file should be placed in, in order to minimize the cost of data migration later [16]. Finally, the data access logs are collected to predict the hotness of data access and proactively move the data between different storage such as tape, disk, and SSD [17].

The LHAASO data storage system effectively meets the needs of massive storage, high-performance access and long-term data preservation by adopting a series of advanced technologies and systems.

4.3 High-Throughput Computing System

LHAASO data processing is a typical data-intensive computing application. Thousands of users submit ten thousands of jobs simultaneously to run concurrently on the computing cluster and read or write terabytes of data, which needs high throughput computing system. Therefore, we deployed an open-source high-throughput computing software system called HTCondor [18] that creates an HTC environment based on the computing resources installed in on-site data center at Daocheng and large data center at Beijing, as shown in Fig. 9.

Users submit and manage jobs through the Login Farm. Our self-developed HEP_JOB tool is deployed on the login nodes, which assigns jobs to the site at Daocheng or the one at Beijing according to the site name specified by user. Two independent HTCondor systems are deployed at the Beijing site and the Daocheng site to manage the resources of each site separately, and at the same time the two systems can work coherently. In order to support the needs of decoding, reconstruction, simulation, analysis and other computing tasks, we have defined different priorities and resource scheduling strategies for them, and finally achieved high-efficient management of tens of thousands of jobs which keeps good balance between resource sharing and urgent tasks execution.

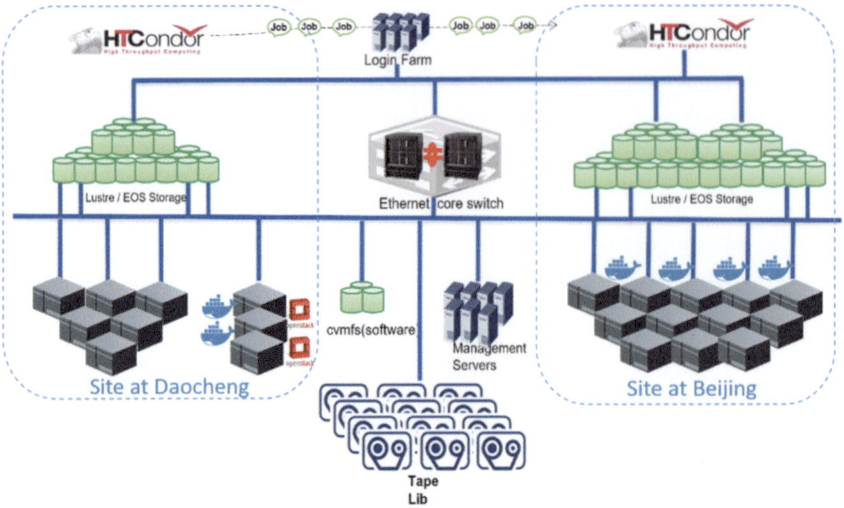

Fig. 9 The HTC cluster for LHAASO

The Daocheng site is located in a national nature reserve at an altitude of 4410 m, and no one is usually on duty, so on-site operation and maintenance is almost impossible. Therefore, we manage the resources remotely based on virtualization technology as shown in Fig. 10. Virtual machines are managed by Openstack [19], and containers run in the virtual machines. All login and management nodes provide services in the form of containers. Kubernetes [20] is used to manage the containers and make full use of the load balancing and fault tolerance mechanisms to improve the availability of the system. The computing tasks are wrapped using singularity containers [21] and are scheduled and run by HTCondor.

4.4 *Metadata Management System*

As mentioned above, the LHAASO produces trillions of events and 10 PB scientific data every year, including raw data, decoded data, calibration data, reconstruction data, analysis data, etc. At the same time, the necessary information related to these data is very rich, such as detector type, data generation time, data type, detector operation parameters, version and parameters of simulation program, version and parameters of reconstruction program, calibration constant, file path, run number, checksum value, etc. Therefore, we have developed a scientific metadata management system to save a large amount of metadata information and also support the feature of dataset management including creation, query, modification and publishing. Its architecture is shown in Fig. 11. Administrators and users can quickly find and access

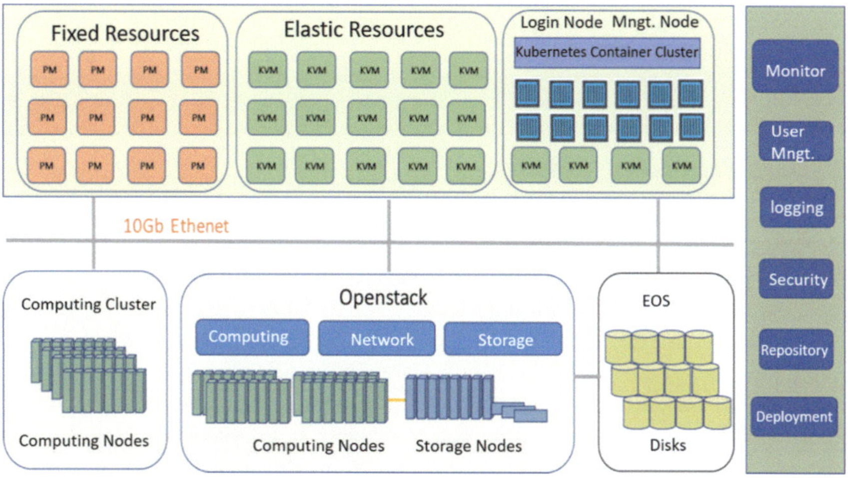

Fig. 10 Remote resources management based on virtualization technology

all kinds of data and their attributes through the scientific metadata management system.

The metadata information is stored in a backend database, and the metadata server publishes the metadata based on HTTP/HTTPS protocol, and the client queries the information using a web page or restful API. Each data or dataset is given a unique identifier ID, which can be flexibly retrieved by conditions such as time, RUN number, and data type and so on. The scientific data management system records the whole processes including data generation, transfer, processing and analysis, so users can retrieve data or trace the whole information of the data by data ID or dataset ID.

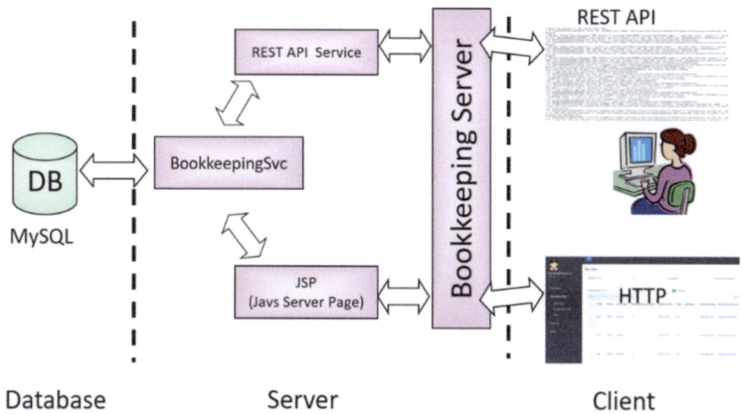

Fig. 11 The Architecture of scientific metadata management system

The metadata management further supports data publishing, opening and sharing combined with data policy.

5 Data Processing Software

The data processing software consists six programs including event simulation software, detector calibration software, data pre-processing software, data reconstruction software, real-time monitoring software for celestial bodies, and data product generation and analysis software, as shown in Fig. 12. The event simulation software mainly is responsible for the Monte Carlo simulation of cosmic ray air shower. The detector calibration software is used to analyze specific calibration data or real experiment data to realize the correction of detector time synchronization, calibration of signal amplitude and calculation of detection efficiency. Data pre-processing software mainly analyzes the raw detector signal and filter the noisy signal, thus reducing the amount of data. The data reconstruction software reconstructs the direction and energy of each event and identifies the particle type. The real-time monitoring software for celestial bodies monitors multiple extragalactic objects and GRB-like outbursts in real time. The data product software extracts the observation data from different sources or partial sky regions according to specific time and conditions to generate the data products of celestial bodies and provide them to users for analysis. Each of the above processes reads in a large amount of data and also generates a large amount of new data, which are managed by data access and job management software provided by the data processing platform.

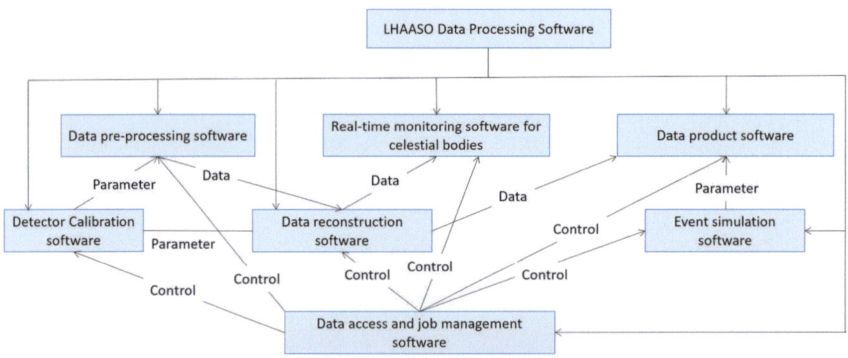

Fig. 12 The sets of LHAASO data processing software and their relationship

5.1 Event Simulation Software

The event simulation software is mainly developed based on CROSIKA – a program for detailed simulation of extensive air showers, and GEANT4—a toolkit for the simulation of the passage of particles through matter. The output data of simulation software is stored in ROOT Tree format [22]. The simulation software can work in four modes: KM2A independent, WCDA independent, WFCTA independent, and three arrays combined. The simulation samples include cosmic rays and gamma rays, which enable the tracking simulation of celestial bodies, as well as the isotropic 26 primary cosmic rays and gamma rays. For example, KM2A focuses on the very-high-energy (VHE) and ultra-high-energy (UHE) range of 10 TeV to 10 PeV; WCDA focuses on the VHE range of 10 GeV to 1 PeV; and WFCTA and the three-array combination focus on the UHE range of 100 TeV to 100 PeV.

5.2 Detector Calibration Software

The detector calibration software is classified into two categories including dedicated programs and general-purpose programs. Dedicated programs are mainly used to analyze data generated from laboratory tests or some dedicated calibration systems, such as waveform sampling data from MD (for calibrating water quality and water level), trigger window edge signal data extracted from ED trigger data (for calibrating detector particle number measurements), single-way data from WCDA (for calibrating PMT gain and efficiency), data from WCDA water decay length measurement system (calibration of water quality), observation data formed by the signals from WFCTA LED and laser calibration systems (calibration of efficiency and atmospheric quality), etc.

The general-purpose program processes the standard experiment data and obtains the time-synchronous differences between different channels of the same detector, or even between different detectors, by reconstructing several iterations of tens to billions of cosmic ray events, using the characteristics of the air shower front surface, and containing the dependence of the time differences on the charge amplitude as well as the air shower core position distance.

5.3 Data Pre-processing Software

The data pre-processing software includes two types of software: raw data decoding software and noise filtering software. The former converts raw data format and generates random-access ROOT Tree format data, which adopts the column storage method and can achieve more than 50% lossless compression based on zlib/xz by using the characteristic of the same type of column data. The noise filtering software is mainly

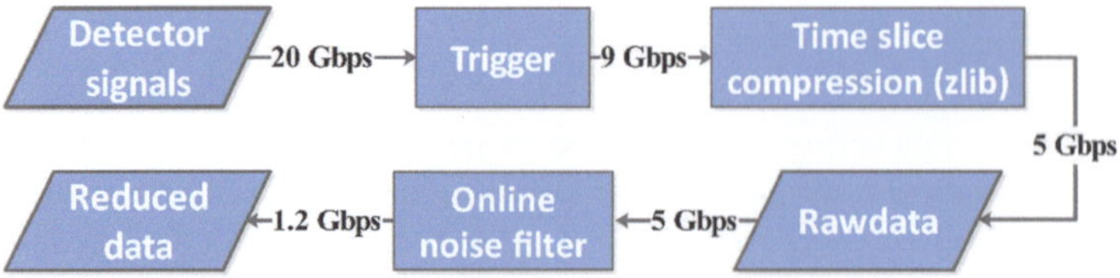

Fig. 13 The data volume changes in each process for LHAASO-WCDA

applied to WCDA data, using a complex event reconstruction method to remove a large number of noise hits from the unit detector, and finally reduce the data volume to less than 1/4 of the original volume. The data volume changes in each process for LHAASO-WCDA are shown in Fig. 13.

5.4 Data Reconstruction Software

The data reconstruction software is used to reconstruct the core position, direction, and energy of the cosmic ray events in extensive air shower, and to perform a comprehensive analysis of the spatial and temporal distribution of HITs to obtain the parameters for the screening of the primary cosmic ray and gamma ray. The main scientific objectives of KM2A and WCDA both focus on gamma astronomy, and the reconstruction algorithms are relatively similar. The core position reconstruction adopts the methods of center-of-gravity, clustering, and the two-dimension Gaussian distribution. The direction reconstruction is obtained by fitting a multi-round linear plane regression on hit points. The energy reconstruction is obtained by analyzing the signal amplitude of each hit and the corresponding two-dimensional lateral distribution of detector positions. In terms of particle classification, KM2A uses the ratio of the number of muon detected by MD to the total number of signals detected by ED to discriminate gamma event, which can exclude most of the cosmic ray backgrounds. WCDA is relatively complex, mainly using the two-dimensional distribution pattern of hit signal time and amplitude to exclude the cosmic ray background, obtaining several parameters related to particle classification, and extracting the category similarity parameters to weight the events to obtain a higher signal-to-noise ratio. WFCTA records the image of the longitudinal development of the air shower in the atmosphere, and then uses an algorithm to remove the noise in the image, followed by the extraction of various parameters of the image such as the size of the image (total signal amplitude), the direction of the long and short axes of the image and its length ratio, etc. It is also necessary to find the corresponding signals from the data of KM2A and WCDA to obtain a variety of information, and to realize the multivariate analysis of particle classification with the help of neural networks and other methods.

5.5 *Real-time Monitoring Software for Celestial Bodies*

The radiation intensity of a large proportion of the celestial bodies is changing, and some unknown sources or astronomical radiation phenomena will suddenly burst. These phenomena can be detected by real-time monitoring of some candidate sources or even the whole sky region, and early warnings can be notified to various International telescopes and detectors, thus realizing multi-band and even multi-messenger observation studies. Therefore, LHAASO has developed real-time monitoring software for celestial bodies to analyze the special signals in real time.

The real-time monitoring software for celestial bodies does this by analyzing the online reconstruction data generated by data pre-processing software. First, a background analysis database is periodically generated containing the acceptance ratio of source-oriented region to the corresponding background region of each selected celestial body in the horizon coordinate system. Then, the experiment data are analyzed in real time to obtain the number of events in source-oriented region and its background region for each source. Meanwhile, if it is found that the ratio of the number of experiment data files that have been processed to the total experiment data exceeds a ratio (e.g. 90%) on the current day (referring to the sidereal day, each source corresponds to a different starting time of the sidereal day), the significance calculation procedure will be started, and the probability of occurrence is decided according to the significance. If the probability exceeds 1 per year, the warning procedure is initiated, and the information is sent out via WeChat and email, and then the web page is updated. This information will eventually be sent to International astronomical monitoring systems such as GCN (Gamma-ray Coordinates Network) [23] to provide early warning for various other International experiments.

5.5.1 Data Product Software

The observation data produced by LHAASO is extremely large, and the number of celestial bodies covered by LHAASO is also very large due to the wide field-of-view and all-day observation. LHAASO meets the needs of astrophysicists by regularly publishing celestial body observation data products.

The data product software divides the whole sky area in the field of view into several windows with approximately equal solid angle, horizontal and vertical scale, and half of the windows in both horizontal and vertical directions are superimposed with each other. A set of observation data of each sky window are published every day, including the events in source-oriented area and the corresponding background area. In order to support energy spectrum analysis, each product includes the time, direction, energy estimation and other parameters of each event in the source-oriented area and background area. The data product also provides a database of detector response for the entire field-of-view sky area with celestial body direction and energy, including parameters such as angular resolution and effective area.

The data product software named LHAASO-PI is built based on a common software package for the astronomy community with the programming language python. The package contains several algorithms for the analysis of celestial bodies, including functions such as position and energy spectrum fitting. For example, it contains the 3ML package [24], which can analyze multiple superimposed sources within the same sky area. In addition, the LHAASO-PI package also incorporates data analysis programs from other experiments (e.g., the Fermi Gamma-ray Space Telescope) to facilitate the joint analysis of multi-band observations.

6 Scientific Application

The LHAASO detectors have been operating stably, has produced a large amount of data. Based on the LHAASO data system, effective data processing and analysis have been carried out, resulting in significant scientific achievements.

6.1 A Large Amount of Scientific Data Has Been Produced

The LHAASO started to operate while it was under construction. The one quarter of detector arrays began to take data in April 2017, which marked the beginning of LHAASO science observation. Then, half and three quarters of detector arrays were put into scientific operation in December 2019 and December 2020 respectively. All LHAASO detectors were completed and put into operation on July 26, 2021, and met all the requirements in the design specifications with duty cycle exceeding 95%. By the end of November 2021, LHAASO has acquired about 20 PB of data and recorded the signals of celestial bodies and cosmic rays in the VHE and UHE ranges using various observation methods and different trigger conditions.

6.2 Several Important Scientific Results Have Been Obtained

The LHAASO is far larger than other similar detectors in International cosmic ray field, and also adopts a number of advanced technologies such as trigger-less and soft-trigger data acquisition technologies, large-scale data processing platform with high-performance massive hierarchical storage, and powerful data processing software. The LHAASO has powerful capability to observe, record and analyze these scientific data due to above technologies. LHAASO has published scientific results in International top journals such as Nature, Science, and PRL, etc. In the first half of 2021, LHAASO discovered 12 gamma-ray sources more than 100 TeV and detecting photons with energies exceeding 1 peta-electron-volt (quadrillion electron-volts or PeV), including one at 1.4 PeV. LAASO also measured precisely the CRAB energy spectrum from 1 TeV to 1 PeV, pushing the traditional theoretical model further. All these discoveries were only based on the data from the half array less one year.

Currently the LHAASO detector array has been completed, and will observe more comprehensive events. More scientific results are expected to achieve along with the accumulation of scientific data.

6.3 Advanced Data Analysis Technologies Are Used to Mine Data Deeply

So far, the LHAASO collaboration has 26 organizations with more than 260 members. With the publication of scientific results, more teams will be attracted to join the collaboration to conduct related research based on observation data. The expansion of the collaboration means that more manpower will be devoted to physics analysis. The rapid development of machine learning and other artificial intelligence technologies provides opportunities for deep mining of the LHAASO data. We will exploit artificial intelligence technologies based on GPU, FPGA, TPU and other heterogeneous computing resources to analyze the LHAASO scientific data, expecting to breakthrough noise filtering, direction reconstruction, particle classification, and significantly improve the observation sensitivity and accuracy. We expect to discover more celestial radiation sources and radiation phenomena, and make higher precision energy spectrum and morphology measurements of celestial radiation sources, explore deeper mechanisms of celestial motion, reveal new laws, and finally solve the century mystery of the origin of cosmic rays.

7 Summary

The LHAASO is one of the highest altitude, largest scale and most sensitive cosmic ray detection instruments in the world. In order to meet the challenge of massive scientific data, the LHAASO has developed and applied advanced information technologies, including: developing clock synchronization technology based on "White Rabbit" protocol in data acquisition, achieving large-scale, multi-node, and high-precision clock synchronization in the wild working conditions above 4,000 m; using lossless data compression, multi-stream and automatic allocation of multiple streams and multiple links, data discovery and verification and other technologies in data transmission, achieving timely data transfer from Daocheng to Beijing; using technologies of in-memory databases, hardware-assisted data redundancy and distributed hierarchical storage in data storage to achieve high performance, large volume and long-term data storage system; adopting the technologies of high-throughput computing, containers and virtualization in computing system to achieve unified management of resources across multiple data centers; developing a modular software framework and artificial intelligence data analysis technology in data processing software to achieve efficient and easy-to-use data analysis. Based on these technologies, the LHAASO data system provides the full-lifecycle management of data acquisition, storage, transmission, processing, publication and sharing, and efficiently handles the massive scientific data of ten petabytes annually to provide high-quality celestial body observation data products and data processing services for astrophysicists at home and abroad.

References

1. Cao Z, Chen MJ, Chen SZ et al (2019) Introduction to large high altitude air shower observatory (LHAASO). Chin Astron Astrophy 43(4):457–478
2. Cao Z, Aharonian FA, An Q, et al (2021) Ultrahigh-energy photons up to 1.4 petaelectronvolts from 12 γ-ray Galactic sources. Nature 594(7861):33–36
3. Cao Z, Aharonian FA et al (2021) Peta-electron volt gamma-ray emission from the Crab Nebula. Science 373(6553):425–430
4. Cao Z, Chen MJ, Chen SZ et al (2019) Introduction to Large High Altitude Air Shower Observatory (LHAASO). Acta Astronom Sinica 60(3):16
5. Gong G, Chen S, Du Q, et al (2011) Sub-nanosecond timing system design and development for LHAASO project. In: Proceedings of ICALEPCS2011, Grenoble, France
6. Moreira P, Serrano J, Wlostowski T, et al (2009) White rabbit: sub-nanosecond timing distribution over ethernet. In: 2009 international symposium on precision clock synchronization for measurement, control and communication. IEEE, pp 1–5
7. Gu M, Zhu KJ, Zhuang J et al (2013) Data acquisition software of LHAASO prototype system. Nucl Electron Detect Technol 33(5):5
8. Schwan P (2003) Lustre: Building a file system for 1,000-node clusters. In: Proceedings of the Linux symposium
9. Peters AJ, Sindrilaru E, Adde G (2015) EOS as the present and future solution for data storage at CERN. J Phys: Conf Ser 664(4):042042
10. Guthmundsson O, Da Silva J, Guomundsson O (1993) The Amanda network backup manager. In: Proceedings of USENIX systems administration (LISA VII) conference
11. Baud JP, Barring O, Durand JD (2000) CASTOR project status. In: CHEP 2000, Padova, February
12. Cano E, Bahyl V, Caffy C, et al. (2020) CERN Tape Archive: production status, migration from CASTOR and new features. In: EPJ web of conferences, vol 245. EDP Sciences
13. Bitzes G, Sindrilaru EA, Peters AJ (2019) Scaling the EOS namespace–new developments, and performance optimizations. In: EPJ web of conferences. EDP Sciences, vol 214. p 04019
14. Ongaro D, Ousterhout J (2014) In search of an understandable consensus algorithm. In: 2014 USENIX annual technical conference (Usenix ATC 14), pp 305–319
15. Gu J, Wu C, Xie X, et al (2019) Optimizing the parity check matrix for efficient decoding of rs-based cloud storage systems. In: 2019 IEEE international parallel and distributed processing symposium (IPDPS). IEEE, pp 533–544
16. Cheng Z, Wang L, Cheng Y, et al (2020) Heat prediction of high energy physical data based on LSTM recurrent neural network. In: EPJ Web of Conferences. EDP Sciences 245, p 04002
17. Cheng ZJ, Cheng YD, Chen G et al (2020) High energy physics data placement strategy based on random forest. Comput Eng Appl 56(21):60–64
18. Fajardo EM, Dost JM, Holzman B et al (2015) How much higher can HTCondor fly? J Phys: Conf Ser IOP Publ 664(6):062014

19. Rosado T, Bernardino J (2014) An overview of openstack architecture. In: Proceedings of the 18th international database engineering & applications symposium. pp 366–367
20. Bernstein D (2014) Containers and cloud: from lxc to docker to kubernetes. IEEE Cloud Comput 1(3):81–84
21. Kurtzer GM, Sochat V, Bauer MW (2017) Singularity: Scientific containers for mobility of compute. PLoS ONE 12(5):e0177459
22. Blomer J, Canal P, Naumann A, et al (2020) Evolution of the ROOT Tree I/O. In: EPJ web of conferences. EDP Sciences, vol 245, 02030
23. Barthelmy SD, Cline TL, Butterworth P, et al (2000) GRB coordinates network (GCN): a status report. In: AIP conference proceedings. American Institute of Physics, vol 526, no 1. pp 731–735
24. Burgess JM, Vianello G (2016) The multi-mission maximum likelihood framework (3ML). In: Eighth Huntsville gamma-ray burst symposium, vol 1962. p 4110

Cao zhen, chief scientist of the LHAASO project, professor of the institute of high energy physics, Chinese Academy of Sciences. He received his BS degree at Yunnan Univ. Kunming, China in 1982, and Ph.D. degree at IHEP Beijing, China in 1994. He worked as a Research Associate in Unvi. Of Oregon, USA on theoretic hadron and heavy ion physics for years before joined the HiRes Fly's Eye Collaboration for ultra high energy cosmic ray physics at Utah, USA in 1998. He was promoted as an Associate Professor in U. Utah in 2003 and as a Professor in IHEP in 2004. His current research focuses are high energy cosmic ray physics above 10^{14} TeV, high energy gamma ray astronomy with LHAASO experiment and ultra-high energy (UHE) tau neutrino searching above 10^{15} eV.

High-Performance Computing Technology Contributes to the Construction of the Earth System Numerical Simulation Device

Guangqing Zhou, Lian Zhao, and Jinrong Jiang

Abstract Numerical simulation is a considerable means to drive the earth science research forward. The earth system model, as the flag of the national science and technology strength, is the fundamental tool for earth system numerical simulation. As one of the national scientific and technological infrastructure construction project during the "Twelfth Five-Year Plan" period, the "Numerical Simulation Device of the Earth System", which is designed based on the Chinese earth system model software with the high-performance computer, is aimed to supply an experimental platform for earth system research and application, which provide guarantees and scientific support for the high-level earth system scientific research and application, national disaster prevention and mitigation, climate and environmental governance and carbon emission and neutrality. Based on a brief introduction to the construction of the Earth system numerical simulation device, this paper will focus on introducing the development of the software, "Earth System Model Numerical Simulation System," and the optimization based on high-performance parallel computing. At last, it will show the trend of the development of high resolution and high complexity of the numerical simulation of the Earth system.

Keywords Earth system model · Device · Parallel computing · Optimization technique · High performance computing

G. Zhou (✉)
Institute of Atmospheric Physics, Chinese Academy of Sciences, Beijing, China
e-mail: zhgq@mail.iap.ac.cn

L. Zhao · J. Jiang (✉)
Computer Network Information Center, Chinese Academy of Sciences, Beijing, China
e-mail: jjr@sccas.cn

China's e-Science Blue Book 2023, https://doi.org/10.1007/978-981-99-8270-7_6

1 Introduction

1.1 Earth System Model

Since the 1970s, climatological research has introduced the "climate system," which considers the four circles of the atmosphere, ocean, snow and ice, and the land surface, mainly regarding the physical processes and their interactions within each circle. In the twenty-first century, the Earth system has become a hotspot and the frontier of earth science research. Based on the climate system, it further considers ecosystems, space weather, and the solid Earth system. It understands the operation of the Earth's energy processes, ecological processes, and metabolic processes by studying the physical, chemical, and geo-biochemical processes within the circles and the energy, momentum, and material exchange between the circles. As shown in Fig. 1. The increasing attention paid to the role of coupled biogeochemical processes such as the carbon and nitrogen cycles in the climate system and the impact of human activities on these cyclical processes. In the past, understanding and knowledge of the laws of the Earth system have relied more on observational experiments. With the enrichment of various kinds of observational experiments, the improvement of understanding, and the development of computers, numerical simulation tools have become an essential experimental tool for the scientific research of the Earth system nowadays. It has penetrated all joints of the scientific research and application of the Earth system, which is one of the essential symbols of the development of the Earth system science to a holistic, integrated, and quantified level [1, 2].

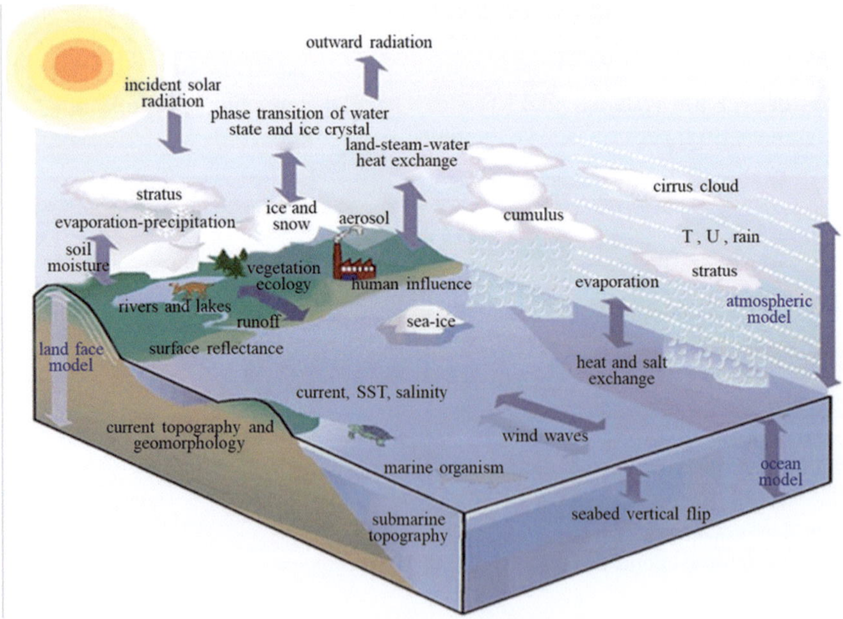

Fig. 1 Schematic diagram of earth system

The Earth System Model (ESM) is the most complex and comprehensive numerical modeling tool and the "Holy Grail" of numerical modeling in Earth system science. The Earth System Model is a mathematical and physical model based on the physical, chemical, and biological processes in the Earth's various circles and the material and energy exchange patterns. It is then solved numerically, compiled into an extensive comprehensive computational program, and put into practice by supercomputers as a scientific tool [3] for the simulation and prediction of the complex behavior and processes of the Earth system. Forecasting the Earth system involves predicting weather and climate and the evolution of the ecological environment and its hazards. The international term "Earth System Model" is limited to the simulation of the Earth's surface layers, mainly consisting of the climate system and the ecosystem [4].

1.2 Significance of Building Earth System Model

The Earth System Model (ESM) provides access to vast amounts of data on the spatial and temporal distribution and evolution of the Earth System's various circles. The Earth system has a vast spatial and temporal tensor. It is physically "inaccessible" (e.g., solid Earth, deep ocean, and near-Earth space). Although Earth observation systems have made significant progress, supported by the development of new satellite and sensor technologies, observations are still far from meeting the requirements for a complete and quantitative description of crucial processes, limiting the understanding of the Earth system. The Earth system model can produce the main components of the Earth system. The Earth system model can generate massive data on the prominent circles of the Earth system by combining data assimilation techniques with observational data. It can provide increasingly complete and realistic data for the quantitative description of the Earth system, laying a solid data foundation for the development of Earth system science.

With an experimental platform for virtual Earth systems such as the Earth System Model, it is possible to design and conduct all types of scientific experiments in Earth system science in the same way as in the laboratory and physics and chemistry. It allows for a detailed differentiation and understanding of the factors and patterns that influence the Earth's spheres and interactions. It answers several significant questions about current global changes, such as the mechanisms and drivers of climate change and how climate and environment will evolve in the future.

Global climate and ecological change are an essential factor in the sustainable development of China and is one of the central issues in international climate and environmental diplomacy. The use of Earth system models to elucidate the mechanisms of global climate and environmental change and to make predictions has become an essential element of international Earth system science research today. Also, it provides a critical scientific basis for international consultation and agreement on climate and environmental change issues (e.g., Intergovernmental Panel on Climate Change (IPCC)). It also provides a critical scientific basis for international consultation and agreement on climate and environmental change issues (e.g., Intergovernmental Panel on Climate Change (IPCC)). It is an essential indicator of a country's overall level of geoscientific research. At the same time, only by mastering advanced and perfect numerical modeling techniques and experimental tools can make China independent simulations and predictions and enhance the right to speak in climate diplomacy negotiations. Major developed countries internationally have taken the development of Earth system models as an essential development direction for Earth system science. China also started to develop Earth system models based on its climate system models in 2007 and has achieved outstanding results [1, 3, 17].

1.3 Earth System Model and High-Performance Computing

The earth system models, defined as computer programs encapsulating many natural laws, are mathematical representations of the Earth system's physical, chemical, and ecological processes and biogeochemical cycle processes such as carbon and nitrogen cycles. The current Earth system model includes mainly physical climate system processes such as atmosphere, ocean, land surface, and sea ice and subsystems of ecological and environmental processes such as aerosols and atmospheric chemistry, terrestrial, marine ecosystems, and biogeochemical cycles. It uses coupler software to transmit information about the material and energy exchanges between the above subsystems, coupling them together. Therefore, the coupler is also often considered an integral part of the Earth system model. The Earth system model is thus a product of understanding the laws of the Earth System based on multidisciplinary interdisciplinary research. It requires understanding physical, chemical, and life processes in the earth sciences and the support of modern high technology facilities such as high-performance computers and mass storage.

High-performance computing technologies play a crucial role in driving the development of Earth system models. The development of 'integrated' and 'fine-grained' Earth system models has led to increasing computation, with high-resolution Earth system models placing great demand on computational power. At the same time, the increase in the number of Physic-chemical processes and the introduction of new component models has led to an order of magnitude increase in computational demand, with the application of assimilation and ensemble simulation techniques pushing the computational demand up another order of magnitude. Thus, the continuous development of high-performance computing technologies is fundamental to the Earth system model.

As the demand for modal computing in the Earth system continues to rise, the efficient use of computing resources has become an inescapable problem. The traditional "serial model parallelization" approach is limited in terms of the efficiency and modularity of model parallelization. New high-performance computing technologies are needed further to improve the scalability and efficiency of model parallelization. Emerging processors and technologies, such as GPUs (Graphics Processing Unit), heterogeneous multicore processors, and variable floating point precision computing technologies, provide new technical routes to improve the computational efficiency of the Earth system model. Therefore, the development of computational middleware applicable to the Earth system model and the collaborative design with the model program to establish a high-performance software support platform has become crucial technical support for developing the Earth system model. In addition, big data and artificial intelligence technologies have been applied to the development of models and forecasting techniques in recent years, such as improved parameterization of cumulus convection and cloud precipitation and ultra-short time forecasting of convective systems help to reduce the uncertainty of complex parameterization schemes in models.

2 Construction of Earth System Numerical Simulation Device

2.1 Construction Background

The Earth System Numerical Simulation Facility is a significant science and technology infrastructure project under China's National 12th Five-Year Plan. The Institute of Atmospheric Physics (IAP) of the Chinese Academy of Sciences (CAS) is the legal entity for constructing the project, and Tsinghua University is the co-construction unit. The facility intends to be the first of its kind in China with independent intellectual property rights. The core software is the numerical simulation software of the Earth system, and the software and hardware indicators are compatible with each other. The device's scale and comprehensive technical level are among the world's leading numerical simulation devices. The core software is the CAS-ESM (Chinese Academy of Science-Earth System Model) [17], developed independently by the Institute of Atmospheric Physics of the Chinese Academy of Sciences after a long scientific research and development period. The Earth System Numerical Simulation Facility can simulate the evolution of the Earth's atmosphere, hydrosphere, cryosphere, lithosphere, and biosphere. It allows for inversion of the Earth's past, observing the present and predicting the future.

2.2 Construction Content

The Earth System Numerical Simulation Facility provides a platform for studying the physical, chemical, and biological processes and their interactions in its atmosphere, hydrosphere, cryosphere, lithosphere, and biosphere. It provides an experimental platform to investigate the impact of these interactions on the Earth System as a whole and regional environment. It integrates simulation techniques with observational data to improve the accuracy of predictions and enables quantitative description and simulation of complex processes in the Earth system at mesoscale resolution. It can provide scientific support for major national disaster prevention and mitigation issues, addressing climate change, and atmospheric environmental management. It promotes interdisciplinary crossover and integration among different disciplines of Earth system science and facilitates the overall leap of China's Earth system science to the international first-class level.

The installation includes an Earth model numerical simulation system, a regional high-precision numerical simulation system, super simulation support and management system, a support database, data assimilation, and visualization system, a high-performance computing system for Earth science (as shown in Fig. 2) and civil engineering facilities.

The Earth System Model Numerical Simulation System is the core part of the whole device, supported by the "Super Support and Management System." It realizes large-scale parallel computation on the hardware system of the device—"High-Performance Computing System for Earth Science." The simulation output provides the side boundary conditions for the regional high-precision environmental simulation system to achieve higher accuracy regional simulation and prediction. In addition, the "supporting database and data assimilation and visualization system" will support the storage and real-time visualization and analysis of the simulation output of the Earth system model numerical simulation system.

2.3 Construction Situation

The installation is in Miyun Economic Development Zone, the eastern group of Huairou Science City, and is an integral part of Beijing's National Comprehensive Science Centre. It is also an essential part of constructing the Beijing Science and Technology Innovation Centre and Huairou Science City. The project has a

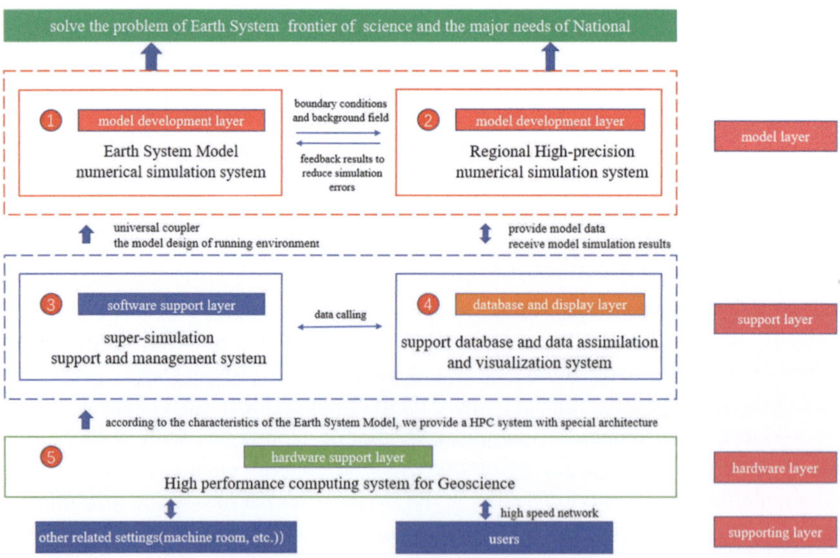

Fig. 2 Device system composition

total investment of CNY 125.52 million, a construction area of 24,310 m^2, and a construction period of four years. The project started in full in November 2018, will be completed in June 2021 and will be ready for acceptance in 2022.

On 28 February 2017, the project proposal received project approval from the National Development and Reform Commission. On 14 June 2018, the National Development and Reform Commission approved the feasibility study report. On 30 July 2018, the joint approval of the Chinese Academy of Sciences and the Ministry of Education on the preliminary design plan of the plant was obtained. On 16 November 2018, the approval of the National Development and Reform Commission on the approved preliminary design budget estimate of the plant was obtained. All the approvals required for the device's construction were completed, and the project entered the full implementation stage.

After more than two years of intensive construction, the "Earth System Numerical Simulation" was inaugurated on 23 June 2021 in Huairou Science City, Beijing, China. The hardware machine of the device is shown in Fig. 3. The Earth System Model Numerical Simulation System is China's first significant science and technology infrastructure with a numerical simulation software as the core construction content. It will serve primary national needs by addressing climate change, ecological environment construction, carbon peaking, carbon–neutral vision, disaster prevention, and mitigation (e.g., weather forecasting). It can provide strong scientific support for international climate and environment negotiations.

In the future, as the solar-terrestrial space environment model, the continental ice cap model, and the solid Earth model are gradually coupled into the Earth system model. The Earth system numerical simulation device will realize fully coupled numerical simulations from the Earth system to space weather. This device can further promote the theorization and numeration of Earth science promote the crossover and integration between various geosciences disciplines. It can promote the overall advancement of China's Earth system science to the international top level.

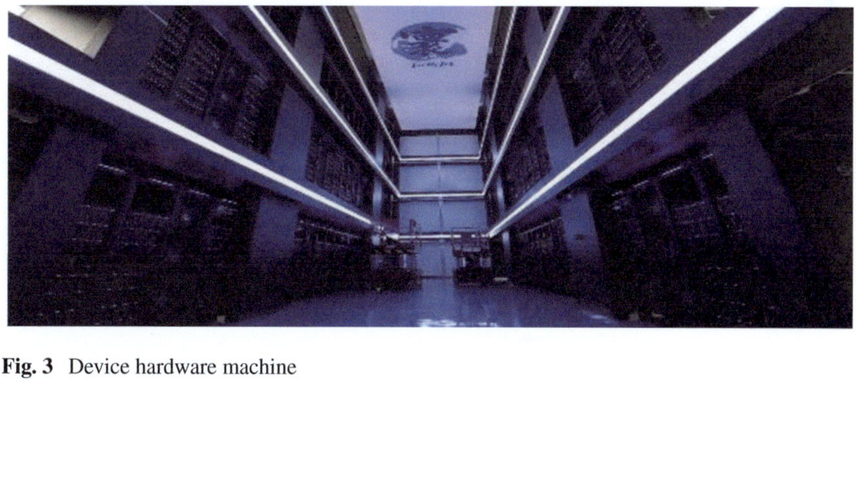

Fig. 3 Device hardware machine

3 Development and Optimization of Numerical Simulation System for Earth System Model

3.1 Core Functions of Numerical Simulation System of Earth System Model

The "Earth System Model Numerical Simulation System" is based on CAS-ESM1, a prototype Earth numerical modeling system developed by the Chinese Academy of Sciences and further developed into CAS-ESM2, include "Atmospheric circulation model" "Ocean (sea ice) circulation model" "Land surface processes model" "Vegetation dynamics model" "Aerosol and Atmospheric Chemical Processes Model," "Marine Biogeochemical Model," "Terrestrial Biogeochemical Model," the "Earth System Model Integration Module," "Solar-Terrestrial Space Environment Model," "Continental Ice Cap Model" and "Solid Earth Model" 11 subsystems (as shown in Fig. 4). The first seven of these subsystems exchange material and energy through the Earth System Model Integration Module, thus forming a unified Earth System Model. In contrast, the three subsystems of the Solar-Terrestrial Space Environment Model, the Continental Ice Cap Model, and the Solid Earth Model are relatively independent due to their characteristic spatial and temporal scales. Currently relatively independent, their response to external forcing is the main consideration at this stage.

(1) The Atmospheric Circulation Model subsystem is developed and refined based on the existing Atmospheric Circulation Model IAP-AGCM4.1 [5] (the Institute of Atmospheric Physics Atmospheric General Circulation Model), which is used to simulate essential atmospheric circulation characteristics such as

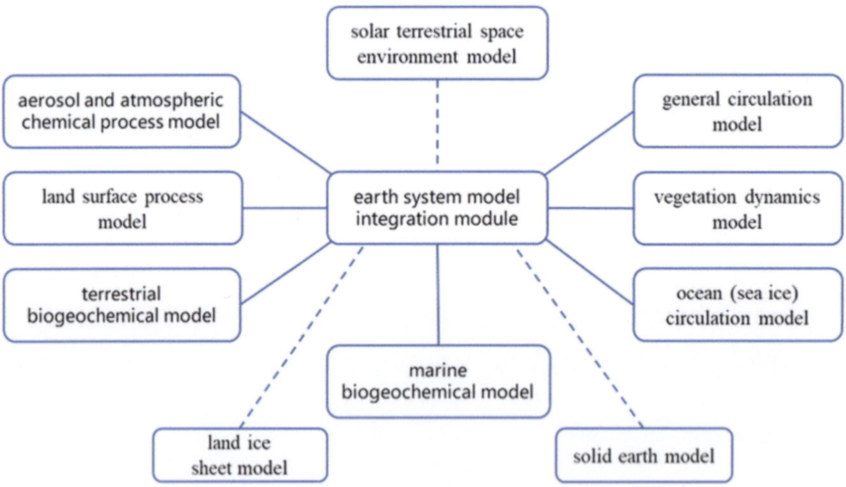

Fig. 4 Structure of earth system model numerical simulation system

atmospheric temperature, wind speed, pressure, precipitation, cloudiness, and water vapor. The results of this simulation can provide upper boundary conditions for land surface models, sea ice models, ocean models, and side boundary conditions for regional atmospheric models. It establishes a subsystem of atmospheric circulation models with a horizontal resolution of about 25 km, enabling high-resolution global climate (atmosphere) simulations. It can provide advanced simulation software for climate simulation research, short-term climate prediction and scientific prediction, and early warning for disaster prevention and mitigation.

(2) The ocean circulation model subsystem is a coupled ocean-sea ice subsystem comprising a separate ocean circulation model and a sea ice model. The flux exchange through the coupler allows the simulation of essential characteristics such as seawater temperature, salinity, current velocity, sea surface height, sea ice current velocity, sea ice thickness, and density under given boundary conditions and initial conditions. The simulation results can also be used as lower boundary conditions for the atmospheric model and provide side boundary conditions for the regional ocean model. The LICOM [6] model of the Institute of Atmospheric Physics of the Chinese Academy of Sciences is used for the ocean circulation model subsystem. The CICE model [7], which is commonly used globally, is used for the sea ice model. This model makes the overall model level of the coupled ice-ocean ocean circulation model subsystem comparable to the mainstream international models and provides an advanced simulation software platform for climate simulation and prediction, ocean and sea ice forecasting, and research in China.

(3) The Land Surface Process Model subsystem is our Common Land Surface Model CoLM [8], which incorporates a range of sub-models such as the Soil Hydrothermal and Vegetation Canopy Double Macro-leaf models. It is a systematic study that explores the physical, chemical, and biological processes and their patterns of change in the Earth's hydrosphere, lithosphere, and biosphere by considering them as a sizeable interacting system. It is an essential platform for researching advanced operational regional/global meteorology-hydrology-ecology forecasting, land use and change, and climate and ecological change. This platform can provide the scientific basis for quantitative inquiry into the role played by land surface processes in the Earth system and the interaction between human activities and the environment.

(4) The vegetation dynamics model subsystem is mainly based on the vegetation dynamics model IAP-DGVM [9] of the Institute of Atmospheric Sciences, Chinese Academy of Sciences. It is mainly based on climatic data, vegetation parameters, and bare soil properties. Based on the data on soil physical environment conditions, photosynthesis and respiration of vegetation given by the land surface model, and based on the basic principles of vegetation dynamics, the global vegetation distribution, and its evolution with climate, as well as the temporal and spatial changes in the characteristics of various vegetation types, are simulated. The effects of human activities (including crop and forestry management) are considered. The fully coupled Earth system model

results provide information on the physiological and ecological characteristics of plants for the land surface process model and the vegetation-surface-atmosphere fluxes (e.g., carbon fluxes, trace gas emissions) for the atmospheric circulation model.

(5) The aerosol and atmospheric chemical process model subsystem is mainly developed based on the aerosol and atmospheric chemical model Global Environmental Atmospheric Transport Model (GEATM) [10] of the Institute of Atmospheric Sciences, Chinese Academy of Sciences. This subsystem is developed by acquiring global the chemical constituent emission source data from anthropogenic activities (e.g., industry, traffic.) and natural processes (e.g., vegetation, volcanoes.). It simulated forecast data of global meteorological elements from the atmospheric circulation model subsystem and upper boundary conditions of the climate mean state. This subsystem integrates internationally advanced mechanisms of atmospheric Physic-chemical processes to achieve global three-dimensional spatial and temporal distribution simulations and predictions of major atmospheric chemical constituents. The established aerosol and atmospheric chemical process model subsystem will provide international advanced simulation software tools for scientific research on the global atmospheric environment, climate change, and interactions. It can carry out global multi-scale air pollution forecasting and quantitative assessment of cross-continental and international transport of pollutants, providing scientific and technological support for China's "Belt and Road Initiative" strategy. The software system will provide scientific and technological support for China's "The Belt and Road" initiative.

(6) The ocean biochemical model subsystem is based on the IAP's global ocean biogeochemistry general circulation model (OBGCM) [11], which mainly describes the biogeochemical cycling of carbon in the ocean and consists of two parts: the ocean biogeo-chemical model and the ocean ecological model. The marine biogeochemical model describes the processes of carbon uptake, transformation, and storage in the ocean, starting from the exchange of carbon at the sea-air interface, and quantifies the regulation of atmospheric carbon concentrations by modeling the response of these processes to different climatic conditions. Marine Ecological Processes (MEP) describes the dynamics of pelagic plankton and provides a process description of the conversion of inorganic to organic carbon in the ocean's upper layer through online simulations of ecosystem growth and extinction. The marine biochemical model subsystem provides the atmospheric model subsystem with sea-air carbon fluxes to calculate the change in atmospheric CO_2 concentration over time. The ocean biochemical model quantifies the ocean's capacity to absorb anthropogenic CO_2 and the response of this capacity to climate change. The interaction between the carbon cycle and climate change can be studied through full coupling with other subsystem models. The future evolution of the marine environment, such as seawater pH, can be predicted.

(7) The terrestrial biogeochemical subsystem is based on the physical environmental conditions of soils given by the land surface model. The subsystem is a database of farmland distribution, agronomic management practices, crop parameters, basic soil properties, constructed by fusing data from multiple sources. It can help to stimulate the carbon and nitrogen biogeochemical processes in soils. It mainly includes the transport and transformation of carbon and nitrogen elements. Its simulation results can provide the atmospheric model, aerosol, and atmospheric chemical process model with the net emission fluxes of CO_2, CH_4, and N_2O from different terrestrial ecosystems, i.e., the lower boundary conditions of material exchange in the atmospheric system. The primary function is to simulate the carbon and nitrogen cycling processes in terrestrial ecosystems and the associated carbon and nitrogen gas emissions based on the biogeochemical variable fields constructed by the environmental factor subsystem simulations and to provide the required spatial and temporal distribution of carbon and nitrogen substances for the analysis of land surface process models. The subsystem will enable the modeling and estimation of meaningful greenhouse gas (CO_2, CH_4, N_2O) emissions from agricultural ecosystems.

(8) The continental ice cover model subsystem uses the polar ice cover and geomorphological data provided by the National Snow and Ice Data Center (NSIDC). A high-resolution numerical continental ice cover software model is established based on the more mature international ice cover models. It can make reliable predictions of continental ice cover trends over hundreds of years, providing a valuable reference for global atmospheric and ocean-related scientific research.

(9) The Solid Earth Model Subsystem aims to develop a global-scale, multi-physical field geodynamic model, which simulates the physical environment and physical motions of the Earth's different layers, or different scales of the same layer, through different subsystems. At the same time, the subsystem will also provide computational data for the simulation of massively parallel visualization systems for the systematic analysis of data obtained from multiple subsystems within the subsystem. This software will enable the scientific study of various regions of the solid Earth under complex conditions. In the future, it will be possible to establish coupling with several model subsystems of different dimensions and scales, such as atmosphere, ocean, ice cover, and vegetation, at multiple temporal and spatial scales, to realize a complete simulation system of the Earth's complex systems.

(10) The solar-terrestrial space environment model subsystem integrates kinematic and 3-dimensional magnetohydrodynamics (MHD) numerical models. The model is driven by observations such as the solar atmosphere and considers physical mechanisms such as bulk heating to simulate the evolution of solar wind propagation in the solar corona to Earth space. The model is used to study the evolution of the background solar wind during different periods of solar activity, the propagation evolution of large-scale solar wind perturbation structures, and the physical processes such as coronal mass ejection

(CME)/interplanetary surge interactions. The solar-terrestrial space environment model outputs the spatial and temporal variations of the fundamental parameters of the solar wind magnetic field and plasma in the vicinity of the Earth's space. It can compensate for the scarcity of current solar-terrestrial space exploration data and can also be used to analyze the causes of geomagnetic storms and space weather prediction experiments.

(11) The Earth System Model Integration Module subsystem is based on the CAS-ESM coupling framework of the Earth System Model of the Chinese Academy of Sciences. The coupling interface automatic generation function module, complex grid mapping, and 3D coupling functions were added to improve the compilation and operation system. The coupling of Earth system models was completed, and the data interaction between the subsystem as mentioned above and the operation, optimization, and diagnosis of Earth system models were realized. Realize easy assembly and operation of the coupled Earth system models.

3.2 Parallel Optimization of Numerical Simulation System of Earth System Model

The main directions in developing Earth system models are increased resolution and increased complexity. To guarantee the computational stability and accuracy of the model during numerical simulations, the time integration step that can be selected decreases with increasing resolution. However, the computational effort of the model grows exponentially with increasing resolution. At the same time, to carry out climate and environmental simulations, it is often necessary to integrate the Earth system model over decades, centuries, or even millennia, and the amount of computation required is enormous. As the demand for Earth system model computation continues to rise, the efficient use of computational resources becomes an unavoidable issue. Due to the limitations of the scalability and parallel efficiency of early models, the model runs slowly, making it challenging to meet the time-sensitive needs of practical applications. Its core algorithms can only use less than 5% of the computational power on general-purpose processors. The adoption of crowdsourced and heterogeneous architectures has made programming difficulties and computational inefficiencies even more pronounced for the highly complex Earth system model. Earth system models often have millions of codes with various computational characteristics between different algorithms, such as different computational methods in the east–west, north–south, and vertical directions. It is impossible to find a standard parallelization method for all segments of the code or all core algorithms. Computation is scattered. There are many variables, many synchronizations, and many branching computations, and these result in low computational access ratios and much data migration communication, making floating-point and parallel computation inefficient. It is conceivable that a million-line code with just 5% of the computation not suitable for parallelism would only achieve a 20 × parallel speedup at the limit, while a 10 × increase in pattern resolution would increase the amount of computation required thousands of times. Therefore, it is impossible to carry out high-resolution simulations without massively parallel computing to improve computational efficiency. To improve the computational efficiency and parallel scalability of the model, researchers have done much work in the areas of model parallel algorithm design, parallel implementation, parallel performance optimization, and floating-point computational efficiency optimization. The main bottleneck facing the atmospheric circulation model and ocean circulation model in the Earth system model is to improve the computational efficiency and exploit the comprehensive computational capability of large computing platforms.

3.2.1 Parallel Optimization of General Circulation Model

Improving the computational efficiency of the atmospheric circulation model has become the key to improving the overall Earth system model. IAP-AGCM uses a finite difference separation method on a spherical meridional grid in China's independent atmospheric circulation model. The method is characterized by high computational accuracy and low computational effort, but there is the problem of poles. In the high north–south latitude (pole) region, an east–west (x-direction) FFT filter is used in the calculation to increase the time integration step and maintain computational stability. It results in a strong correlation of the east–west calculations, which cannot be dissected. In turn, the calculation of physical processes such as vertical convection and turbulence calculations strongly correlates in the vertical direction (z-direction). There is why the IAP-AGCM was the first to use the one-dimensional region profiling parallelization method. To improve the scalability of the model, the researchers developed a hybrid two-dimensional area-profile parallelization method, i.e., without FFT filtering in the east–west direction (x-direction) at low and medium altitudes. It is only strongly correlated in the vertical direction when horizontal two-dimensional (x–y direction) region profiling is used.

In contrast, strong correlations are computed at high latitudes in both the vertical and east–west directions, where a north–south-vertical two-dimensional (y–z direction) area profile is used. It increases parallel scalability. By implementing MPI + OpenMP computing, the parallel scalability and computational speed of the IAP-AGCM are increased by a factor of nearly 10 [12]. However, the communication overhead of data conversion between multiple region profiles also increases. The researchers developed adaptive local Gaussian filtering to improve scalability instead of FFT filtering, turning global communication into proximity communication. A parallel algorithm for three-dimensional dissection of the dynamical framework of atmospheric modes was implemented. Instead, horizontal two-dimensional area profiles are used in the physical process calculations. Some of the computational processes with vertical solid or east–west correlation are specially treated, and the overall performance of the atmospheric model is substantially improved overall [13]. At the same time, for heterogeneous calculations, GPU heterogeneous acceleration of some physical process modules (long and short-wave radiation module,

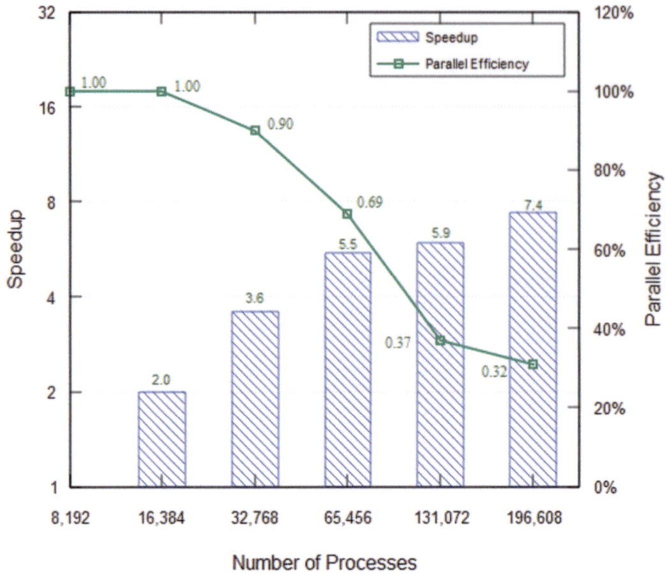

Fig. 5 Experimental results of large-scale scalability test

cloud microphysics module) has been implemented [14, 15], and good results have been achieved. Heterogeneous acceleration of the atmospheric circulation model dynamical framework is also underway.

The IAP-AGCM has been tested on a large scale on a domestic X86 processor-based supercomputer. The model's scalability and performance acceleration ratios were tested at a horizontal resolution of 1/4° × 1/4° (around 25 km), using 8,192 to 196,608 processes, respectively. The results are shown in Fig. 5 and show that the program scales to nearly 200,000 cores, with a 7.4 × speedup compared to 8,192 cores (23.6 × compared to 1,024 cores) and parallel efficiency of 32% (based on 8,192 cores).

3.2.2 Parallel Optimization of Ocean Circulation Model

The ocean circulation model is also an essential part of the Earth system and a hot spot for computation. Our ocean circulation model, LICOM3.0, uses a three-polar coordinate grid and finite difference discretization to solve the pole problem and uses horizontal two-dimensional area dissection parallelism. To further improve its computational speed, we designed a GPU heterogeneous computational method for LICOM3.0 and implemented a heterogeneous computational version of LICOM3-HIP. LICOM3-HIP first uses Horizontal 2D region dissection parallelism. The computation on each computational grid after dissection is given to the GPU to compute; each GPU uses 1D, 2D, or 3D dissection to assign grid points to different GPU cores to compute according to the computational characteristics.

LICOM3-HIP [16] uses traditional optimization techniques such as shared memory, rematerialization, loop unswitching, loop expansion, and the Single Instruction, Multiple Data (SIMD) architecture of GPU accelerated cards to adapt to the single instruction, multiple data (SIMD) architecture. In addition to the Single Instruction, Multiple Data (SIMD) architecture of GPU-accelerated cards, this is optimized by redesigning specific numerical algorithms or implementations to decouple data dependencies, avoid memory write conflicts, and rewrite logic to optimize branch divergence.) It allows the computation to be dissected in three dimensions on each GPU, increasing its parallel scalability. To reduce communication overhead, LICOM3-HIP uses a "scalable pseudo-boundary" strategy to optimize communication, replacing it with computation to reduce communication. Finally, the parallel scalability and computational speed of LICOM3.0 were significantly improved. The second-highest resolution global ocean long time simulation in the world was achieved on the domestic heterogeneous supercomputing system (26,200 DCU accelerated cards), with a computational speed of 5.37 times faster than that of 382 cards and computational efficiency of 8% (Fig. 6) [16].

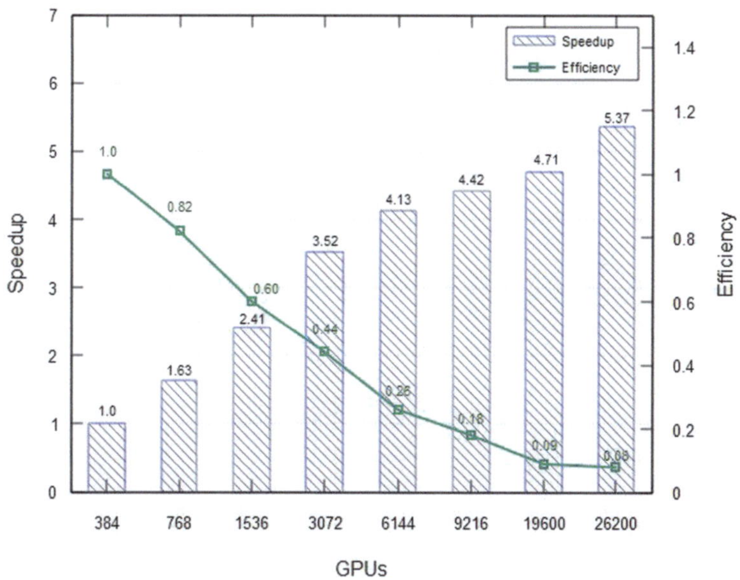

Fig. 6 Acceleration ratio and efficiency test

Table 1 Component mode combination

	ATM	OCN	LND	ICE
Name	IAP-AGCM	LICOM-HIP	CoLM	CICE
Resolution ratio	25 km	10 km	25 km	10 km
Use node	2880	768	384	768

3.2.3 Parallel Optimization of Coupled Integration Mode

China's independent high-resolution climate system coupled model was further developed based on parallel optimization of atmospheric and oceanic component models. The million-core parallel computing and high-resolution simulation applications realized on the domestic heterogeneous supercomputing system Zhongke Pioneer I. The model resolution index transformed from a follower to a runner-up in international. The combination of component models used and the number of nodes used is shown in Table 1.

A set of performance tests was performed on the above-coupled models of the high-resolution climate system on a cluster of domestic X86 processors. The computational process was performed concurrently using the atmospheric and oceanic component models based on the estimated computational effort for each component model. The ocean component model uses 768 computational nodes, and the

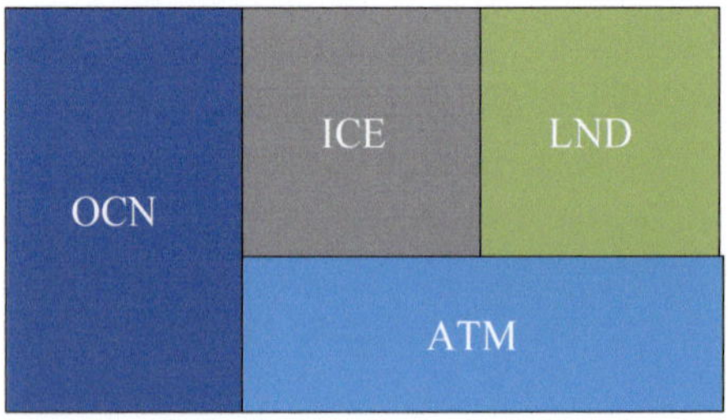

Fig. 7 Processor allocation scheme

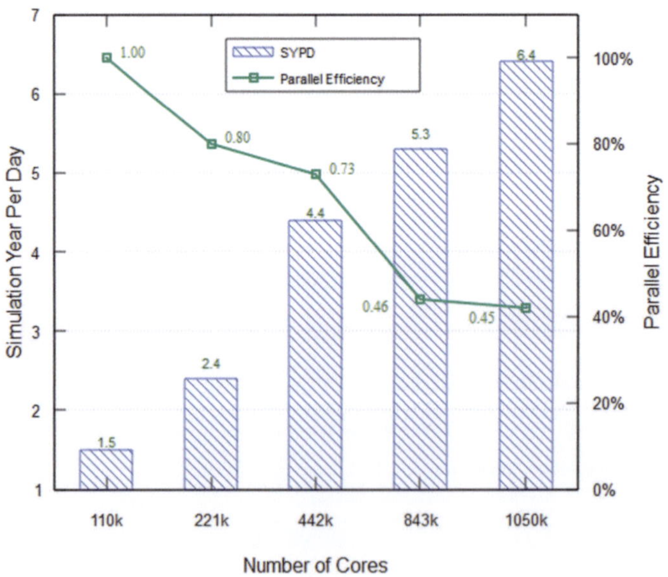

Fig. 8 Simulation speed

atmospheric component model uses 2880 computational nodes for 3648 nodes. The land surface component mode and sea ice component mode were executed serially using 2880 nodes and the atmospheric component model. The processor allocation scheme is shown in Fig. 7, and the test results show in Fig. 8. Using 3648 nodes and 1.05 million cores, the coupled model can compute at a 6.4 model years/day rate. There is over 40% efficiency relative to 100,000 cores in parallel.

3.3 Earth System Model Coupling Integration and Simulation Results

The Earth System Model Integration Module subsystem is responsible for the data transfer between the component models and the operation of the Earth System Model. The coupled Earth system modes are easily assembled for operation. The Earth System Model Integration Module (flux coupler) consists of four subsystems, the general control program, and the upper, middle, and bottom layers, see Fig. 9.

(1) The Earth system mode integration module's master control program is responsible for the Earth system model's start, run, and end. In the start phase, the master control program constructs the parallel environment, allocates processors, calibrates the various clocks, prepares the I/O, generates the data structure, reads the data mapping configuration file, initializes the mapping program, and starts each component mode of the Earth system mode in turn. In the run phase, the master control program sets the start, end, and run time points of each component mode according to the various clock alarms while migrating, mapping, fusing, and rearranging the data. In the end phase, the master control program is responsible for the recovery and destruction of the storage.
(2) The upper subsystem of the Earth system model integration module includes the flux calculation module, the side boundary module, and the coupling module. These modules are called the general control subsystem.
(3) The middle subsystem of the Earth System Model Integration Module includes modules for time management, 3D data structure, parallel control, complex grid mapping, I/O, and module statistics. These modules are invoked by the general control subsystem and the upper-level subsystem.
(4) The underlying subsystem of the Earth System Model Integration Module includes script management, extended MCT, interpolation routines, and automatic code generation modules, with MCT being invoked by the general control subsystem and the middle and upper subsystems. The script management is used for the configuration of the operational parameters of the Earth system model and the interpolation procedure used for the generation of the mapping matrix. It is followed by automatic code generation for the Earth system model extensions, all of which are supported by the Earth system model integration module. Once all three have completed their functions, the Earth System Model Integration Module General Control Subsystem then calls each of the other subsystem

modules in a running instance to complete the start, run and finish of the Earth Model System.

Eight subsystem models—atmospheric, oceanic, sea ice, land surface, vegetation dynamics, aerosol and atmospheric chemistry, and terrestrial and oceanic biogeochemical processes—have been coupled and integrated. A 'fully coupled' Earth system model CAS-ESM2 has been developed. Simulations of global CO_2 and terrestrial and oceanic (natural) carbon processes driven by anthropogenic emissions since the Industrial Revolution (1850) (Fig. 10) [17].

In addition to the fully coupled Earth system model described above, CAS-ESM2 also enables high-resolution climate simulations with a coupled atmosphere/land surface of 25 km and ocean/sea-ice of 10 km. Figure 11 shows the simulation results

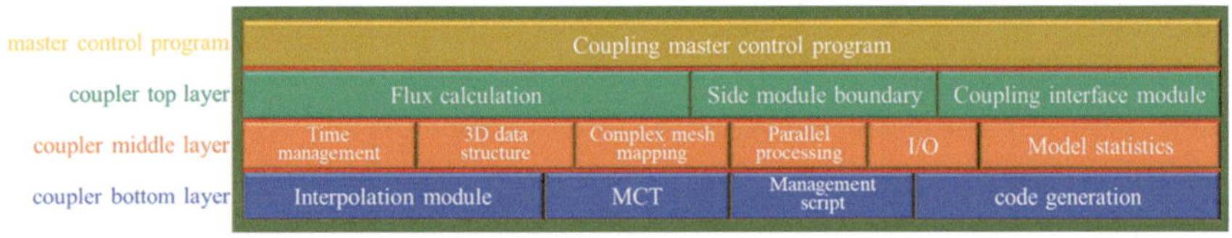

Fig. 9 Subsystem composition of earth system mode integration module

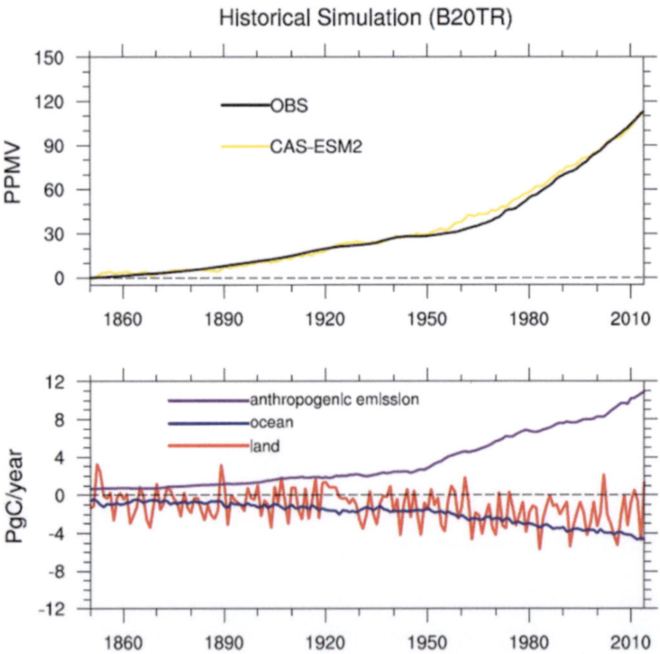

Fig. 10 Global CO_2 and anthropogenic emissions, terrestrial and marine (NATURAL) carbon changes [18]

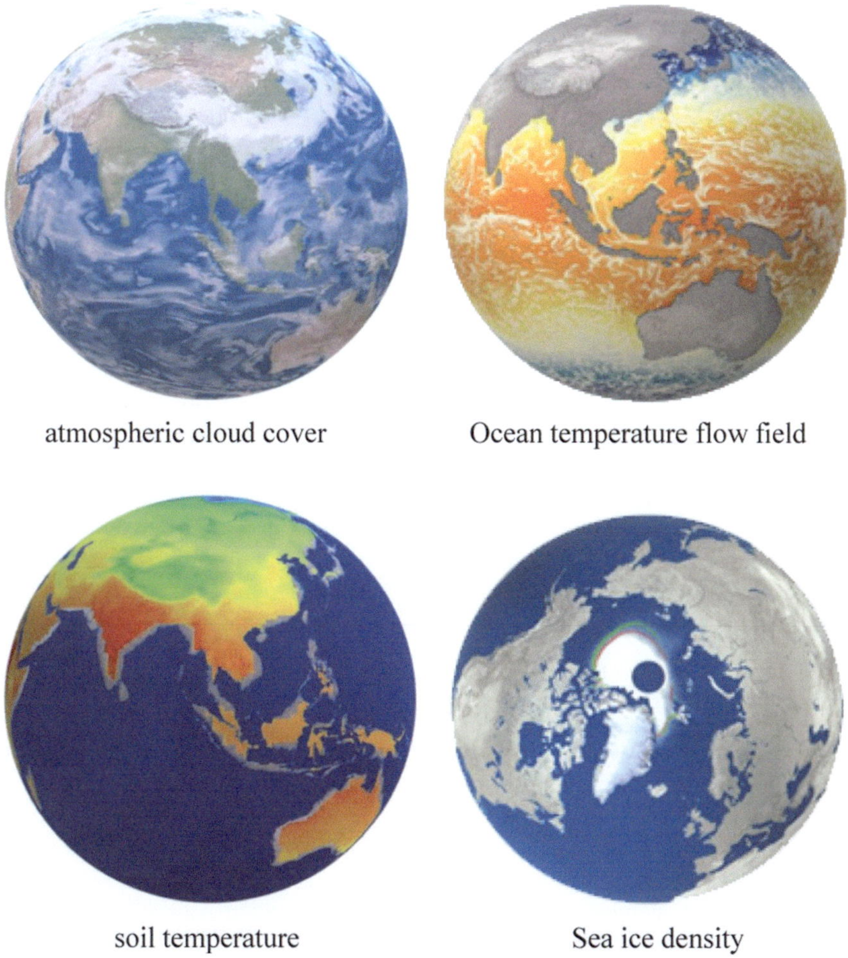

Fig. 11 High resolution coupling simulation of atmosphere/land surface and ocean/sea ice

for atmospheric cloudiness, soil temperature, ocean temperature flow field, and sea ice density.

4 Summary and Prospect

The Earth System Model is an important tool and experimental platform for conducting scientific research on the Earth system. It is a complex, large-scale software system involving multiple disciplines and is closely related to the development of computational mathematics, a high-performance computing software, and hardware. Our Earth system model and high-performance computer system development have laid the foundation for constructing an internationally advanced Earth system modeling device. We have carried out extensive work on high-performance parallel computing methods and the parallel optimization of CAS-ESM2, the core software system of the installation. We have achieved fully coupled Earth system simulations of climate and ecosystems. CAS-ESM2 participated in the sixth phase of the Coupled Model Intercomparison Project Six (CMIP6), carrying out experiments in several categories of CMIP6 and generating a large amount of experimental data on climate change simulations, providing a milestone in the development of CAS-ESM2 and in the development of the system. The successful development of the system is also a successful practice in constructing advanced high-performance computing hardware and software in China, which facilitates the sustainable development of the Earth system model and provides an example for the construction of software-centered large scientific devices in China.

In the future development direction, CAS-ESM will further improve the resolution of the model based on the autonomous high-resolution climate system model (25 km atmosphere + 10 km ocean). A fully coupled high-resolution Earth system model will also be developed. More subsystem/subsystem models will also be coupled, especially for anthropogenic processes, continental ice caps, middle and upper atmosphere, etc.

Thank

Thanks for the supported by the National Key Scientific and Technological Infrastructure project "Earth System Science Numerical Simulator Facility" (EarthLab).

The numerical calculations in this paper have been done on ORISE computing environment.

Thanks to the project participants for their work for the completion of the "earth system numerical simulation device", include the Institute of Atmospheric Physics of the Chinese Academy of Sciences, the computer network information center of the Chinese Academy of Sciences, the Institute of computing technology of Chinese Academy of Sciences, Zhongshan University and other units.

References

1. Zhou GQ, Zhang YQ, Jiang JR et al (2020) Earth system model: CAS-ESM. Front Data Comput 2(1):38–54
2. Wang B, Zhou TJ, Yu YQ (2008) A perspective on earth system model development. Acta Meteor Sin 66(6):857–869
3. Zeng QC, Lin ZH (2010) Recent progress on the earth system dynamical model and its numerical simulations. Adv Earth Sci 25(01):1–6
4. Zhou TJ, Chen ZM, Zou LW et al (2020) Development of climate and earth system models in China: past achievements and new CMIP6 fesults. Acta Meteor Sin 78(03):332–350
5. Zhang H, Zhang MH, Zeng QC (2013) Sensitivity of simulated climate to two atmospheric models: interpretation of differences between dry models and moist models. Mon Weather Rev 141(5):1558–1576

6. Liu HL, Lin PF, Yu YQ et al (2012) The baseline evaluation of LASG/IAP climate system ocean model (LICOM) version 2. Acta Meteor Sin 26(3):318–329
7. Hunke EC, Lipscomb WH (2002) CICE: the Los Alamos sea ice model documentation and software user's manual version 4.0 LA-CC-06-012. Los Alamos National Laboratory, Los Alamos NM
8. Dai Y, Zeng X, Dickinson RE et al (2003) The common land model. Bull Am Meteor Soc 84(8):1013–1023
9. Bonan GB, Oleson KW, Vertenstein M et al (2002) The land surface climatology of the community land model coupled to the NCAR community climate model. J Clim 15(22):3123–3149
10. Zeng XD (2010) Evaluating the dependence of vegetation on climate in an improved dynamic global vegetation model. Adv Atmos Sci 27(5):977–991
11. Xu YF, Li YC, Chu M (2013) A global ocean biogeochemistry general circulation model and its simulations. Adv Atmos Sci 30(3):922–939
12. Wang YZ, Jiang JR, Zhang H et al (2017) A scalable parallel algorithm for atmospheric general circulation models on a multi-core cluster. Future Gener Comput Syst 72:1–10
13. Wu BD, Li SG, Zhang YQ et al (2018) AGCM3D: a highly scalable finite-difference dynamical core of atmospheric general circulation model based on 3D decomposition. IEEE, Singapore
14. Wang YZ, Guo MX, Zhao Y et al (2021) GPUs-RRTMG_LW: high-efficient and scalable computing for a longwave radiative transfer model on multiple GPUs. J Supercomput 77(5):4698–4717
15. Wang YZ, Zhao Y, Jiang JR et al (2020) A novel GPU-based acceleration algorithm for a longwave radiative transfer model. Appl Sci 10(2):649
16. Wang P, Jiang J, Lin P et al (2021) The GPU version of LASG/IAP climate system ocean model version 3 (LICOM3) under the heterogeneous-compute interface for portability (HIP) framework and its large-scale application. Geosci Model Dev 14(5):2781–2799
17. Zhang H, Zhang MH, Jin JB et al (2020) Description and climate simulation performance of CAS-ESM version 2. J Adv Model Earth Syst 12(12):1–35
18. Gao XF, Zhu JW, Zeng XD et al (2021) Changes in global vegetation distribution and carbon fluxes in response to global warming: simulated results from IAP-DGVM in CAS-ESM2. Adv Atmos Sci 2022:1–14

Guangqing Zhou born in 1965, Doctor of Science, doctoral supervisor. Researcher fellow of Institute of Atmospheric Physics, Chinese Academy of Sciences, director of Information Science Center, chief technologist of earth system numerical simulation device. He is mainly engaged in the research and development of numerical model of earth system, numerical simulation and short-term climate prediction. He graduated from the Department of Meteorology of Nanjing meteorological college in 1988 and received a doctorate from the Institute of Atmospheric Physics of the Chinese Academy of Sciences in 1996. The leader of the leading group of "earth system dynamics model research" of the important direction project group of "phase III" of the knowledge innovation project of the Chinese Academy of Sciences, and presided over the project of "IAP ocean model improvement and coupling research with atmospheric and marine biochemical processes".One of the winners of the second prize of the National Natural Science Award for "Research on climate system model, simulation and climate predictability" in 2005.

Development Status and Prospects of Biodiversity Informatics in China

Cui Xiao, Congtian Lin, Zheping Xu, and Keping Ma

Abstract Driven by computer science, artificial intelligence and other technologies, biodiversity informatics has gradually developed into a data intensive scientific research paradigm. Internationally, biodiversity informatics has made great progress in the construction of large data platform (species checklist, specimen digitization, literature, color photos, public scientific data), online data analysis and modeling methods, interdisciplinary integration and data products. In recent 20 years, China's biodiversity informatics has developed rapidly. By combing the concept, development process and discipline frontier of biodiversity informatics, this paper focuses on the achievements of China's biodiversity informatics in database construction, special platform and international cooperation. The rapid development of biodiversity informatics in China has changed the weak situation of digital biological data in Asia. Nevertheless, the development of biodiversity informatics in China still needs to be strengthened: (1) Attach importance to the integration of biodiversity big data resources to form a more powerful comprehensive information platform; (2) Increase data sharing efforts. At present, the sharing level of data in the whole Asia is very different from that in Europe and the United States; (3) Establish the connection between big data platform and data source. Only continuous data source can keep the data updated; (4) Strengthen capacity-building in biodiversity informatics, especially in data mining. (5) Strengthen regional and international cooperation. Only through the accumulation of theoretical knowledge and accurate grasp of the

C. Xiao (✉) · K. Ma (✉)
State Key Laboratory of Vegetation and Environmental Change, Institute of Botany, Chinese Academy of Sciences, Beijing, China
e-mail: xiaocui@ibcas.ac.cn

K. Ma
e-mail: kpma@ibcas.ac.cn

C. Lin
Key Laboratory of Animal Ecology and Conservation Biology, Institute of Zoology, Chinese Academy of Sciences, Beijing, China

Z. Xu
Resource Construction Department, Documentation and Information Center, Chinese Academy of Sciences, Beijing, China

China's e-Science Blue Book 2023, https://doi.org/10.1007/978-981-99-8270-7_7

problem, together with the collection, integration and analysis of data, can we really answer scientific problems and realize the development of biodiversity informatics.

Keywords Biodiversity · Data standard · Specimen digitization · Data platform · International cooperation

The development of information technology and the advent of the era of big data led scientists to apply informatics technology to the field of biodiversity, including taxonomy, biogeography and ecology. Researchers used modern computer technology to view and analyze existing information, developing new predictive models. This integration of biodiversity research and information technology resulted in biodiversity informatics (BDI) [1]. Biodiversity informatics refers to the use of information technology to integrate, manage, analyze and interpret the basic data of biological organisms, especially at the species level [2, 3]. Focusing on individual organisms, populations, taxa and their interrelationships, BDI covers a wide array of disciplines, including systematics, evolutionary biology, population biology, behavioral science, pollination biology, parasitism, and plant community [4].

Biodiversity informatics is an emerging discipline that provides information for decision-making on biodiversity conservation and sustainable use [5] . Over the past decades, key software tools for managing, normalizing and publishing data—such as Lifewatch [6] and The Integrated Publishing Toolkit have been developed [7]. Concurrently, a number of authoritative databases and large-scale biological information networks have been produced [8], such as the Global Biodiversity Information Network (GBIF). These systems have the potential to promote development in industries such as agricultural science, biomedicine, biotechnology, environmental management, pest control, public health, popular science education and biodiversity conservation (GBIF Secretariat 2019).

China began exploring BDI in the 1980s. Scholars across different fields followed BDI's international development and discussed the necessity, problems, research methods, construction contents and possible support for the construction of biodiversity informatics in China [4, 9–14]. In 1983, the Chinese Academy of Sciences proposed the "Scientific Database and Its Information System," beginning to build special databases for a number of scientific disciplines. Some of these included biodiversity-related databases, like the China Animal Database, China Plant Database, Southwest China Resources Plant Database, and Freshwater Algae Seed Bank. This amalgamation of databases have gradually developed into today's Scientific Data Center of the Chinese Academy of Sciences. Since 2003, the Ministry of Science and Technology has been building a national platform to track and store basic scientific and technological information, including the "standardization, integration and sharing platform construction of specimen resources". In 2019, in order to improve the resource sharing service system and promote the opening and sharing of scientific and technological resources to the public, the Ministry of Science and Technology and the Ministry of Finance formed 20 national scientific data centers

Table 1 List of 19 national science and technology resource sharing service platforms related to biodiversity informatics

Serial number	Country platform name	Relevant department
1	National Genome Science Data Center	Chinese Academy of Sciences
2	National Microbial Science Data Center	Chinese Academy of Sciences
3	National Qinghai-Tibet Plateau Scientific Data Center	Chinese Academy of Sciences
4	National Ecological Science Data Center	Chinese Academy of Sciences
5	National Earth System Science Data Center	Chinese Academy of Sciences
6	National Basic Discipline Public Science Data Center	Chinese Academy of Sciences
7	National Forestry and Grassland Science Data Center	Forestry and Grass Bureau
8	National Ocean Science Data Center	Ministry of Natural Resources
9	National Important Wild Plant Germplasm Resource Bank	Chinese Academy of Sciences
10	National Horticultural Germplasm Bank	Ministry of Agriculture and Rural Affairs
11	National Tropical Plant Germplasm Bank	Ministry of Agriculture and Rural Affairs
12	National Forestry and Grassland Germplasm Bank	Forestry and Grass Bureau
13	National Marine Aquatic Germplasm Resource Bank	Ministry of Agriculture and Rural Affairs
14	National Freshwater Aquatic Germplasm Resource Bank	Ministry of Agriculture and Rural Affairs
15	National Parasite Resource Bank	Health Committee
16	National Strain Resource Bank	Ministry of Agriculture and Rural Affairs
17	National Pathogen Resource Bank	Health Committee
18	National Herbarium Repository	Chinese Academy of Sciences
19	National Taxidermy Resource Bank	Chinese Academy of Sciences

and 30 national biological germplasms (Table 1). The goal of this project was to provide high-quality scientific and technological resource sharing services for scientific research, technological progress and social development.

Under the encouragement of the Biodiversity Committee of the Chinese Academy of Sciences and related units, China has actively cooperated with internationally

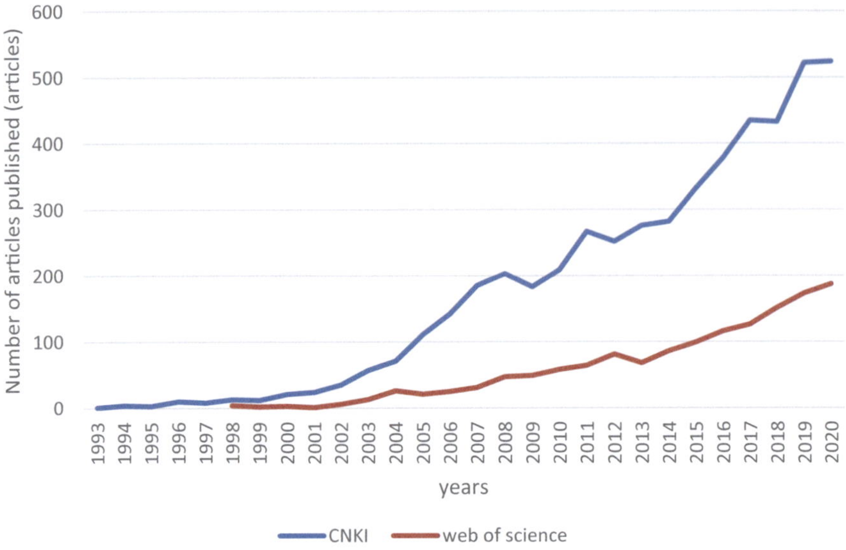

Fig. 1 The number of articles retrieved by the combination of "biodiversity" and "data"

renowned biodiversity informatics projects (such as the Species 2000 Global Inventory of Biological Species, EOL, Biodiversity cooperation with the Historical Documentation Library, the Global Biodiversity Information Network, and others). Beginning in 2009, a National Symposium on Biodiversity Informatics has been organized every two years to foster a professional academic exchange platform and promote the development of biodiversity informatics in China. An analysis of the annual number of articles searched in CNKI and Web of Science with the combination of "biodiversity" and "data" (Fig. 1) suggest that biodiversity informatics is gaining more and more attention.

1 Database Construction

The foundation for the development of biodiversity informatics involves specific information in a number of fields. Species names and distribution information are perhaps the most important. Chinese species data are centered on biodiversity cataloging, specimen data, color photos, bibliography, and research projects. In terms of genetic diversity, information mainly includes germplasm resources, DNA barcoding, and omics data. In terms of ecosystems and landscapes, collected information mainly includes sample plot data from field surveys, and data obtained by means of satellite remote sensing and near-ground remote sensing.

1.1 Data Standards

Biodiversity informatics has extensive and complex data sources, and data standards are particularly important during data fusion and sharing. Darwin Core is widely used as a biodiversity data standard, and it can extend into other biological fields (molecules, genes, paleontology, herbarium management). With more diverse sources of observational data and increases in the degree of integration, the Darwin Core has come to involve a basic text collection file. The files are organized by a simple descriptor, the format is pre-defined according to the rules of the Darwin Core text guidelines. It is no longer limited to species record data, but is tightly linked to Dublin Core. It is used by the Global Biodiversity Information Network (GBIF) to code and link biological names, taxa, species information, distributions and publications. The key idea of the archive format is the organization of the structure through the association of a core data file and multiple extension files. Each extension record points to a record in the core file. Thus, multiple extension records can exist in the form of each individual core record, which can greatly reduce data redundancy and compress the size of the data set. GBIF Integrated Publishing Tool (IPT) is based on this standard, aggregating 1.9 billion data records worldwide.

1.2 Species Diversity Data

1.2.1 Catalogue of Life China

In order to track the progress of taxonomic research and integrate the Chinese biological species list, the Biodiversity Committee of the Chinese Academy of Sciences has cooperated with the International Species 2000 organization since 2008 to initiate the construction of the "Chinese Biological Species List" (Fig. 2). This integrates the lists of all biological species distributed in China, conducts database construction in accordance with the International Species 2000 standard and norms, and provides free and open services to the public. The China Biological Species List releases an updated version annually. As of 2021, the list included more than 120,000 species and subspecies, covering animal, protozoa, plant, fungi, pigment, bacteria, viruses and other biological groups. The data content includes species scientific name, Chinese name, classification system, alias, distribution and related literature. It is the most comprehensive and widely used catalog database of taxa in China.

In order to protect and develop biodiversity-related research, Chinese scholars have adopted the classification and standards of the IUCN Red List, and have brought together more than 600 experts to assess the threatened status of various flora and fauna, excluding fish and macrofungi. The results indicate that the protection status of birds, reptiles, and amphibians in China is deteriorating, while the protection status of mammals, gymnosperms, and angiosperms is improving [15] . The release of the China Biodiversity Red List has laid a scientific basis for formulating biodiversity

Fig. 2 Cover of the 2008–2021 Inventory of Biological Species in China

conservation policies and plans, provided a data basis for scientific research on biodiversity, and created the necessary conditions for public participation in biodiversity conservation. This list has inspired specific actions, including the implementation of the China Biodiversity Conservation Strategy and Action Plan (2011–2030) and the Convention on Biological Diversity.

In February 2021, the State Forestry and Grassland Administration and the Ministry of Agriculture and Rural Affairs publicly released a new version of the National Key Protected Wildlife List. A total of 980 species in 8 categories of wild animals are included in the new list. One category contains 234 species of national first-class protected wild animals. The remaining seven categories account for 746 species of national second-level protected wild animals. This is the first major adjustment to the checklist in 32 years in China [16]. In September 2021, with the approval of the State Council, the State Forestry and Grassland Administration and the Ministry of Agriculture and Rural Affairs released the adjusted List of National Key Protected Wild Plants, accounting for 40 categories and 455 species of wild plants. This is the first update since 1999, when the original [17].

Fig. 3 Annual increments of specimen digitization data

1.2.2 Digitization of Biological Specimens

Biological specimens are based on physical objects, and have the characteristics of being "identified by professionals, covering a wealth of information, and spanning a large scale of time and space" [18]. A complete biological specimen includes multiple fields such as collection and identification information. With the development of photographic technology, scanning technology and computer technology, the digitization of specimens has become simple and feasible, thus overcoming many inconveniences related to the use of the Science and Technology Basic Conditions Platform of the Ministry of Science and Technology started the digitization of biological specimens in 2003 [19] . In the past ten years, the digitized data of specimens has increased year by year (Fig. 3). By the end of 2020, more than 15.6 million specimens had been digitized. Specimen digitized data includes plants (over 10.96 million), animals (over 4.35 million), fossils, rocks, minerals, ores, sediments, meteorites, ice and snow samples, and fungi. Plant and animal specimen data accounted for 70.7% and 28% of total data, respectively. All digitized specimens are shared in real time on the National Specimen Resource Sharing Platform, and users can obtain data by browsing online or filling in the data application form [20].

1.2.3 Literature Digitization

The digitization of Chinese biodiversity literature relies on the BHL China node (Biodiversity Heritage Library-China) of the China Biodiversity Digital Library. The BHL China node comprehensively collects, scans, extracts and systematically organizes important biological diversity documents; establishes a network platform for consultation and exchange of Chinese biodiversity information; and opens up interfaces to provide information on biodiversity information. Platforms, such as Encyclopedia of Life China Node, NSII and China Digital Herbarium, provide literature data services. The BHL China node currently contains 18,000 volumes of Chinese literature and 108,000 volumes of English literatures, including flora books, species lists, scientific research reports, maps, and other documents. The sharing of English documents addresses an issue that once plagued Chinese universities, which was that some institutions could not directly view foreign documents. At present, the bibliography of the BHL China node adds 1,000 volumes annually.

Bibliography digitization is the basis for extracting species distribution data. The names and distribution of species are critical information for biodiversity research. Since the published literature and published chronicles are highly accurate in content, data collection through literature and chronicles has become an important aspect of biodiversity data construction. For example, the current Chinese plant species distribution database based on bibliographical records includes 5,789,603 species distribution records.

1.2.4 Multimedia Data

The database of species color photographs was created due to the popularity of digital cameras. Color photos allow for the analysis of species type, location, specimen health, and specimen characteristics. The construction of China's plant color photo library, for instance, and the digitization of specimens developed almost concurrently. After 18 years of accumulation and construction, the China Natural Herbarium (CFH) and China Plant Image Database (PPBC) are the main body of China's plant color photo data, accounting for 68.9% of China's plant species color photos collection.

With the development of artificial intelligence, the past five years have seen a number of plant recognition software that can identify flower shape and color, among other characteristics. When users use software that links to a larger database, latitude and longitude coordinates are automatically generated with species data.

1.3 Genetic Diversity Data

Genetic diversity is one of the core components of biodiversity. Conservation of biodiversity is ultimately about preserving its genetic diversity, and the stability and evolutionary potential of a species depends on its genetic diversity, while the economic and ecological value of a species depends on its unique genetic composition [21]. Genetic diversity data of species include germplasm resources, barcodes, and DNA sequence data, among others.

1.3.1 Germplasm Resources

Crop germplasm resources. Crop germplasm resources are an important part of biodiversity and the most valuable and strategic resource of a country [22]. China was the first to domesticate and cultivate crops such as soybean, corn, plum, peach, and apricot, and is also one of the native origins of rice, flax, eggplant, banana, and sweet orange. Chinese crops involve a total of 574 taxonomically cultivated species and 1930 wild relatives. There are 20 species (classes) of food crops, 45 species (classes) of vegetables and spices, 53 species (classes) of fruit crops, and 26 species (classes) of cash crops that originated in China. The China Crop Germplasm Information Network has information on more than 340 crops with 470,000 germplasm, including grain, fiber, oilseeds, vegetables, fruit trees, sugar, tobacco, tea, mulberry, pasture, green manure, and heat crops. 18 genetic resources, and more than 30,000 new aquaculture germplasm resources were collected [23].

Forest tree species. There are more than 8,000 species of woody plants in China, including about 2,000 species of trees. The variety of timber species provides the basis for the cultivation of timber forests and the selection and breeding of good species. Among the timber species, there are 10 species and 1 variety of larch, 53 species of poplar, 51 species and 14 varieties and 1 variant of oak. Economic tree species are about 1000 species, mainly woody oil and grain, medicinal, chemical raw materials, fruit trees, woody vegetables and other tree species. There are more than 200 species of woody oil and grain trees, of which 50 species are edible, more than 100 species of woody grain trees, about 140 species of fruit trees, and about 1,000 species of woody medicinal plants. Protective tree species mainly include camphor pine, ash, Xinjiang poplar, dry willow, and purple acacia. For example, there are about 256 species and 63 varieties of willows in China, and seven species and four subspecies of the genus Buckthorn. There are more than 1200 species of ornamental trees in the garden, mainly seven-leaved trees, gooseberry, cedar, white pine, begonia, hedgehog, hibiscus, etc. For example, there are 11 genera and 140 species of Chinese magnolia family trees, more than 20 species of lilac genus; 90 species of rose genus. There are many species of energy trees, including 60 species of fast-growing and high-quality charcoal trees, and more than 10 species of major woody oil trees. Bamboo and rattan species are abundant, with more than 500 species of bamboo in 37 genera in China, including 10 genera and 48 species of endemic bamboo taxa; rattan resources are abundant, with 3 genera and 42 species and 26 varieties of palm rattan, and 37 species and 26 varieties of provincial rattan [24]. The National Forest Germplasm Repository shows that a total of 761 new varieties of forestry plants were authorized in 2021, with the authorized varieties mainly in the genera of rose (159 in total), plum (105), peony (95), rhododendron (49), magnolia (23), and apple (23). In 2021, 22 new species were approved as forest tree species, including poplar, sea buckthorn, thin-shelled pecan, oil tea and 12 other tree species.

By the end of 2021, China has announced two batches of 99 national forestry and grass germplasm resource repositories, preserving 240 species of trees, mainly preserving native species such as oak, *Camellia oleifera*, *Eucommia ulmoides*,

Gleditsia sinensis, Camellia japonica, Acer truncatum, Sinojackia xylocarpa, etc., and preserving about 50,000 copies of rare and endangered tree species and tree species resources with cultivation and utilization value or potential utilization value. Construction of national forestry and grass germplasm resources facilities preservation library 4 sub-repositories.

Flower resources. China has a wide variety of flowers, with 1,331 genera and 4,708 species of flowers, including more than 1,000 kinds of herbaceous flowers, including more than 400 kinds of flowering plants, more than 100 kinds of foliage plants, more than 100 kinds of aquatic ornamentals, nearly 300 kinds of grasses and more than 200 kinds of succulents. Camellias, azaleas, moonflowers, peonies, osmanthus, primroses, etc. are mostly distributed in China as the center of resources. There is a lack of a national floral resource website to unify and display national floral germplasm resources. As of the end of 2021, there are more than 300 varieties of plum blossoms, 150 varieties of osmanthus, 3000 varieties of chrysanthemums, 800 varieties of peonies, 400 varieties of peonies, 300 varieties of camellias. In addition, there are many famous varieties of moonflowers, roses, lilacs, violets, azaleas, peach blossoms, apricots, and cherry blossoms [24]. Taking camellia as an example, by the end of 2021, 144 varieties of Camellia spp. have obtained the national certificate of new plant variety rights. The sequencing of floral resources such as plum, Chinese lotus, reeling flowers of the genus Rosa, Rhododendron macrophylla, chrysanthemum brain of the genus Chrysanthemum, and peony has been completed, especially the resequencing of 333 plum varieties and 15 wild species of plum.

Forage resources. A total of 604 new forage grass varieties have been validated in China, including 227 breeding varieties and 83 native grass varieties, 32 grass varieties, including 'Xiuzhu Muqin' sheep grass, have been validated twice, including 25 old mangosteen, mountain wheatgrass and wolftail Grass species, mostly used for both pasture and ecological restoration grasses. The cumulative identification and evaluation of more than 13,000 grass germplasm resources, screening out a number of excellent germplasm resources such as high protein alfalfa, long spike alfalfa, forage high-yielding oats, etc., discovering 6 genes for cold resistance of lentil, screening more than 500 excellent germplasm resources, and providing breeding materials for new variety selection and breeding.

Wild plant resources. The National Wild Plant Germplasm Resource Center is based on the Wild Plant Germplasm Resource Center in Southwest China and cooperates with relevant units nationwide to collect and preserve wild plant germplasm resources. With national strategic needs as a framework, standardized collection, sorting, and preservation will continue. By the end of February 2022, 18,390 plant resources compposed of 307 families, 2893 genera and 18,390 species were shared through the website.

Wildlife resources. China has carried out genetic diversity determination, genetic management and genealogical archiving of artificial populations of endangered wildlife, such as giant pandas, elk, South China tigers, golden monkeys, ibis and Yangtse crocodiles. Taking the red panda as an example, the red panda is divided into two subspecies, the Himalayan subspecies (A.f. fulgens Cuvier, 1825) and the Chinese subspecies (A.f. styani Thomas, 1902), and the whole genome, Y chromosome and mitochondrial genome can comprehensively distinguish species, subspecies and populations. Data from 65 whole genomes, 49 Y chromosomes and 49 mitochondrial genomes of red pandas completed in China provided the first comprehensive genetic evidence for species differentiation in red pandas [25]. China has carried out genetic exchange of captive populations of wild animals to avoid inbreeding, maintain genetic diversity, and use wildlife genetic resources to improve the genetic quality of domesticated animals.

Edible mushrooms. China is one of the richest countries in the world in terms of large fungal resources, with about 1020 known species of edible mushrooms, about 90 species domesticated, more than 50 species commercially cultivated, and an annual production of more than 30 million tons of edible mushrooms, accounting for more than 70% of the global total annual production. Geographically, southwest China is the richest region of mushroom resources in the country, with about 161 genera and 882 species of edible mushrooms in 54 families found in Yunnan province only. Some widely cultivated edible mushrooms, such as shiitake, enoki, fungus and chrysalis, are widely distributed and have high genetic diversity in China.

1.3.2 DNA Barcode Data

DNA barcoding (DNA barcoding) is a tool for accurate species-level identification by base sequence differences using one or several short and standardized gene fragments that are commonly found in organisms as universal fragments. Compared with traditional taxonomy, DNA barcoding technology has the advantages of not relying on morphological features and developmental stages, digital identification, rapid and accurate, and simple and standardized operation. It has been successfully applied in many fields such as biodiversity monitoring, customs, herbal medicine authenticity identification, forensic identification, animal and plant quarantine, biological invasion, food and drug market supervision, etc. After 15 years of development, DNA barcoding research has basically determined the use of different universal fragments for different biological taxa, while a standard global digital database of vital barcodes has been constructed [26]. At present, DNA barcoding research is mainly focused on how to improve the resolution of closely related taxa species and construct a regional barcode database [26].

China has made great contributions in promoting the development of DNA barcoding technology, especially in the field of plant barcoding. Through extensive data analysis, Chinese scientists recommended ITS (Internal Transcribed Spacer) as the core fragment of plant barcoding and proposed technical standards and specifications suitable for plant barcoding research [27, 28]. The Chinese plant DNA barcode reference sequence database constructed by the Institute of Botany, Chinese Academy of Sciences covers more than 185,000 species in 1058 families and 16,854 genera, including more than 5.03 million international common DNA barcodes and taxon-specific DNA barcodes, of which the data volume of four common barcodes of Chinese domestic taxa accounts for about one-third of all data. The database management platform has built a high-throughput DNA barcode acquisition technology system based on second-generation sequencing technology, focusing on sample labeling methods and automated data processing methods.

In 2012, Kunming Institute of Botany, Chinese Academy of Sciences led the launch of a new generation of intelligent phytobotanical journal (iFlora) research program [29]. The program is based on the Chinese e-Phytobotanical Journal, which integrates elements of DNA barcode data, next-generation sequencing technology, geographic information data, and computer technology to achieve a digital platform for rapid and accurate species identification and information sharing. As of January 20, 2022, the platform has completed the integration of core data, basic data and expanded data of 39,971 species (including subspecies rank) of 300 families and 34,434 genera at three levels of information.

China is the country and region that submitted the most plant barcode data in The Barcode of Life Data System (BOLD). In addition, for Chinese herbal medicines, the Chinese Academy of Medical Sciences has established a DNA barcode identification system for Chinese herbal medicines, and the Chinese University of Hong Kong has constructed a DNA barcode database for herbal medicines.

1.3.3 Genomic Data

The rapid development of high-throughput sequencing technologies has provided us with powerful genome sequencing capabilities, thus enabling biological data to grow at an unprecedented exponential rate at different histological levels (including genomics, transcriptomics, proteomics, epigenomics, metabolomics, etc.). The National Genome Science Data Center was established in June 2019 on the basis of the Beijing Institute of Genomics (BIG), established in 2003, and the BIG Data Center (BIGD), established in 2015. Facing the strategic development needs of national big data, the National Genome Science Data Center focuses on the construction of data resources and database system around human, animal, plant and microbial genomic data, and carries out a series of work on data services, system operation and maintenance, technology research and development, and data mining. At present, the National Genome Science Data Center has the genomic data remittance, management and sharing system with independent intellectual property rights, which guarantees data security and supports and serves the data storage, management and sharing of more than 300 scientific research projects, such as the National Key Research and Development Program, National Natural Science Foundation of China, and CAS pioneering special projects. As of September 28, 2021, the National Genome Science Data Center has reached an archive of over 10 petabytes of raw sequence data.

The Shenzhen National GeneBank was approved for construction by four ministries: National Development and Reform Commission, Ministry of Finance, Ministry of Industry and Information Technology, and National Health and Wellness Commission (formerly Ministry of Health), and was constructed by Shenzhen

Huada Institute of Life Sciences (formerly Shenzhen Huada Gene Research Institute), and was completed and used in 2016. It was completed and used in 2016. Shenzhen National Genebank has 10 million samples storage capacity, 691 trillion times/sec computing capacity, petabyte-level data output capacity, million bases/year genome synthesis support capacity, 30+ science-specific databases, and 10+ analysis tools. The Shenzhen National Genebank stores, reads, synthesizes and uses biogenetic resources, and builds a public welfare, open, leading and strategic science and technology platform for mining genetic resources and supporting life science research and bio-industry innovation and development, which is the world's leading comprehensive genebank for storing, reading and writing biogenetic resources.

1.4 Vegetation Ecological Data

In 2020, the "Vegetation Map of China (1:1,000,000)" was updated to 866 groups and subgroups of 12 vegetation types, with approximately 3.3 million km^2 of vegetation types changed compared to the "Vegetation Atlas of China (1:1,000,000)" published in 2001 and the "Vegetation Map of the People's Republic of China (1:1 000 000)" published in 2007 [30]. The "1:1,000,000 China Vegetation Map Database" on the NSII website contains 60 submaps and their data, indicating the geographical distribution of 11 vegetation type groups, 55 vegetation types, 960 group systems and subgroup systems, and more than 2,000 community dominant species, major crops, and cash crops in China. The "1:500,000 China New Generation Vegetation Map Database" will further refine the 1:1 million China vegetation map and contain 40 TB of satellite remote sensing and near-ground remote sensing data and more than 200,000 ground records.

2 Information Platform Construction

The data platform is the display window and output carrier of data. The good or bad platform construction is closely related to the quantity and quality of data. In recent years, various data platforms have emerged in the field of biodiversity informatics. From national to provincial and municipal, and even protected areas, various comprehensive and application-oriented platforms have been established one after another. Relevant government agencies, research institutes, universities and colleges have built biodiversity-related data and information platforms according to their needs.

2.1 Integrated Biodiversity Platform

2.1.1 Biodiversity and Ecological Security Big Data Platform

The Biodiversity and Ecological Safety Big Data Platform (BioONE), relying on the Institute of Botany of the Chinese Academy of Sciences, aims to comprehensively integrate the biological and ecological information resources of the Chinese Academy of Sciences, improve data correlation and fusion, quality control and data mining capabilities, improve data management standard development and data visualization tool development, and provide the most core scientific BioONE aims to provide the core scientific support for the construction of ecological civilization, ecological function zoning, ecological security pattern construction, biodiversity and regional ecological protection, biological resources industry innovation and sustainable development goals. At present, BioONE has integrated 2.6 billion biodiversity and ecological security-related data, with a total capacity of 5133TB, including paleontological data, species diversity data, biogenetic data, vegetation map and ecological security data, covering from paleontology to extant organisms, including genetic data, species data and ecosystem data, providing multi-dimensional data retrieval and visualization at different temporal and spatial scales. It provides multi-dimensional data retrieval and visualization at different temporal and spatial scales.

2.1.2 Strategic Biological Resources Programme of Chinese Academy of Sciences

In the 12th Five-Year Plan, the Chinese Academy of Sciences (CAS) launched the Strategic Biological Resources Programme (BRP), with the basic function of serving social development and supporting scientific research. The BRP integrates resources from botanical gardens, herbaria, repositories, biodiversity monitoring networks, and experimental animal platforms to build a holistic resource system that is geared to the major needs of the country and the main battlefield of the national economy. On the basis of long-term collection and preservation of resources, we have realized the analysis, evaluation and utilization of resources, promoted the digitization and informatization of resources, and provided resource support, talent support and technical support for the completion of important scientific research tasks.

The Strategic Bioresources Program of Chinese Academy of Sciences has formed a "5 + 3 + 1" network structure, i.e., five resource collection and preservation platforms, three resource evaluation and transformation platforms, and one strategic bioresources information center, and established a scientific steering committee and a management committee to strengthen the guidance and management of the Strategic Bioresources Program. As of January 22, 2022, the Strategic Biological Resources Program of Chinese Academy of Sciences has 6,735,941 biological specimen data, 139,945 plant records, 158,375 biogenetic resources, 19,822 model and experimental animals, and 384,624 biodiversity monitoring networks.

2.2 *Thematic Data Platforms*

2.2.1 National Specimen Resource Sharing Platform

The National Specimen Resource Sharing Platform (NSII) relies on the Institute of Botany, Chinese Academy of Sciences, and is mainly for sharing digital specimens, providing services such as specimen query, specimen data download, specimen data cleaning, visual display of specimens, and specimen resource promotion (Fig. 4). There are six sub-platforms under the specimen resource sharing platform: plant specimen sub-platform (to be developed into National Plant Specimen Resource Center in 2019), animal sub-platform (to be developed into National Animal Specimen Resource Center in 2019), protected area sub-platform, rock and mineral fossil sub-platform (to be developed into a National Rock and Mineral Fossil Specimen Resource Center in 2019), teaching sub-platform, and polar resource sub-platform, each sub-platform or Each sub-platform or resource repository has a corresponding website and provides services to the public. As of December 1, 2021, the NSII website has 16.36 million specimen records, 6.74 million specimen photos, 16.1 million color photos of species, 100,000 volumes of literature and more than 200 topics. It has become the largest data sharing platform for biological specimens in Asia.

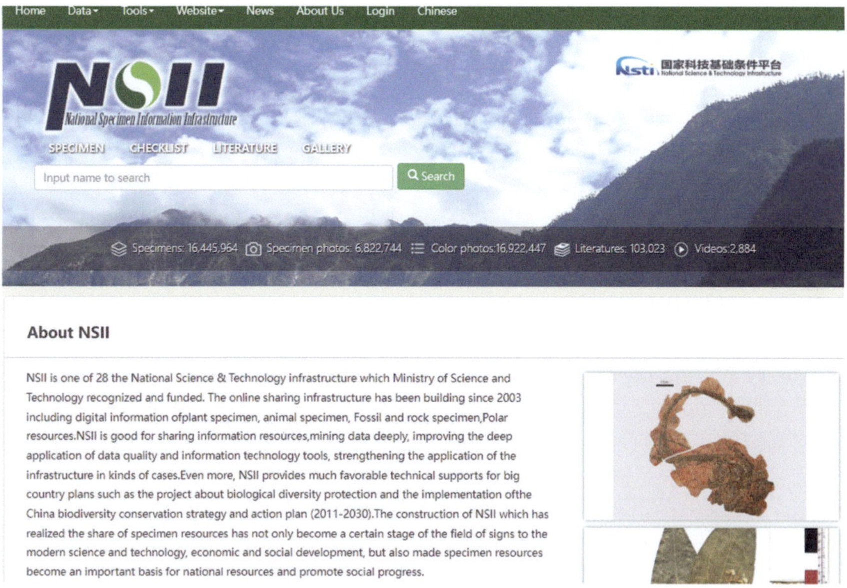

Fig. 4 Website home page of NSII

2.2.2 China Plant Subject Database

Relying on the Institute of Botany of the Chinese Academy of Sciences, China Plant Subject Database is based on the Species 2000 China node and the Chinese Flora List, integrating color photos of plants, plant literature records and different types of plant data to serve scientific research, government decision-making and education and science popularization. As of December 1, 2021, the data capacity of Chinese plant subject database exceeds 2.74 TB, covering 90% of the species in the Chinese plant list, plant pictures cover 80% of Chinese plant species, specimen data cover 60% of Chinese plant species, there are 29 datasets, 7 software tools.

2.2.3 China Animal Scientific Database

China Animal Scientific Database covers all vertebrate taxa, some insects and other invertebrate taxa, with over 40,000 species, over 350,000 visits per year, and over 1TB of downloads per year, with users from research institutions, universities, government departments and public science teams worldwide. It is the most comprehensive and widely used scientific database on zoology in China.

2.2.4 National Microbial Resource Platform

The National Microbial Resource Platform is jointly undertaken by 9 research institutes and universities, with 9 national specialized microbial strain management and preservation centers in China for agriculture, medicine, medicine, industry, veterinary, general, forestry, typical cultures and marine as core units respectively, to carry out the collation and integration of microbial resources and shared operation services, and 104 resource-advantaged units in different fields are organized to carry out standardized collation of resources. The platform has systematic description specifications, technical protocols as well as operation mechanism and management system, and has established a perfect microbial resource information database, physical library and sharing system, and the library resources have increased from 40,000 to 162,000 strains.

2.2.5 National Plant Specimen Resource Center

National Plant Specimen Resource Center was established in 2020, relying on the Institute of Botany, Chinese Academy of Sciences. It is one of the national science and technology resource sharing service platforms approved by the Ministry of Science and Technology and the Ministry of Finance. The National Herbarium has 8,181,464 specimen data, 66,688 type specimens and 66,113 specimen photos. The National Herbarium aims to comprehensively improve the collection, representativeness, management level and sharing service capacity of the physical library and digital platform through macro layout and precise collection, and serve domestic and foreign plant classification and evolutionary biology research, plant resource conservation and utilization.

2.2.6 National Animal Collection Resource Center

The National Animal Collection Resource Center is a national scientific and technological resource sharing service platform approved by the Ministry of Science and Technology and the Ministry of Finance. It is one of the 30 national biological germplasm and experimental material resource banks. At present, the collection has more than 20 million kinds of animal specimen physical resources. The collected specimens cover all provinces, regions, sea areas and typical ecosystems in China, accounting for nearly 70% of China's animal specimen resources and nearly 90% of known animal species. The National Animal Specimen Resource Bank aims to digitize the collection, arrangement, production, preservation, and research of animal specimens by integrating China's animal specimen resources. By disseminating digital and physical resources, the Resource Bank promotes the preservation, management and construction of animal specimen resources.

2.2.7 Plant Science Data Center

The Plant Science Data Center is based on the existing data, platforms and facilities of the plant science-related institutes (gardens) of the Chinese Academy of Sciences (CAS), and focuses on the idea of "supporting scientific research and promoting national development". Based on the idea of "supporting scientific research and promoting national development", we will comprehensively integrate the data resources of the Chinese Academy of Sciences in the areas of plant species, vegetation ecology and relocation conservation, and realize a new model of centralized management of plant science data resources of the Chinese Academy of Sciences. A new model of centralized management of plant science data resources. As of the end of January 2022, a total of 1,138,157,858 plant data with 266,121,886 MB file volume have been aggregated.

2.2.8 Joint Conservation Plan of Chinese Botanical Gardens

The Joint Conservation Plan of Chinese Botanical Gardens, formerly known as the Chinese Botanical Gardens Alliance, is a coordinated management mechanism among botanical gardens (arboretums and medicinal botanical gardens) in China, jointly established with the support of the Chinese Academy of Sciences, the State Forestry and Grassland Administration, the Ministry of Housing and Urban-Rural Development and the Ministry of Ecology and Environment. It aims to promote the standardized construction and orderly development of botanical gardens nationwide, gradually improve the layout of botanical gardens, promote the sharing of species resources, information and technical exchanges among botanical gardens, promote the establishment of China's botanical garden system and the enhancement of innovation capacity, and serve the development of ecological civilization and the construction of an innovative country. At present, we have effectively promoted the conservation of national strategic plant resources, the conservation of plant diversity in China and the construction of the national botanical garden system through the implementation of the "Native Plant Protection Plan", the construction of the national botanical garden standard system, the public science popularization program and the capacity program, and achieved remarkable results; at the same time, we have built and continuously optimized the relocation conservation Plant big data platform (covering 7 sub-systems such as PIMS plant information management system, botanical garden image library and full-coverage conservation database of native plants), and officially released the global biological image search engine with 3.4 million name data, which can search for 350,000 species of organisms in 43,031 genera of 4470 families.

3 International Cooperation

The international collaborations related to biodiversity informatics initiated or participated by China are shown in Table 2.

3.1 China's Involvement in Global Projects

Species 2000 is an international organization dedicated to building an electronic global list of biological species. The Species 2000 China Node is a regional node of the international Species 2000 project, proposed by the Species 2000 International Secretariat on February 7, 2006, and officially launched on October 20 of the same year. The Biodiversity Committee of the Chinese Academy of Sciences, together with its partners, supports and manages the establishment of the Species 2000 China Node. The main tasks of the Species 2000 China Node are to organize and collate taxonomic information of all biological species distributed in China according to the Species 2000 standard data format, to establish and maintain the Chinese biological species list, and to provide a free service for users worldwide.

The Global Biodiversity Information Network (GBIF), an international network and data infrastructure built in cooperation with several countries and organizations worldwide, is the largest basic database on biodiversity.GBIF aims to provide common standards and open source tools for data depositories around the world, providing open access to all types of vital data on Earth.GBIF Asia region includes Northeast Asia, South Asia Southeast Asia, and West Asia subregions. A total of nine

Table 2 Introduction to the main international cooperation institutions related to biodiversity informatics in which China is involved or initiated by China

Serial number	Name	Target	Introduction	China action
1	Convention on Biological Diversity (CBD)	To protect the sustainable use of biological diversity and its components, and to share the benefits arising from the utilization of genetic resources in a fair and equitable manner	An international convention for the protection of the earth's biological resources. As of November 2021, the CBD had a total of 196 parties	China entered on December 29, 1993, and CBD is applicable to Hong Kong and Macao SAR. On October 11–15, 2021, the fifteenth CBD Conference of the Parties (COP15) was held in Kunming, China, with the theme of "Ecological Civilization: Building a Community of Life on Earth"
2	International Union for Conservation of Nature (IUCN)	Major missions include saving endangered flora and fauna, establishing national parks and nature reserves, and assessing the conservation status of species and ecosystems	It is the largest and oldest global non-profit environmental protection organization in the world. Active in more than 160 countries, it brings together more than 1,400 member organizations and more than 18,000 experts	In 1996, the Chinese government joined the IUCN and China became a national member. The China Liaison Office was established in 2003, and the IUCN China Representative Office was officially established in 2012. IUCN has 32 Chinese member units, including 4 in Hong Kong.

(continued)

Table 2 (continued)

Serial number	Name	Target	Introduction	China action
3	Convention on International Trade in Endangered Species of Wild Fauna and Flora (CITES)	Designed to ensure that international trade in species does not endanger the survival of wildlife.	It is an intergovernmental international trade convention with 183 contracting parties as of December 2019.	CITES officially came into force for China on April 8, 1981. In 1982, the Scientific Committee on Endangered Species of the People's Republic of China was established in the Chinese Academy of Sciences as a scientific organization for the implementation of the treaty, and the State Forestry and Grassland Administration established the Endangered Species Import and Export Administration of the People's Republic of China
4	Species 2000	Aims to create an inventory of all known species on Earth as a base dataset for global biodiversity research	Membership in Species 2000 is open to any organization, project or individual interested in advancing the organization's goals	Species 2000 China Node was officially launched on October 20, 2006. The Biodiversity Committee of the Chinese Academy of Sciences, together with its partners, supports and manages the establishment of the Species 2000 China Node
5	Global Biodiversity Information Infrastructure (GBIF)	Aims to provide common standards and open-source tools for data custodians around the world, providing open access to data on all types of life on Earth	It is constructed in cooperation with many countries and organizations around the world.	The Chinese Academy of Sciences joined GBIF in 2013 and released 1.6 million pieces of data.

(continued)

Table 2 (continued)

Serial number	Name	Target	Introduction	China action
6	Biodiversity Heritage Library (BHL)	Aims to build a repository of biodiversity literature	The world's largest open access digital library of biodiversity documentation and archives.	The Institute of Botany, Chinese Academy of Sciences contributed 1,033 volumes of books with a total of 327,000 pages.
7	Botanical Garden Conservation International (BGCI)	Aims to mobilize botanical gardens and their partners to participate in the conservation of plant diversity	Founded in 1987, it currently has more than 800 botanical gardens, research institutions and individual members in more than 120 countries.	The BGCI China Project Office was established in 2007 to use technological expertise in the global botanical garden network to support China's plant protection.
8	Marine Biodiversity Information System (OBIS)	Aims to create a global open access data and information clearinghouse	More than 20 OBIS nodes around the world connect 500 institutions from 56 countries	China joined OBIS in May 2004
9	Asia Plant Digitization Program (MAP)	It aims to establish a network platform for the distribution of Asian plant diversity by collecting and sorting data sets	At present, the sorting of plant lists of 48 countries in Asia has been completed, and the plant lists of Northeast Asia, North Asia and Central Asia will be published and released soon	Established in November 2015

(continued)

Table 2 (continued)

Serial number	Name	Target	Introduction	China action
10	"China-ASEAN Environmental Information Sharing Platform"	The interconnection and exchange of environmental information between China and ASEAN, improving environmental data collection, sorting, analysis, processing and application; and promoting regional sustainable development	Participating countries include China and ten members of ASEAN	Construction started in 2016 by the China-ASEAN Environmental Protection Cooperation Center
11	"One Belt One Road Biodiversity Monitoring and Protection Research Center"	Strengthen cooperation in ecology and biodiversity in investment and trade	Promote the "Digital Belt and Road Initiative" biodiversity protection work	In May 2018, the China Biodiversity Conservation and Green Development Foundation and experts from the Institute of Remote Sensing Earth, Chinese Academy of Sciences formally established a cooperative relationship

countries or institutions are members of GBIF. The Chinese Academy of Sciences joined GBIF in 2013 and established a node in China, which has published 1.6 million digitized plant specimen data.

The Biodiversity Heritage Library (BHL), the world's largest open access digital library of biodiversity literature and archives, aims to develop the most comprehensive, reliable, and reputable repository of data-rich biodiversity literature and other primary materials, create services and tools that facilitate the discovery of BHL services and tools that help discover BHL content, etc. The Institute of Botany of the Chinese Academy of Sciences contributed 1,033 volumes, totaling 327,000 pages.

The Ocean Biodiversity Information System (OBIS), a global open access data and information clearinghouse whose mission is to build and maintain a global consortium working with the scientific community to promote free and open access to biodiversity and biogeographic data and marine biological information and applications. Currently, there are more than 20 OBIS nodes worldwide connecting 500 institutions from 56 countries, and China joined OBIS in May 2004.

3.2 An International Collaboration Initiated by China

The Mapping Asia Plants (MAP) digitization program was proposed by the Committee on Biodiversity of the Chinese Academy of Sciences (CAS) in November 2015 at the annual working meeting of the Asian Biodiversity Conservation and Information Network (ABCDNet), which it leads, with the support of the Bureau of International Cooperation of CAS, the One Belt, One Road The aim of MAP is to establish a network platform for the distribution of plant diversity in Asia by collecting and organizing data sets on plant species and distribution in Asia, and to provide a comprehensive platform for plant MAP aims to provide comprehensive basic information and interdisciplinary data mining environment for plant diversity conservation and research in Asia. At present, we have completed the initial combing of the plant lists of 48 Asian countries, and will soon publish and release the plant lists of Northeast Asia, North Asia and Central Asia.

The "China-ASEAN Environmental Information Sharing Platform" was initiated by the China-ASEAN Environmental Protection Cooperation Center in 2016. The main objectives of the platform construction are to promote the interconnection, interoperability and mutual use of environmental information between China and ASEAN, to improve the capacity of environmental information and data collection, collation, analysis, processing and application, and to promote regional green and sustainable development, etc., including various environmental themes such as biodiversity and ecological conservation and environmental sustainability. Participating countries include China and ten ASEAN member countries.

In the programmatic documents, guidelines and planning documents of the "Belt and Road" construction, the need to strengthen cooperation on ecology, environment and biodiversity in investment and trade is emphasized, and the need to establish biodiversity databases and information sharing platforms is expressed. in May 2018, the Institute of Remote Sensing and Earth Sciences of the Chinese Academy of Sciences took the lead in promoting "Digital Belt and Road" (DBAR) construction, including information on biodiversity.

4 Outlook

The availability of biodiversity data and information, and our ability to use this information effectively, reflects the evolving value of biodiversity informatics. Developing the capacity to effectively mobilize, manage, publish, and use biodiversity data can support national strategies that will incorporate international initiatives, including the Convention on International Trade in Endangered Species (CITES), the Intergovernmental Platform on Biodiversity and Ecosystem Services (IPBES), the Convention on Biological Diversity (CBD) Aichi Targets, the United Nations Convention to Combat Desertification (UNCCD), the United Nations Framework Convention on Climate

Change (UNFCCC), and the Sustainable Development Goals (SDGs), among others. The development of these strategies requires relevant, reliable and accurate data [31].

In April 2019, the seventh plenary session of the Intergovernmental Science-Policy Platform on Biodiversity and Ecosystem Services (IPBES) was held in Paris, France, to take stock of the achievement of the targets of the first round of the work program (2014–2019) on the one hand, and to adopt the second round of the work program (2019–2030) on the other. 2030). The second round of the work program will scientifically assess the progress of the 2030 Global Biodiversity Targets, the 2030 Agenda for Sustainable Development and the Paris Agreement. In discussing the results of the first round of the work programme, some key issues were also identified. "Some of the data used for assessments are relatively old, with reports generally assessed on a 3–4 year cycle, and data even 10 years old, reducing the relevance of the findings for reference and policy guidance....... Cultural differences in the background of assessment authors, language limitations, etc., resulting in low confidence in the assessment evidence such as data and information" [32]. One of the key elements of the latest conceptual framework proposed by IPBES is the Nature's Contributions to People (NCP, Nature's Contributions to People), which builds on the concept of ecosystem service proposed by the Millennium Ecosystem Assessment (MA) and encompasses all of nature's contributions to people [33, 34]. In 2019 IPBES completed its first global assessment exploring the overall status and trends of nature's contribution to humans [35], but modeling spatio-temporal patterns of change in NCP remains a major challenge, especially in terms of data credibility and computational power [36]. Therefore, data from biodiversity informatics are the basis for conducting global assessments of spatial and temporal patterns of change in NCP.

For China, there is a lack of local research base in related fields or the research results are not internationally competitive because of incomplete assessment data [37], and even some assessment topics (e.g., pollinators, pollination) are in a state of background ambiguity [38]. Given that IPBES-related assessments are receiving more and more attention from governments and international organizations and can influence the negotiation process of relevant conventions, China needs to actively carry out the second round of work program-related thematic assessments, develop discipline construction and basic research in response to research vacancies, and accelerate the construction of a comprehensive domestic biodiversity information platform to further improve data storage, management, and application for future assessment work To provide support for future assessments. The development of biodiversity informatics in China will lead to the construction, sharing and application of data in neighboring countries and their Belt and Road countries, thus promoting the accumulation, mining and service of global biodiversity data.

The Kunming Declaration and the establishment of the Kunming Biodiversity Fund adopted at the first phase of the 15th Conference of the Parties to the United Nations Convention on Biological Diversity (COP 15) in Kunming in 2021 demonstrate China's future international role in biodiversity conservation. Biodiversity informatics in China has been developing rapidly in recent years, and the number of cases supported by scientific research and governmental decisions is increasing.

Nevertheless, the development of biodiversity informatics in China still needs more efforts.

First, we should pay attention to the integration of biodiversity big data resources and form a more powerful integrated information platform. In particular, we need to strengthen the integration and sharing efforts of multi-source data, and further break the isolation of data between different departments and institutions on the basis of joint cataloguing of data and API interaction.

Secondly, we need to increase the sharing of data. The current level of data sharing throughout Asia differs greatly from that of European and American countries. Explore mutually beneficial multi-party cooperation mechanisms in China (including academic institutions, government agencies, publishing institutions, citizen science platforms and social media, etc.) to promote the construction of a "soft environment" of mutual recognition among data stakeholders.

Third, we need to establish a good connection between the big data platform and the data sources, as only a continuous flow of data sources can keep the data up-to-date. Although 16 million plant and animal specimens have been digitized and shared online in China, the degree of digitization needs to be significantly increased compared to the size of China's specimen collections [39]. In addition, the documentary resources in the collection contain a large amount of survey, observation and distribution data to be organized and mined.

Fourth, we should strengthen capacity building in biodiversity informatics, including technical training and project implementation to develop the relevant capacity of practitioners in all aspects of the data life cycle, especially capacity building in data mining. Encourage the utilization of new technologies and methods in the whole life cycle processes of data collection, management and mining, such as infrared camera technology, audio and video recording technology, remote sensing technology, environmental DNA technology, artificial intelligence technology and research workflow technology.

Fifth, strengthen regional and international cooperation. In the Asian region, through channels such as CAS overseas science and education centers, the national "One Belt, One Road" cooperation network and the new "Biodiversity Fund" established by COP 15, we will actively "go out" and expand China's presence in the Asian region. We will expand the scale and influence of China's cooperation in the Asian region. In terms of global international cooperation, through international platforms such as GBIF, IUCN, BHL and BGCI, we actively participate in international projects and affairs, and contribute China's biodiversity data and cases toward the UN Sustainable Development Goals (SDGs).

References

1. Xu Z, Chen B, Wang L, et al (2013) Progress and development trend of biodiversity informatics research. Ann New Biol:290–312

2. Soberón J, Peterson T (2004) Biodiversity informatics: managing and applying primary biodiversity data. Philos Trans R Soc London Ser B: Biol Sci 359(1444):689–698
3. Johnson NF (2007) Biodiversity informatics. Annu Rev Entomol 52:421–438
4. Lisong W, Bin C, Liqiang Ji et al (2010) Advances in biodiversity informatics research. Biodiversity 18(5):429–443
5. Hardisty A, Roberts D (2013) A decadal view of biodiversity informatics: challenges and priorities. BMC Ecol 13(1):1–23
6. Fuentes D, Fiore N (2014) The LifeWatch approach to the exploration of distributed species information. ZooKeys 463:133–148
7. Robertson T, Döring M, Guralnick R et al (2014) The GBIF integrated publishing toolkit: facilitating the efficient publishing of biodiversity data on the internet. PLoS ONE 9(8):e102623
8. Bisby FA (2000) The quiet revolution: biodiversity informatics and the internet. Science 289(5488):2309–2312
9. Xu HG (1998) A study on the construction of biodiversity information system in China. Rural Ecol 14(4): 11–15
10. Yang Z, Liang Z, Wenwei R et al (2000) Biodiversity informatics: an emerging direction and key technologies. Biodiversity 8(4):397
11. Ma K (2014) Rapid development of biodiversity informatics in China. Biodivers Sci 22(3):251
12. Shao GZ, Li H, Lin YC et al (2014) Marine biodiversity information resources. Biodiversity 22(3):253
13. Wang X, Zhang F, Zhang J (2017) Biodiversity information resources. I. Species distribution, cataloguing, phylogeny and life history traits. Biodiversity 25(11):1223–1238
14. Jian Z (2017) Biodiversity science and macroecology in the era of big data. Biodiversity 25(4):355
15. Mi X, Feng G, Hu Y, et al (2021) The global significance of biodiversity science in China: an overview. Natl Sci Rev 8(7):nwab032
16. The List of National Key Protected Wildlife (revised on February 1, 2021). J Wildl 42(02):605–640. https://doi.org/10.19711/j.cnki.issn2310-1490.2021.02.045
17. Lu Z, Qin HN, Jin X, Zhang Z, Yang Q, Hong D, Li DB, Li K, Yuan L, Zhou Z (2021) Necessity, principles and procedures for the adjustment of the List of Wild Plants of National Importance. Biodiversity 29(12):1577–1582
18. He P, Chen J, Man QG (2019) The present and future of the biological herbarium (museum) of the Chinese Academy of Sciences. Proc Chin Acad Sci 34(12):1359–1370. https://doi.org/10.16418/j.issn.1000-3045.2019.12.005
19. He Y, Zhang Y, Zhang Z (2017) Relevance and reflection of digital construction of plant herbarium. South Agric Mach 48(09):43–44+48
20. Xiao C, Luo HR, Chen TM, Mark P (2017) Analysis of the progress and status of digitalization of national specimen resource sharing platform. Res Inf Technol Appl 8(04):6–12
21. Hong-Xin WANG, Zhi-Ang HU (1996) Breeding systems, genetic structure and conservation of genetic diversity in plants. Biodiversity 02:32–36
22. Cao YS, Fang WEI (2010) Establishment and application of the national crop germplasm resource platform[J]. Biodiversity 18(05):454–460
23. Ji L (2021) The census of agricultural germplasm resources has been progressed in stages. Economic Daily. https://doi.org/10.28425/n.cnki.njjrb.2021.008192
24. Xue DY, Zhang YY (2022) Biogenetic diversity and conservation in China. Henan Science and Technology Press, Zhengzhou
25. Hu Y, Thapa A, Fan H, et al (2020) Genomic evidence for two phylogenetic species and long-term population bottlenecks in red pandas. Sci Adv 6(9):eaax5751
26. Liu Juan Hu, Dongnan ZZ et al (2019) Application of DNA barcode technology in forestry science research. For Sci Res 32(3):152–159
27. China Plant BOLG et al (2011) Comparative analysis of a large dataset indicates that internal transcribed spacer (ITS) should be incorporated into the core barcode for seed plants. Proc Natl Acad Sci U S A 108(49):19641–19646

28. Gao L, Liu J, Cai J et al (2012) Technical specifications for plant DNA barcoding studies. J Plant Taxon Resour 34(6):592–606
29. Li DZ, Wang YH, Yi TS et al (2012) The next-generation flora: iFlora. Plant Divers Resour 34(6):525–531
30. Mark P, Guo QH (2021) Progress and trends of vegetation ecology research in China. Chin Sci: Life Sci 51(03):215–218
31. Parker-Allie F, Pando F, Telenius A, et al (2021) Towards a post-graduate level curriculum for biodiversity informatics. perspectives from the global biodiversity information facility (GBIF) community. Biodivers Data J 9
32. Pan Y, Zhang B, Wu Y et al (2020) The latest developments of IPBES and China's countermeasures. Biodivers Sci 28(10):1286–1291
33. Russo K, Smith Z (2013) What water is worth: Overlooked non-economic value in water resources. Springer
34. Díaz S, Pascual U, Stenseke M et al (2018) Assessing nature's contributions to people. Science 359(6373):270–272
35. Scholes R, Montanarella L, Brainich A, et al (2018) Summary for policymakers of the assessment report on land degradation and restoration of the Intergovernmental Science-Policy Platform on Biodiversity and Ecosystem Services
36. Chaplin-Kramer R, Sharp RP, Weil C et al (2019) Global modeling of nature's contributions to people. Science 366(6462):255–258
37. Pan Y, Tian Y, Xu J et al (2018) Methodological assessment on scenarios and models of biodiversity and ecosystem services and impacts on China within the IPBES framework. Biodivers Sci 26(1):89–95
38. Tian Y, Lan C, Xu J et al (2016) Assessment of pollination and China's implementation strategies within the IPBES framework. Biodivers Sci 24(9):1084–1090
39. Cui X, Mingyuan Li, Fang Ye, Mingxue F, Ling Y, Xue F, Ping M (2018) Exploration of the development direction of NSII based on ten million specimen records. Res Inf Technol Appl 9(05):7–26

Cui Xiao Ph.D. in Wildlife Conservation and Utilization, Institute of Botany, Chinese Academy of Sciences. She has long been engaged in the construction of plant data resources, data platform construction, data management, data mining and services in China, and has published more than 10 relevant papers.

Data Engineering-Based Research on the Disasters Along the China–Pakistan Economic Corridor

Yaonan Zhang, Jianfang Kang, Minghao Ai, Yufang Min, Hongxing Li, Keting Feng, Guohui Zhao, Xirong Li, and Yamin Wu

Abstract As a result of the natural geographical features and climatic conditions along the China–Pakistan Economic Corridor (hereinafter referred to as "CPEC"), the CPEC is a hazard-prone region frequently inflicted by various geological hazard, such as ice-fall, avalanche, freeze–thaw, landslide, debris flow, rock-fall, flood and glacial lake outburst, which pose great threats to the infrastructure construction in the region. In the light of the scarcity of observation systems and a lack of observation data along the corridor, we introduced the engineering idea and established a data engineering technological architecture based on available data resources and finite problem-solving goals in this paper. Instead of pursuing the process mechanism of hazards along the CPEC, we strive to figure out the hazard laws behind the data, so as to reveal the typical natural disasters along the corridor according to these laws,

---

Y. Zhang (✉) · J. Kang · M. Ai · Y. Min · H. Li · K. Feng · G. Zhao
Northwest Institute of Eco-Environment and Resources, Chinese Academy of Sciences, Lanzhou 730000, P.R. China
e-mail: yaonan@lzb.ac.cn

J. Kang
e-mail: kangjf@lzb.ac.cn

M. Ai
e-mail: aimh@lzb.ac.cn

Y. Min
e-mail: myf@lzb.ac.cn

H. Li
e-mail: lihongxing@lzb.ac.cn

K. Feng
e-mail: fengkt@lzb.ac.cn

G. Zhao
e-mail: zhgh@lzb.ac.cn

Y. Zhang · J. Kang · M. Ai · Y. Min · H. Li · K. Feng · G. Zhao · X. Li · Y. Wu
National Cryosphere Desert Data Center, Lanzhou 730000, P.R. China
e-mail: lixirong@lzb.ac.cn

Y. Wu
e-mail: wuyamin@lzb.ac.cn

China's e-Science Blue Book 2023, https://doi.org/10.1007/978-981-99-8270-7_8

and analyze the evolution of typical disasters. For this purpose, we established data production systems for studying the conditions of the hazard-inducing environmental factors along the CPEC, built online models of disaster identification and analysis along the corridor, carried out the vulnerability assessment of debris flow and gravel-sliding slope, typical flood prediction and the simulation and prediction of extreme low temperature, strong wind and rainstorm in the coming 7 days in Gaizi Valley. In addition, we explored the application scenarios of data engineering in disaster analysis and prediction along the CPEC, and verified the feasibility of data engineering in disaster prevention and mitigation.

Keywords Data engineering · China–Pakistan economic corridor · Natural disaster · Disaster prevention and mitigation · Vulnerability assessment

1 Introduction

In September and October, 2013, President Xi Jinping proposed the Belt and Road Initiative, which aims to build a community of shared interests, shared responsibilities, and a shared future with countries along the road, featuring political mutual trust, economic integration, and cultural inclusiveness. In this way, China unveiled the Belt and Road Initiative, including six economic corridors and one 21st-century Maritime Silk Road. As a flagship project of the Belt and Road Initiative, the CPEC was officially launched on April 20, 2015. The 3,000-km-long corridor starts from China's Kashgar and ends at Pakistan's Gwadar. It is a pilot project under the Belt and Road Initiative [1, 2]. Crossing the intersection of the Himalayas, Karakoram and Hindu Kush mountains, the CPEC has the most complex natural environment and engineering geological conditions in the world [3]. Due to the huge topographic changes, the climate is characterized by vertical climate difference and regional differentiation along the corridor. The steep topography is a hotbed of surface processes and hazards, resulting in extremely active natural hazards in this area [4]. It has become an important factor for the regional sustainable development and the safety of major projects along the corridor.

Many scholars have studied the CPEC hazards. In 2016, Yao Xin et al. summarized 17 major engineering geological problems of 4 categories; they divided the corridor into 8 remarkable sections, and made a brief analysis of the main engineering geological problems in the western section of the CPEC [5], and performed an analysis of InSAR identification of creep-slip landslides in the Karakoram Mountains [6] and the deformation characteristics and distribution laws during 2007–2011. In 2018, Deng Ensong et al. studied the risk assessment and prevention of debris flow in Aoyitage-Blulunkou Section of Karakoram Highway (KKH) in terms of glacial debris flow, rainfall debris flow, etc. [7–10]. In 2018, Ding Peng et al. adopted the box-counting method in fractal theory to analyze the spatial development characteristics of various types of slope hazards by taking the slope hazards along the KKH of the CPEC as an example [11]. In 2021, Cui Peng et al. focused on the spatial pattern of landslides and debris flows along the CPEC, assessed hazard risks, predicted their future development trends, and put forward corresponding countermeasure [12]. In 2021, Zhang Yaonan compiled the CPEC thematic data articles; Min Yufang, Kang Jianfang, AI Minghao et al. respectively carried out spatio-temporal dynamic monitoring and research on the desertification degree of CPEC by reorganizing and compiling basic datasets, including the data on basic geographic information, glacier, frozen soil, snow cover, desertification, extreme climate, gravel-sliding slope and surface deformation [13–28]. However, in view of the scarcity of the overall data on the CPEC hazard research, it is necessary to establish a new study method architecture for the further hazard research, evolution analysis, prediction and early warning of the hazards disaster prevention and reduction of linear projects (e.g. highways) along the corridor, so as to form a practical technical method architecture.

Starting from the establishment of the scientific big data of the cold and arid areas in 2013, the National Cryosphere Desert Data Center (hereinafter referred to as "NCDC") pitched data engineering as the main subject development direction in the NCDC. The NCDC set up the Scientific Data Engineering Technology Research Center of Resources and Environment in Gansu Province in 2014, focusing on the theory and application of the scientific data engineering data in resources and environment. On the basis of many filed surveys of the CPEC organized and participated in by the NCDC, we further improved the data engineering technology architecture. The NCDC Center came up with data engineering methods architecture and technology systems, and produced the catalog data of various environmental factors about the geological hazards along the corridor. Then we completed the thematic data production of glaciers, ice lakes, frozen soil, snow cover, deserts, drought index, surface temperature, surface deformation, desertification distribution and precipitation on the annual, monthly and daily scales, 3km $\times$ 3km spatial scale. Moreover, we assessed the vulnerability of debris flow and gravel-sliding slope, specific flood prediction, the simulation and prediction of extreme low temperature, strong wind and rainstorm in the coming 7 days in Gaizi Valley. Our products are expected to provide supports for the identification of geological hazards and economic development of the CPEC, and form a set of automatic generation systems of data-driven online decision-making maps and reports. The data engineering method systems have been applied to the online monitoring of offshore Enteromorpha prolifera, the analysis of temporal and spatial distribution and evolution of the hazards along Wenzhao Highway, and the monitoring and analysis of snow cover hazard along Duku Highway. In addition, the systems can support the national defense and security, provide a new idea of using data engineering to hazard prevention and reduction, and worked effectively in data-driven for disasters monitoring and early warning in the early stages.

2 Demonstration Scenarios and Data Engineering

2.1 Demonstration Scenarios and Disaster Features

The CPEC extends from the Himalayas, Karakoram Mountains, the northern edge of the Pamir Plateau to the transitional zone of the southern plain in Pakistan, across the area at the highest altitude in the world with geographic the locations ranging from 23°47′ N–41°55′ N and 60°20′ E–80°16′ E. The terrains of CPEC is high in the north and low in the south, and it with complex geomorphologies and landforms, active neotectonic movements, distinct vertical zones, changeable climate environment, special natural environment and volatile climatic conditions, which form a peculiar landscape of high mountains and deep valleys [29, 30]. The study area across the Karakoram Mountains and part of the Pamir mountains, where modern glaciers are widely distributed. This is the second largest land glacier belt, only second to the polar regions. There are 1,050 glaciers, 80% of which are mountain glaciers [31]. Most part of the corridor is located in Indus River Basin, with a lot of turbulent rivers [32, 33]. The climate along the corridor varies greatly. The south is hot and humid, and the northern mountainous area is dry and cold in winter. The region is characteristic of harsh climate and a large temperature difference between day and night. Moreover, there is rare rainfall in most part of the study area, and the temporal and spatial distribution of rainfall is quite different. Due to the complex and changeable terrain as well as the wide variations of hydrological and climatic conditions in time and space along the corridor, the ecological environment here is vulnerable, and the disasters inducing by factors (such as glaciers, earthquakes, droughts, etc.) are prevalent, which makes the region more prone to hazards. For example, the fluctuation and melting of glaciers can easily lead to various glacier hazards, and the flood and its secondary hazards caused by glacial lake outburst are one of the typical hazards in the study area [26]. Frequent earthquakes can easily cause avalanches. Geological hazards such as debris flow and rockfall occur frequently in the corridor [28]. The arid areas along the corridor are extremely exposed to drought and geological hazards [14, 34]. The special natural geographical and climatic conditions along the CPEC result in 13 main disasters, including avalanche, ice-fall, freeze–thaw (frost heaving and thawing), glacial lake (glacier-dammed lake) outburst, landslide, debris flow (glacier and flood), rock-fall, gravel-sliding slope (gravel flow), flood (snow melting and rainfall), earthquake, ice-sliding, desertification, blizzard and gale. As a result, the research on geological hazards is a tough challenge for the CPEC construction [35].

2.2 Data Engineering and Features

Data Engineering originated from the IEEE Data Base Engineering Bulletin [36] in the United States in 1977. It was not until the foundation of the IEEE Technical

Committee on Data Engineering by the First International Data Engineering Conference in 1984 that the definition of data engineering was officially announced by the committee. According to the committee description, the data engineering is the engineering of information system about data production and data use [37]. With the need of data as an asset to promote productivity, data engineering is becoming more and more important. It has been expanded in concept, connotation and extension, and has become a systematic support method of transforming data to into knowledge, decision-making strategy, service and products. We proposed a preliminary definition for data engineering: it is a data-based product of interdisciplinary integration, introducing engineering idea into the data field and combining the engineering theory and technology of data research with the discipline theory of various fields. It is a subject about data application practice, engineering implementation and innovative application, and provides systematic solutions for creating data value and exploring laws behind data. In terms of the history of scientific development, data engineering is a necessary process from data science to data technology and data engineering. The engineering process, namely performing engineering practices on data is a prerequisite for a digital economy by transforming data into decision-making strategies, products and services.

To our minds, the problem-solving method with the help of data engineering is based on available data resources and finite goals. It does not purposely pursue the process mechanism and universality of problems or research objects, but lays stress on discovering laws behind data and solving problems according to the laws. Based on available data resources, data engineering puts forward solutions and methods to solve problems under the condition of saving resources for sustainable development[38]. The most prominent feature is to creatively reveal the special relation and data value behind data. Therefore, data engineering science makes full use of research results in mathematics, data science, information technology, computer technology and other various fields on the basis of available data resources. Targeting the demands or problems in certain fields and focusing on the whole life cycle of data, it aims to establish specialized data engineering to tap data value, innovatively bring forth data derivatives, implement data value transformation, and fulfill the problem-solving purpose for finite goals, appropriate management decision and data empowerment to some extent. Driven by the further acquisition of more sound data and in the light of the laws behind data, it is expected to form a business of digital economy by gradually approaching problem solving, scientific decision-making and efficient empowerment through iterative steps.

Data engineering research mainly involves the research on the engineering implementation, such as the innovation of data products, the creation of data derivatives, the building of data application model, the representation of data characteristics and the construction of data standards and specifications. It covers data acquisition, standardization, management, analysis, mining, production, application, security, as well as the effective combination of data and related fields. It includes the research on the innovation of data products and the value analysis of derivatives, and the measures of data value transfer and transformation, targeting at models, algorithms, deduction rules, application scenarios, etc. The important means of data engineering technology comprise statistical analysis, data processing, data-driven model development, artificial intelligence model and parameter optimization, as well as uncertainty analysis, visual analysis and other technologies. The key of data engineering lies in the model construction of data application analysis in certain fields in combination with many important technologies, such as data algorithm analysis, data quality control, uncertainty analysis and data deduction technology.

Data engineering takes the process of problem solving as a data flow, and decomposes the data-driven work process into the following engineering systems: available data preparation, data driving, data production (data reanalysis), data mining, the production of data derivatives (or data decision-making), etc. The engineering system of data preparation involves the selection of data type, granularity, specification and quantity as well as data extraction and other engineering work. The data analysis engineering system involves the data analysis, model application in various fields, the production of driving data, model parameter optimization, the simulation of driving models, etc. The data preparation engineering system involves the construction of spatial–temporal filed elements, data integration, data evaluation and so on. The data mining engineering system involves the building of data characterization models, artificial intelligence models, special charting, data features assessment, etc. The data decision-making application engineering system involves the construction of decision-making platform, the analysis of application field scenarios, the map and report production of prediction and early warning, the evaluation of iteration optimization and other decision-making work.

3 The Data Engineering Construction of Disasters Analysis

The NCDC carried out the researches on the natural disasters along the CPEC in 2016, and organized the scientific investigation for scientific data collection, reorganization, preparation and online production of current data for the purpose of "understanding the features and laws of natural hazards and improving the accuracy of prediction and early warning" for disasters in 2017. According the principal line of "data driven–knowledge discovery–information service" to establishing a system of "intelligent perception, intelligent identification, intelligent prediction and intelligent decision-making" on the base of data-driven, we sorted out the tasks of each section, including target design, data acquisition, data preprocessing, data storage, data modeling, data analysis and data application. The work of target design selects the target, locates the target hazard, looks into the current situation and development at a deeper level, analyzes the features of the target disasters and determines the indicators of hazard research. The work of data acquisition retrieves data, collects various forms of data, including remote sensing data and meteorological data, and unifies the standard requirements to solve the problem of missing data. The work of data preprocessing deals with the non-standard and non-uniformity resulted from the data aggregation multi-source, which has a great impact on the data analysis and application later. Therefore, the data shall be pre-processed in accordance with the unified specifications through the supplement of the missing value and the elimination of outlier data, so as to determine the data standard and improve data quality. When it comes to data storage, there are some disadvantages about the multi-source data aggregation; for example, the relation between data cannot be revealed; the data in the past cannot be used now, and the data life cycle is blurred. Large-scale and long-term storage forms and technologies can endow data with time attributes and store data in multiple dimensions through special data organization and storage structure, so as to lay a solid foundation for multi-dimensional statistics, analysis and application of hazards. The work of data modeling counts up data scope and the relation between data, establish the corresponding database, systematically analyze the relation between data; the work converts the conceptual model abstracted from the conclusion into physical models, and uses professional tools to perform machine learning for the data and model in the database to build appropriate data models. The work of data analysis taps the laws behind data, performs multi-dimensional analysis of data relation, explores different values and laws, and develops data products about different data; the work predicts the development of data to provide reference for the future application. Finally, the work of data application provides intuitive analysis results through data visualization tools based on stable data platforms, and expand the data application according to different needs, so as to provide base support for the development of professional tools.

According to the task analysis of each stage as well as the data engineering framework system and specific process, the research carries out disaster identification and analysis, disaster vulnerability assessment, and realized disaster prediction and early warning, so as to learn more about the natural disaster along the CPEC.

The 4 data engineering systems (as shown in Fig. 1.) about the natural disaster application scenarios of the CPEC have been built as following: the engineering system of data aggregation, the engineering system of data analysis and production, the engineering system of model building and simulation, and the engineering system of decision-making support, The multi-source remote sensing data aggregation project mainly involves the data aggregation, integration and fusion of Gaofen-1 and Gaofen-6, Sentinel, MODIS 1 and MODIS 2 and Landsat-8. The special data production project of environmental factors mainly integrates a small amount of observation data, prepares basic geographic information, and completes the online production of special data, such as surface temperature, glacier, snow cover, frozen soil and meteorological element precipitation. The data analysis project of environmental factors mainly involves the analysis of meteorological and hydrological factors. The model building and simulation project mainly involves the building of disaster analysis models of typical landslide, gravel-sliding slope, debris flow and flood to realize the data-driven analysis of landslide, debris flow, gravel-sliding slope and flood. The data engineering of disaster vulnerability assessment mainly introduces machine learning to build the vulnerability assessment of landslide, debris flow, gravel-sliding slope and other disasters, so as to form the thematic maps and reports of online vulnerability assessment of these hazards along the CPEC.

There are two systems for the data engineering research along the CPEC.

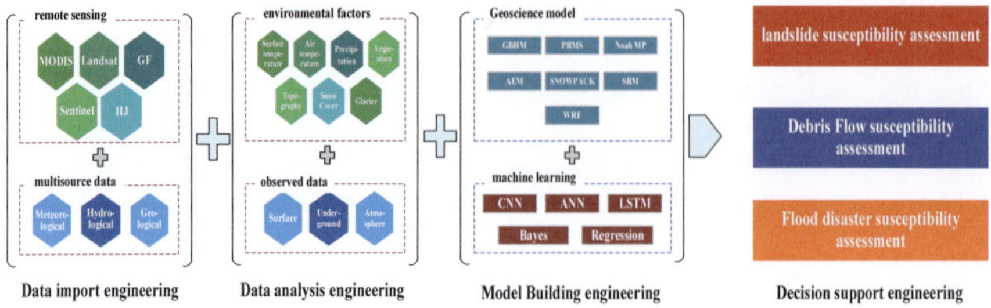

Fig. 1 The data engineering system framework of hazard vulnerability assessment along the CPEC

(1) Automatic generation system of present situation data. We collected and aggregated basic geographic data, domestic GaoFen series satellite images, MODIS remote sensing data products, Landsat, Sentinel, QuickBird and other satellite images as well as global reanalysis data, geological data (e.g. ice, snow, frozen soil and desert), meteorological and hydrological station data, calculation simulation results and other data resources. We analyzed the aforementioned data specifications, established a data resource platform for the CPEC based on metadata management, and formed the data cloud of more than 70,000 pieces of data. The remote sensing data are characterized by small time-spatial constraints and strong current representation. On this basis, we integrated multi-source data (e.g. meteorology and topography), made a breakthrough in the key technology of the inversion of multi-source remote sensing and multispectral characteristic parameters and thematic data integration to prepare the current data for special environmental elements. Moreover, we developed multi-source heterogeneous data processing method, remote sensing data inversion automation, the production system of current data and the analysis system of element situation evolution. Under the support of supercomputing platform, we optimized the existing processing methods of remote sensing data in parallel, realized the automatic downloading of remote sensing data, automatic data splicing and cutting, data processing calculation and automatic generation of map and report results, developed the inversion method of characteristic parameters, such as ice and snow, vegetation, water body and surface temperature for different underlying surfaces and different application fields in alpine and steep mountainous areas. In this way, we can produce the data production in the inversion of glacier, ice lake, snow cover, frozen soil, land surface temperature, desertification, and build the GBHM-driven hydrological models (e.g. radiation and precipitation). The production system of current data is shown in Fig. 2.

(2) The analysis system of environmental situation evolution. In view of the distinctive environmental along the CPEC, we built data engineering-based geoscience models corresponding to special geoscience factors such as glacier, frozen soil, snow cover and ice lake, the models of natural hazards in connection with ice, snow and frozen soil (e.g. landslide, debris flow, flood and freeze–thaw), as well as various physical analysis models and statistical analysis models, so

Fig. 2 The production system of current data along the CPEC

as to form a model base suitable for situation analysis for special elements in the study area. The models can be used as the basis for situation evolution analysis and situation prediction. 147 models have been collected by now. Based on the model coupling of special environmental factors (as shown in Fig. 3), breakthrough can be made in the spatial–temporal barrier of model coupling, so that the data generated by the meteorological prediction model can be used for hydrological simulation and hazard early warning, and the realization of meteorology-hydrology coupling and meteorology-hazard models. We carried out model simulation, data production and future prediction with the key elements and spatial parameters necessary for the situation evolution analysis of 9 geoscience factors (i.e. glacier, frozen soil, snow cover, ice lake, debris flow, landslide, flood, meteorology, freeze–thaw), so as to form a comprehensive prediction model of situation evolution in the study area.

According to the idea of data engineering, we have engineered the whole process of data production, monitoring and analysis, realized the data production and analysis of key hazard-inducing factors driven by multi-source data, and completed the current data production of special environmental factors along the corridor. In addition, we have made great progress in hazard vulnerability assessment and the simulation prediction of hazard factors in the future. We have obtained thematic data about glaciers, ice lakes, frozen soil, snow cover, deserts, drought index, surface temperature, surface deformation, desertification distribution and precipitation on the annual, monthly and daily scales, covering 3 km in space. Moreover, in the system, we performed the vulnerability assessment of debris flow and gravel-sliding slope, flood prediction, the simulation and prediction of extreme low temperature, strong wind and rainstorm in the coming 7 days in Gaizi Valley.

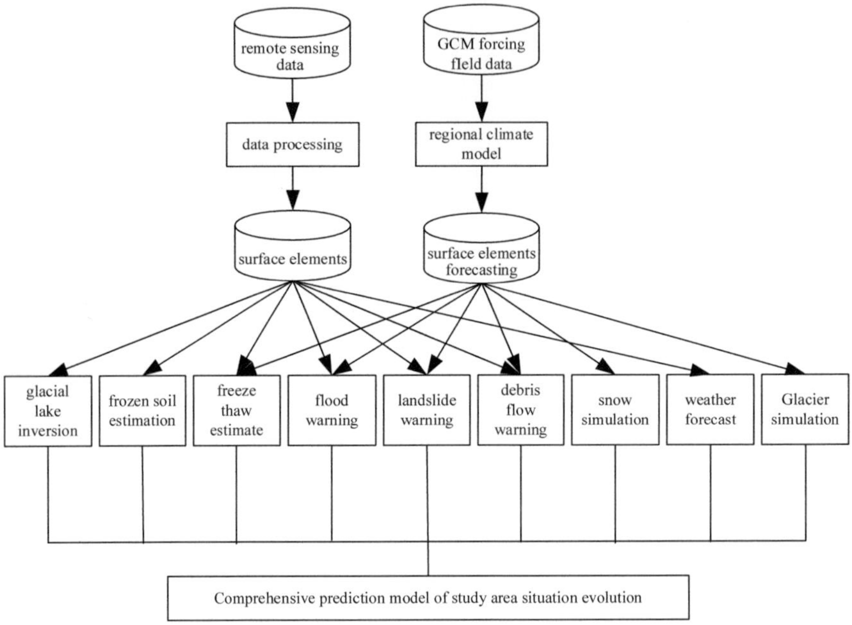

Fig. 3 The model coupling analysis framework of special environmental elements

4 The Application of Data Engineering in Different Scenarios

4.1 *The Analysis of Glacier Evolution Along the CPEC*

Taking the completely open and shared Climate Research Unit (CRU) reanalysis data of the Climate Research Center of University of East Anglia in the UK as the climate-driven dataset, we combined the Second Glacier Inventory Dataset of China (2009) and the Inventory of glaciers in Pakistan (2003) with 90m DEM data as parameters, and adopted the completely open source OGGM model of glacier simulation (Parks, Goosse, et al. simulated the length changes of 421 glaciers in the world and compared them with the available datasets in 2019, proving that the accuracy of OGGM model was 81%), simulated and calculated the length change and material balance change of glaciers with an area of more than 10 km^2 in the CPEC from 1901 to 2020, and perform simulation and prediction for the change of glacier thickness under the scenario of an increase of global temperature by 1.5 °C. The data results are shown in Fig. 4.

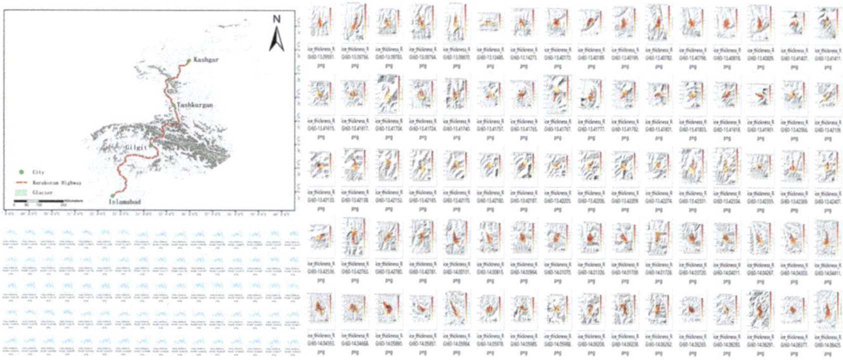

Fig. 4 The result analysis diagram of glacier evolution

4.2 The Distribution Pattern of Frozen Soil Along the CPEC

Taking the 5-year product of MODIS global surface temperature as the driving data and combining with the Second Glacier Inventory Dataset of China (2009), the Inventory of glaciers in Pakistan (2003) and the global soil data as parameters, we used the Top Temper of Permafrost (TTOP) to simulate and estimate the distribution of permafrost along the CPEC with a spatial resolution of one kilometer. The calculation and simulation results are consistent with the results of the field survey performed by Zhu Yanying et al. along the highway, with the conclusion that there is no permafrost along the CPEC highway except Khunjerab section. The field survey shows that the permafrost in Khunjerab section affects about 4.3km of the highway. According to model estimation, the permafrost in this section is about 3.7km, with an error accuracy of about 13%. The calculation and simulation of the permafrost area in Khunjerab section is shown in Fig. 5.

4.3 Snow Cover Evolution Analysis Along the CPEC

We took the ground element data from the reanalysis data—Global Land Data Assimilation System (GLDAS) for weather forecast data as the driving data, adopted the elevation, slope gradient and slope aspect of the CPEC as the parameters, and the observation data from Pishan Station in Xinjiang and Shiquanhe Station in Tibet as the verification data. Then, we adopted the Utah Energy Balance (UEB) snowmelt model to simulate and predict the regional snow distribution, snow thickness and snow water equivalent along the CPEC. The UEB model has been applied in Irtysh River in Xinjiang and Qinghai Lake Basin, and received positive feedbacks in comparison with local meteorological observation stations. Comparing the snow cover days calculated and simulated by the UEB model with the snow cover days of MODIS snow products observed by satellites, the MODIS data has higher resolution and finer

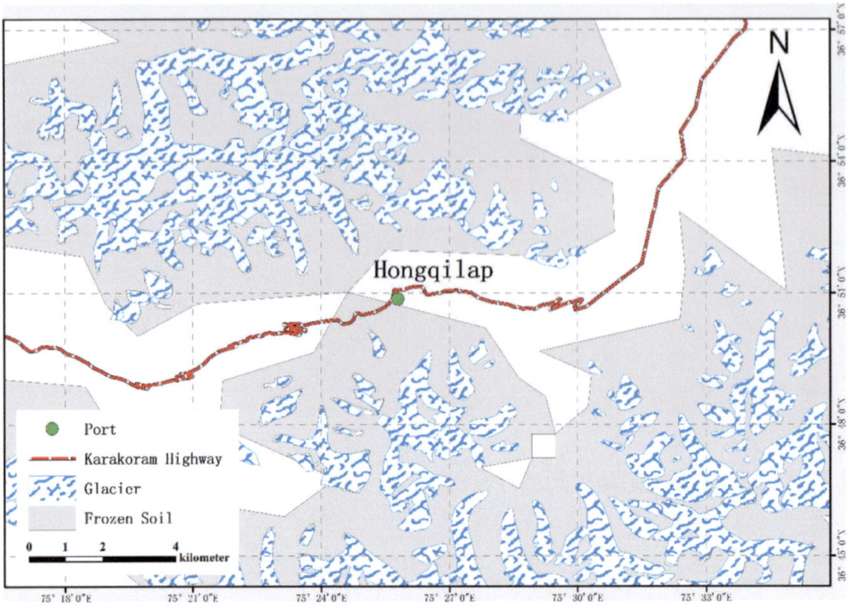

Fig. 5 The diagram of permafrost simulation results in Khunjerab section

description, but the calculation results of the UEB model are highly consistent with the spatial distribution of MODIS snow products. The spatial resolution of reanalysis data plays an important role in the comparison. The comparison results are shown in Fig. 6.

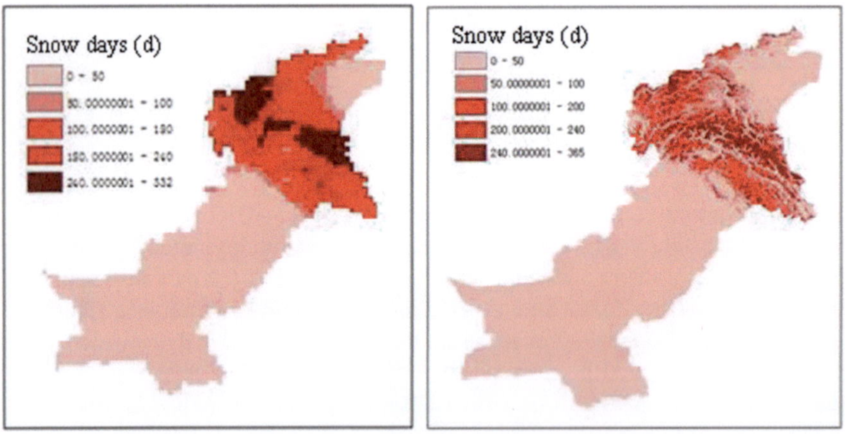

Fig. 6 The comparison between the calculated and simulated snow cover days (left) and remote sensing data (right)

4.4 The Evolution Analysis of Debris Flow Along the CPEC

We took the temperature and precipitation data of available meteorological stations in the study area as the historical driving data, and the meteorological forecast data as the future prediction data. First, we made spatial–temporal integration of these two kinds of data. In the following, in reference to many resources, including the 7-day cumulative temperature, 15-day cumulative temperature, cumulative precipitation, as well as cumulative hydrothermal factors of the year, debris flow gully gradient and the geomorphic information entropy described in previous literature on debris flow in the Caucasus as the indices, we made linear regression on the risk degree of debris flow and obtained the risk rating of debris flow outbreak along the CPEC. Compared with the results with the records of debris flow outbreak events in the Aoyitage-Blulunkou section of Karakoram Highway in 2013 and 2017, the risk of debris flow in the next 48 h can be predicted. Moreover, we have completed the distribution map of typical periglacial geological hazard areas, as shown in Fig. 7.

4.5 The Analysis of Landslide Evolution Along the CPEC

Taking time series Landsat remote sensing images and slope gradient data in the regions of both China and Pakistan as data sources, we assessed the spatial distribution of landslides through the detection of remote sensing image changes. After comparing the pixel perturbation-based Behling method, the time series mining method based on Recurrent Neural Networks (RNN) and the classification method based on vector machines (SVM), we selected RNN as the prediction algorithm of landslide spatial distribution. Due to a long distance between the predicted landslide sites and the highway along the CPEC, it is difficult to directly perform survey and verification. Therefore, we selected Ludian County of Yunnan Province with similar image characteristics as the verification area. Through the comparison between the deep learning simulation results and the expert visual interpretation results, the simulations and predictions of landslide are highly consistent on the scale of space, covering 94.96% of the landslide areas in this region. The comparison results are shown in Fig. 8. The left is the simulation prediction results, and the right is the visual interpretation and field survey results.

5 Conclusion and Suggestion

Data engineering is a new data-driven means to solve practical problems with specific data-driven objects, such as patterns, models, algorithms, deductive rules and application scenarios. It is aimed to solve problems on the basis of available data resources

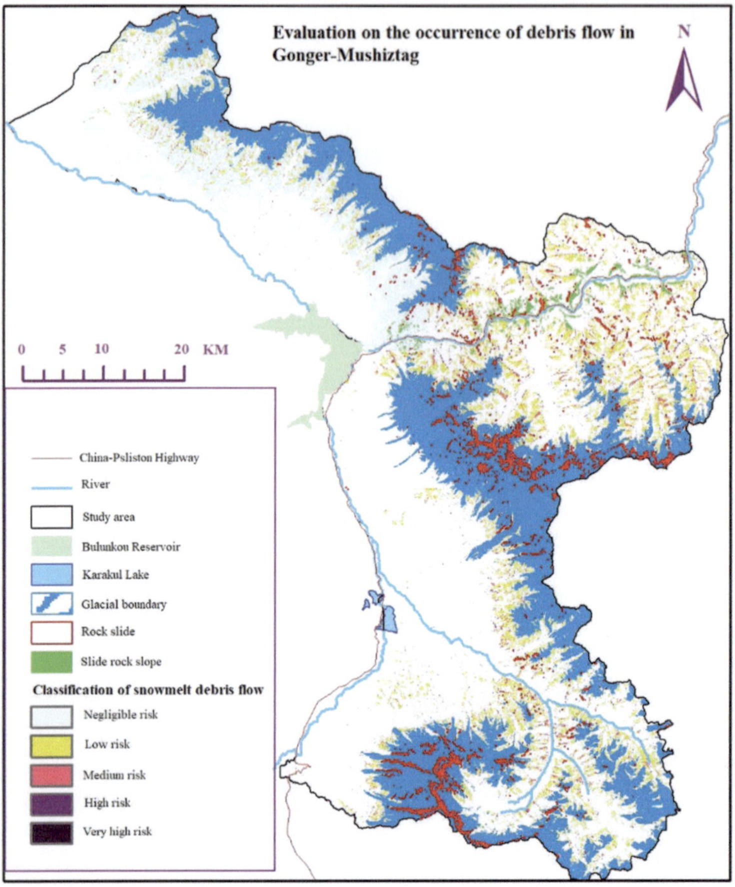

Fig. 7 Vulnerability zoning of snowmelt debris flow of GaiZi valley

and finite goals. Instead of purposely pursuing the process mechanism and universality of problems or research objects, it lays stress on discovering laws behind data and solving problems according to the laws. Data engineering revolves around data knowledge system, and is comprehensively supported by mathematics, computer science, artificial intelligence science and disciplines in various fields. With data-drive support, field model support, the calculation support of artificial intelligence and the iterative optimization technology support, it gives rise to procedural measures for data for decision-making support, product innovation, data application and data empowerment of the assessment indicators of finite goals. The basic data of hazard-inducing environmental factors are the foundation for the implementation of major projects, highway route selection and geological hazard prevention along the CPEC. Therefore, based on the data engineering, we have obtained the production technology for the background data of special environmental factors and the current data of hazard-inducing environmental factors, and created the systems of hazard

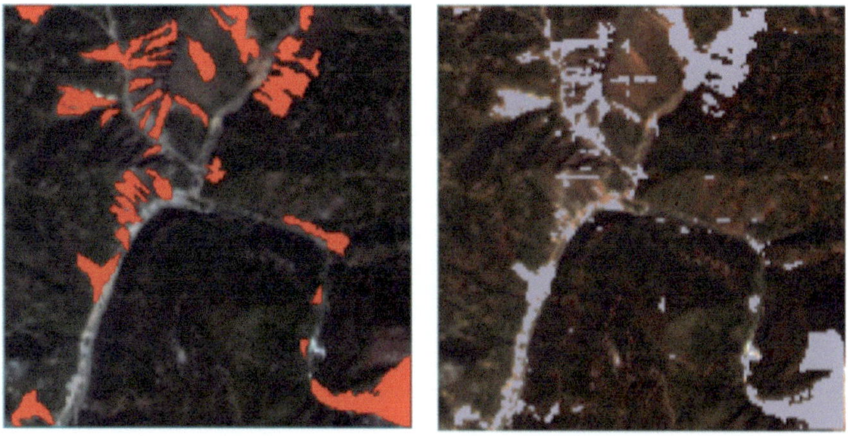

Fig. 8 The comparison between landslide prediction results

zoning, prediction and early warning for the vulnerability of flood, debris flow and gravel-sliding slope. The systems are expected to support the research on vulnerability assessment, prediction and early warning of typical hazard along the CPEC. The identification of natural hazards and prediction data engineering along the CPEC also show that the engineering work on data through data engineering can create a digital economy by providing a new idea for transforming data into decision-making strategies, services and products.

Acknowledgements Thanks to the support from the project of the "National Service Center for Speciality Environmental Observation Stations" by the National R&D Infrastructure and Facility Development Program of China (Y719H71001), the project of the "Construction and Application of Cold and Arid Region Environmental Evolution Research Cloud" by the Information Program of the Chinese Academy of Sciences (XXH13506 and XXH13505), and the project of the National Cryosphere Desert Data Center (E01Z790201).

References

1. Yanqian P (2019) Study on the CPEC geo-hazard characteristic and vulnerability analysis. Northwest University, Xian
2. Qiao L, Yong N, Xin W, et al (2021) An inventory dataset of glacier and glacial lake related hazards (events) along the China Pakistan economic corridor. China Sci Data (Chinese-English Online Version) 6(1):172–181
3. Yong Y, Yongxing W, Jinfeng L et al (2018) Mountain disasters and countermeasures of traffic engineering along China-Pakistan economic corridor. High Speed Railw Technol S2:38–42
4. Zhang F (2021) Analysis of Temporal and Spatial Variation of Snow Cover along China-Pakistan Economic Corridor during 2001–2019. Lanzhou:Lanzhou Jiaotong University
5. Xin Y, Yongshuang Z, Lingjing L (2016) Brief analysis of the engineering problems in west syntaxis section of China-Pakistan Economic Corridor. J Eng Geol (10):909–912

6. Xin Y, Sheng L, Lingjing L et al (2016) Study of slow-moving landslide recognition by InSAR in the Karakoram mountain section of China-Pakistan Economic Corridor. J Eng Geol 10:1300–1306
7. Ensong D, Xueli W, Zhixin Z et al (2018) Hazard evaluation of glacial debris flows in Aoyitake-Bulunkou section of Karakoram highway. Highw Transp Res Dev 35(05):16–23
8. Ensong D, Xueli W, Bin L et al (2018) Hazard evaluation of rainfall-induced debris flows in the Aoyitake-Bulunkou section of Karakoram highway. Sci Technol Eng 18(03):1–8
9. Wengong L, Xueli W, Baocheng C, et al (2018) Analysis of debris flow damage along the China-Pakistan Economic Corridor: taking Aoyitake-Bulunkou Section of the Sino-Pakistan Highway as a case. J Glaciol Geocryol 40(4):773–783
10. Xueli W, Baocheng C, Bin L, et al (2018) Hazard assessment and prevention analysis of debris flows along Aoyitage-Blulunkou section of Sino-Pakistan Highway. Highway 63(11):14–21
11. Pengpeng D, Zongji Y, Jianping Q et al (2018) Fractal characteristic analysis on slope disasters along the KKH of China-Pakistan economic corridor. China Sci Pap 13(15):1685–1689
12. Qiang Z, Peng C, Xiaojun G et al (2021) Spatial pattern and response of landslide and debris flow risks in China-Pakistan economic corridor. Bull Chin Acad Sci 36(2):160–169
13. Yaonan Z (2019) Preface to the China–Pakistan economic corridor dissertation. China Sci Data (Chinese-English Online Version) 4(3):1–3
14. Keting F, Yaonan Z, Deyu T, et al (2019) A dataset of monthly temperature vegetation dryness index along the China–Pakistan economic corridor during 2000–2017. China Sci Data (Chinese-English Online Version) 4(3):97–111
15. Minghao A, Yaonan Z, Jianfang K, et al (2019) A dataset of permafrost distribution along the China-Pakistan economic corridor in 2017. China Sci Data (Chinese-English Online Version) 4(3):8–17
16. Guohui Z, Yaonan Z, Jianfang K (2019) A dataset of high-resolution land surface temperature inversion for the China-Pakistan economic corridor (2013–2018). China Sci Data (Chinese-English Online Version) 4(3):63–74
17. Deyu T, Yaonan Z, Liqin H, et al (2019) Data of stone-slide investigation and vulnerability distribution in Gaizi Valley. China Sci Data (Chinese-English Online Version) 4(3):53–62
18. Weiyi M, Junqiang Y, Jing C (2019) A dataset for the extreme temperature-rising process over Eastern Pamirs Plateau in Northern China–Pakistan Economic Corridor (1961–2017). China Sci Data (Chinese-English Online Version) 4(3):32–41
19. Zehua F, Hui T, Jinyu C (2021) A dataset of SPEI in the China-Pakistan Economic Corridor from 1961 to 2015. China Sci Data (Chinese-English Online Version) 6(4):102–110
20. Yufang M, Keting F, Jianfang K, et al (2019) A Dataset of Desertification distributions along the China–Pakistan economic corridor during 2000–2017. China Sci Data (Chinese-English Online Version) 4(3):75–85.
21. Jiali X, Changzhen Y, Cun C (2019) A dataset of land cover in Northwest China from 1990 to 2010. China Sci Data (Chinese-English Online Version) 4(3): 86–96.86–96
22. Junqiang Y, Weiyi M, Wenfeng H, et al (2019) A dataset of drought indices based on the standardized precipitation evapotranspiration index (SPEI) over Xinjiang, China (1961–2015). China Sci Data (Chinese-English Online Version) 4(3):112–121
23. Liqin H, Yaonan Z, Deyu T, et al (2019) A high-resolution DOM dataset of the China–Pakistan economic corridor (Kashgar to Islamabad). China Scientific Data (Chinese-English Online Version) 4(3):122–132
24. Xia F, Chi Z, Yaonan Z, et al (2018) A dataset of organic carbon storage and carbon dynamic in arid and semiarid China from 1980 to 2014. China Sci Data (Chinese-English Online Version) 3(3):64–67
25. Xia F, Chi Z, Yaonan Z, et al (2018) Dynamic spatial datasets of ecological carbon pools and carbon sources and sinks in arid regions of Western China under the Impact of Climate Change from 1980 to 2014. China Sci Data (Chinese-English Online Version) 3(3):68–78
26. Yanrun R, Yaonan Z, Jianfang K (2019) A dataset of glacier and glacial lake distribution in key areas of the China-Pakistan economic corridor during 2013–2017. China Sci Data (Chinese-English Online Version) 4(3):18–31

27. Dandan C, Yaonan Z, Michel K, Piégay H (2019) DOM and DEM dataset of the Drôme River remotely sensed with a UAV (2005–2009). China Sci Data (Chinese-English Online Version) 4(3):133–140
28. Yanping B, Jianfang K, Meng L, et al (2018) A dataset of surface deformation along the China-Pakistan economic corridor from 2014 to 2018. China Sci Data (Chinese-English Online Version) 4(3):47–55
29. Tandong Y, Fahu C, Peng C et al (2017) From Tibetan Plateau to third pole and pan-third pole. Bull Chin Acad Sci 32(9):924–931
30. Lailin Z, Zhenmin J, Guitang P, et al (2004) A geological comparison between the Eastern and Western Himalayan Syntaxes. Earth Sci (03):269–277
31. Jingrong W (1987) Glaciers and debris flow in the Kashgar–Taxkorgan section of Karakoram highway. J Glaciol Geocryol (01):87–94+102
32. Jie L, Aimin M, Libo W et al (2015) Debris flow disaster and prevention on the Aoyitage-Bulunkou section of Karakoram highway from China to Pakistan. Highway 60(12):8–14
33. Jin H, Yingyan Z, Zhiquan Y et al (2014) Comparison of riverbed sediments and debris flow sediments along Karakoram highway. Chin J Geol Hazard Control 25(3):1–8
34. Yufang M, Yaonan Z, Jianfang K, et al (2021) Study on spatial-temporal dynamic monitoring of degree of desertification in CPEC based on MODIS image. Remote Sens Technol Appl 36(4):827–837
35. Xiaomin L, Xiao H (2019) On the natural disaster risk and management involving Chinese citizen in the contribution of the "belt and road initiative": taking the China-Pakistan economic corridor as an example. J Gansu Radio Telev Univ 20(2):31–35
36. Vincent L (1977) Data base engineering bulletin. IEEE Tech Comm Data Eng 1(4):16–24
37. Nirenburg S, Attiya C (1984) Toward a data model for artificial intelligence application. In: First international conference on data engineering. IEEE, Los Angeles, CA, USA 1(1):446–453
38. Mengsun Y (2014) Learn QIAN Xuesen's thought of "engineering science" and implement engineering research on human health. J Med Postgrad 27(1):1–7

Yaonan Zhang, researcher, doctoral supervisor, director of the National Cryosphere Desert Data Center of Northwest Institute of Eco-Environment and Resources, Chinese Academy of Sciences, dedicated to the long-term research of scientific data and application. He has successively presided over more than 40 projects, including the key fund projects of the National Foundation of China, the projects of Basic Science Talents Training Fund (Cryosphere), the projects of the Ministry of Science and Technology, the key deployment projects of Chinese Academy of Sciences and the projects of local government, with over 60 papers published on the basis of his research. At present, he focuses on the development of data engineering, builts the hazard prevention and reduction system of data engineering, and carries out the research on data driving, coupling model, data engineering integrating with artificial intelligence, and multi-source data fusion.

Construction and Application of Data Intensive Knowledge Service Platform for Agricultural Scientific Research

Ruixue Zhao, Hua Zhao, Jianhua Zheng, and Lihua Jiang

Abstract With the accelerated penetration of new-generation information technologies such as big data and artificial intelligence applied in various fields, scientific research has entered a new paradigm of data-intensive scientific research, and higher requirements have been put forward for scientific research informatization. The development of modern agriculture in China is inseparable from the support of agricultural technological innovation, and agricultural technological innovation is inseparable from professional, efficient and precise knowledge services. Under the new environment of data-intensive scientific research, agricultural technological innovation has put forward new demands for knowledge discovery and knowledge services. Based on agricultural sci-tech big data and the new-generation information technology, an agricultural intelligent knowledge service platform with sci-tech big data was built, which can store, manage, analyze, mine and release agricultural sci-tech information resources and release the value of knowledge, it can also provide research data management services for scientific research institutes and researchers through building of collaborative environment for scientific research, and assist in improving the efficiency of scientific research, thereby promoting the development of agricultural scientific research innovation. The objective needs and construction background of the agricultural intelligent knowledge service platform with sci-tech big data were analyzed, the key technologies and construction results of the platform were introduced in detail, and the future service scenarios and applications of agricultural intelligent knowledge services were prospected.

Keywords Sci-tech big data · Data governance · Knowledge organization and calculation · Intelligent knowledge services · Data intensive

---

R. Zhao (✉) · H. Zhao · J. Zheng · L. Jiang
Agricultural Information Institute, Chinese Academy of Agricultural Sciences, Beijing 100081, China
e-mail: zhaoruixue@caas.cn

R. Zhao · H. Zhao · J. Zheng
Key Laboratory of Knowledge Mining and Knowledge Services in Agricultural Converging Publishing, National Press and Publication Administration, Beijing 100081, China

Key Laboratory of Agricultural Big Data, Ministry of Agriculture and Rural Affairs, Beijing 100081, China

China's e-Science Blue Book 2023, https://doi.org/10.1007/978-981-99-8270-7_9

1 Introduction

Agricultural science and technology information is a strategic resource for promoting scientific and technological progress [1]. With the advent of information era, emerging technologies such as big data, artificial intelligence, and digital twin technology are rapidly developing and being applied, pushing scientific research into the fourth paradigm of data-intensive scientific discovery. In addition to the changes caused by digitization, networking and intelligence, the environment for the production, dissemination, management and utilization of scientific and technological information is also undergoing revolutionary changes [2]. It is a huge challenge for researchers to accurately discover and quickly obtain the required information resources, due to the huge number, diversity, complexity and multidisciplinary. The researcher survey results made by Elsevier showed that researchers spend over four hours searching articles a week on average, and their reading amount of articles was reduced by 10% while the searching time of articles was increased by 11% in recent ten years, which indicated the serious asymmetry and imbalance between the supply and the demand of scientific and technological information for scientific researchers. Information search and delivery as traditional information services, have been unable to meet the needs of knowledge innovation, and the user-oriented and knowledge content-based knowledge services with knowledge value-added and problem-solving are the development direction of information services.

Knowledge services have been explored and practiced by academies, professional information service institutions and even commercial companies, such as knowledge graph based intelligent searching service launched by Google, the knowledge discovery services which could help users discover the inner connection of knowledge launched by Harvard University Library based on Yewno concept network [3], the new generation of open knowledge service system integrating knowledge organization and cognitive computing designed by the National Agricutural Library of China [4], the big data based knowledge service system [5], and the knowledge service framework of big data based on knowledge fusion process [6], which showed that knowledge services have made great progress. However, there are still problems in the balance and connection of knowledge supply and demand, especially in the face of the challenges of data-intensive scientific research. To strengthen the guarantee of information resources, it is not only necessary to provide users with large-scale, diverse, rich content, reliable and high-quality agricultural scientific research resources, but also it is urgent to use emerging technologies such as big data and artificial intelligence to build an intelligent knowledge service platform in the agricultural field, which can strengthen the comprehensive guarantee and accurate service of agricultural science and technology information, and it can also improve its efficient dissemination, exchange and utilization.

In this paper, the technical framework, key technologies and core functions of the constructed data intensive knowledge service platform were introduced based on its objective needs, and the construction results and future application scenarios of the platform were summarized.

2 Characteristics of Data-Intensive Scientific Research and Knowledge Services Demand

Scientific research has gone through three interrelated paradigms of empirical, theoretical and computational science, and comes to the fourth paradigms of data-intensive science, which is seen as a data-driven, exploration-centered style of science [7]. Under this paradigm, the driven mode of science has changed, data has become the direct driving force of scientific research activities. During the process of data collection, storage, management and analysis, the proportion of intelligent instruments, computers, and efficient data processing technology in the entire scientific research activities is constantly rising, and data-centered computer technology, data research tools and data management and analysis methods have become an indispensable and important part of scientific activities in the era of big data. Scientific researchers pay more attention to the communication and cooperation in multidisciplinary fields in the scientific research organization model, which emphasizes breaking the barriers between various innovation entities, focusing on the common innovation goal, giving full play to the advantages and characteristics of different innovation entities, and effectively gathering innovation resources and innovation elements [8]. Currently, the solution of scientific research problems in agricultural field presents new characteristics, such as interdisciplinary research, team collaboration and the systematic integration of disciplinary knowledge. The new generation of agricultural scientific research has entered the era of data-intensive scientific research represented by technologies such as cloud computing and big data. Agricultural scientific research has gradually showed the development trend of 'data' - 'mining analysis' - 'field application', such as crop computational breeding, plant disease and insect pest monitoring and early warning, genetic engineering vaccines research and development, and intelligent prediction of agricultural green development, which reflect the profound impact of the fourth paradigm of data-intensive scientific discovery.

With the continuous breakthrough of agricultural scientific research and the rapid development of technological innovation, agricultural scientific research models and industrial formats have begun to enter a leap-forward development stage of transformation and upgrading [9], thus the new scientific paradigms and industry status have emerged. In order to adapt to the new characteristics of data-intensive agricultural scientific research, such as huge amount of data, strong data correlation, more reliance on tools and instruments, and frequent exchange of academic information, agricultural scientific researchers pay more attention to collaborative innovation with other disciplines, such as ecological and environmental governance in the agricultural field, biosafety, and the complex industrial issues, which all require multidisciplinary joint research. With the further application of emerging technologies such as biotechnology and information and communication technology in agricultural scientific research, it is necessary to create a new knowledge service environment suitable for the data-intensive open science environment for researchers, so that researchers can easily use scientific literature, scientific data, analysis tools, etc., while at the same time promoting the sharing of scientific research results in researching process. Researchers need to use data resources in this field and obtain cross-domain data resources more easily, which makes higher demands for knowledge services in a new situation.

3 Construction of Data-Intensive Agricultural Research Intelligent Knowledge Service Platform

National agricultural science and technology innovation needs to be driven by data resources. On the basis of comprehensively grasping the layout of knowledge resources construction in the field of agriculture and strengthening the construction of characteristic resources of agricultural disciplines, our research strives to form a national agricultural science and technology knowledge resources infrastructure with reasonable layout, rich types, strong authority, high activity, safe, reliable and shared, and provide strong support for knowledge services. Focusing on the demand of knowledge discovery in the agricultural field for scientific research information, the "data-intensive intelligent knowledge service platform for agricultural scientific research"was developed by comprehensively utilizing the technologies related to big data and knowledge service. Through professional knowledge management and knowledge calculation, the accurate expression and intelligent connection of field knowledge were promoted, and the new ecology of scientific and technological information exchange, including scientific research papers and scientific data was constructed, so as to strengthen the comprehensive guarantee and accurate service of national agricultural scientific and technological information.

3.1 Overall Framework

Based on the above construction ideas, the intelligent knowledge service platform for data-intensive agricultural research was constructed from three aspects: data resource construction, key technology research and development, and intelligent knowledge service. The technology framework of the platform included knowledge resources layer, key technology layer and application system layer from bottom to top, as shown in Fig. 1.

(1) Knowledge resources layer. Knowledge resources are important basis for knowledge services. In the data-intensive scientific research environment, knowledge resources are not only traditional resources dominated by literature, but also various types of scientific data, scientific and technological achievements data, reports and Internet open resources. By means of self-construction, purchasing, sharing and integration, the effective integration and connection of multi-source heterogeneous knowledge resources were realized, and a standardized big data

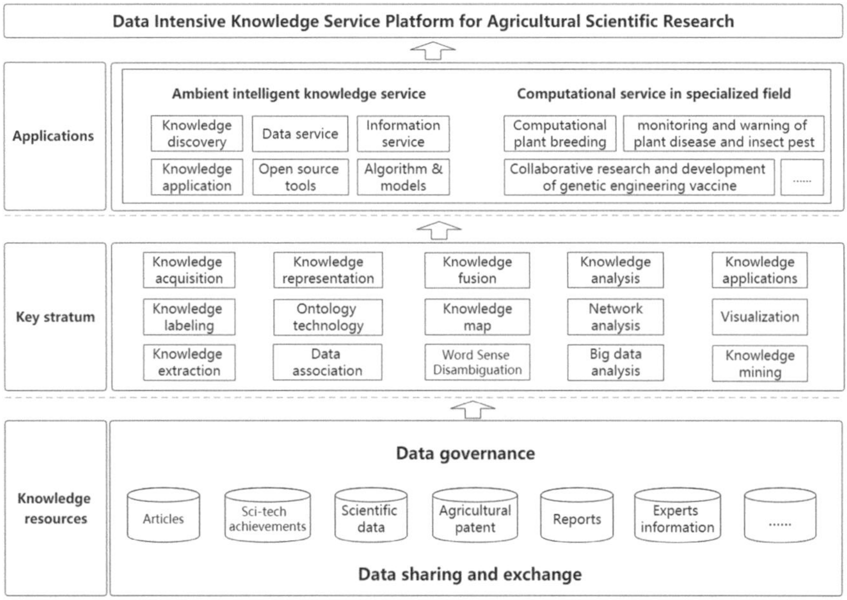

Fig.1 Overall framework of data-intensive agricultural research intelligent knowledge service platform

warehouse and management platform for scientific and technological knowledge in agriculture was constructed to provide reliable infrastructure support for knowledge services.

(2) Key technology layer. Under the data-intensive research paradigm, big data, artificial intelligence and other technologies are impacting the traditional knowledge service mode. The key technologies of the new generation of knowledge service will focus on the integration of traditional knowledge organization technology and new technologies such as big data and artificial intelligence. Knowledge acquisition and organization, knowledge association calculation, knowledge map, big data analysis, and knowledge mining are the key technologies to support data-intensive knowledge services, which can support accurate knowledge discovery and intelligent recommendation in data-intensive scientific research environment, and realize the value transformation and appreciation of information resources from data to knowledge and then to wisdom.

(3) Application system layer. The application system layer was a multi-form knowledge service product for end users, which integrated resources, technologies and services through a unified service portal. System-based knowledge services need to solve the problem of balancing ubiquitous knowledge services and professional services. The data-intensive agricultural intelligent knowledge service platform for agricultural research created a multi-scenario integrated service model for computing services such as inclusive knowledge discovery services, data services, information services, open tools, algorithm models, and crop

computational breeding for professional needs, which solved the adaptability and agility of system services.

3.2 Data Resource Construction

The multi-dimensional, multi-granularity and multi-modal big data resources are the basis for the agricultural intelligent knowledge service platform to provide knowledge services for all kinds of users. Through the systematic collection, aggregation and comprehensive mining of multi-source and multi-type basic data, the aggregation, management, mining and utilization of agricultural scientific research big data were realized, a new ecology of scientific and technological information exchange was built, including scientific research papers, and the comprehensive guarantee and accurate service of national scientific and technological information was strengthened. The content of data resource construction included the formulation of metadata standards and specifications, the exploration of resource construction mode, and the construction of data governance platform.

3.2.1 Metadata Specification

In order to guide and support the standardized description, exchange and reuse, open sharing and interconnection of multi-type digital resources in agriculture and other disciplines, and realize the normalization and standardization of data resources construction in agricultural field, following the principles of modularization, reusability and scalability, the metadata description standard system of scientific research big data resources was studied and formulated by widely referring to the existing domestic and foreign famous metadata standards, such as the Dublin Core Metadata Element Set, JATS of the National Center for Biotechnology Information (NCBI), Document Resource Processing Specification of the National Science and Technology Library (NST), the Science and Technology Bibliography Ontology (BIBO), the Research Community Semantic Web Ontology (SWRC) and the Exchangeable Image File (EXIF). At the same time, the application requirements of human and computer scenarios were taken into account. The metadata description specification included general metadata and characteristic metadata, covering 24 types of resources such as journals, books, papers, scientific data, experts and scholars, scientific research institutions, scientific research projects, scientific and technological achievements, patents and standards.

3.2.2 Data Resource Construction Mode

After long-term exploration and practice, according to the types, sources, carriers, forms, timeliness, accessibility and other characteristics of different resources, five data resource construction modes were established, including digital processing of paper literature, dynamic collection of network resources, open academic resource harvesting, distributed collaborative crowd-sourcing construction, and multi-source heterogeneous big data fusion on the premise of following the above metadata standards.

The digital processing of paper documents was carried out, the main processing object was the characteristic collection of literature from the National Agricultural Library. During digital processing of agricultural scientific and technological literature resources, an efficient digital processing method of paper documents was formed. Based on the methods of workflow, process monitoring and meticulous management concept, and breaking through the key technologies of intelligent batch processing of data, the technical platform for digital intelligent processing and meticulous management of literature resources was independently researched and built. The standardization of data processing and collaborative work environment were integrated, which realized the whole process tracking management, multi-person collaborative processing, quality control and process monitoring of document digital processing under the network environment.

Dynamic collection of network resources mainly refered to the collection of domestic and foreign policies, regulations, and agricultural technology trends publicly released on the Internet. After demand investigation, resource selection and collection source directory determination, different collection and processing methods were adopted according to open resources from different sources, and unstructured resources were automatically identified, classified and indexed. The collection method included collecting with commercial collection tools and developing collection tools independently, and completing the regular harvest of data through programming and calling the open access API interface. After clarifying the goal, users' needs were analyzed, collection sources were determined, collection tools and formulating collection rules were selected, and finally more than 1000 websites were finally selected as sources for information monitoring and collection.

Open data harvesting mode was adopted for open academic storage and open academic resources. Using the network communication and data interoperability protocols followed by open data access, such as OAI-PMH (Open Archives Initiative Protocol for Metadata Harvesting) and restful API interface, a batch collection process system was established based on Kettle tool to realize the dynamic harvesting of open access journal metadata.

Relying on agricultural science and technology innovation alliance and other distributed collaborative crowd-sourcing, factual tool resources were built, such as scientific research institutions, experts and scholars, scientific and technological achievements. According to the characteristics of those resources, the resources were deeply processed through the distributed collaborative crowd-sourcing mode. On the basis of following the unified metadata specification, a man–machine integrated collaborative crowd-sourcing process of task release, task execution, task handover and data quality verification was established.

Multi-source heterogeneous big data were fused. In order to increase the confidence level of data, enhance the complimentarity of data, improve data reliability, reduce data redundancy and inconsistency, and solve the problems of structural heterogeneity and semantic heterogeneity of data, the unified metadata standards and processing specifications were used to classify and integrate multi-source heterogeneous data, realize data fusion at different levels, and seek the most effective data governance means to improve data utilization, enhance data integrity and correctness, and realize data content enhancement and knowledge appreciation.

3.2.3 Data Governance

In order to realize the aggregation, integration, unified storage, classification organization and centralized management of multi-source heterogeneous big data resources, a big data governance platform was built, which was mainly used by data administrator. Its functions included data aggregation, data processing, data management, value-added data association, data visualization, and open interface service management. Data aggregation was carried out by using existing acquisition systems, open academic resource harvesting tools, distributed collaborative crowd-sourcing and other processing methods. After standardized data processing, data warehouse was completed. Data management mainly involved metadata and main data. The value-added function of data association was mainly based on the knowledge organization system such as thesaurus and classification system in the agricultural field, and the data in the platform were automatically annotated to realize the multi-granularity knowledge indexing of subject concepts, subject classification and entity objects. Data visualization was mainly used to realize the dynamic real-time monitoring and management of data resources construction and services.

3.3 Key Technology

The key technologies of this platform included knowledge organization and knowledge computing, such as knowledge indexing, knowledge extraction, knowledge ontology, knowledge association, knowledge fusion, and knowledge map. It also included knowledge service technologies, such as big data analysis and knowledge application. This paper mainly focused on knowledge indexing, linked data construction, and knowledge map construction technology.

3.3.1 Knowledge Indexing

Knowledge index is the process of extracting and identifying of knowledge elements according to the content characteristics of data resources, and it is the core of knowledge organization and knowledge retrieval. Based on the constructed agricultural knowledge organization system, word segmentation and terminology selection on literature topics (title, keywords, abstracts, full text, etc.) were carried out, and the automatic concept indexing was completed. Based on the mapping between subject concept indexing and classification system, the subject classification of literature was completed. At the same time, the entity objects such as institutions, scholars, journals, animal and plant species names and pests were identified and annotated to realize the automatic identification, extraction and annotation of knowledge units, which effectively supported knowledge organization of big data warehouse resources and knowledge discovery in the process of knowledge service.

3.3.2 Linked Data Construction

Massive information is difficult to be intelligently processed by computer because of its discrete isolation and lack of semantics, while linked data can solve this problem. Based on the service scenarios, the semantic association relationship between multiple types of resources such as scientific and technological literature, scientific data, experts and scholars, scientific research institutions, fund projects and scientific research achievements in data warehouse was constructed. Based on the open source tool D2R, the semantic mapping, automatic transformation and relevance construction from multiple types of resources in relational database to multidimensional semantic association data were completed. At the same time, based on the semantic middleware Jena, a special tool was independently developed to dynamically associate and publish the resources stored in the relational database in the form of associated data, which realized the fine revelation, deep integration and knowledge organization of massive multi-source heterogeneous data resources.

3.3.3 Knowledge Maps Construction

The general knowledge map of scientific research and the knowledge map of agricultural thematic areas were constructed. In terms of general knowledge map of scientific research, aiming at scientific and technological resources such as journals, journal papers, conferences, conference papers, patents, scientific research projects, and scientific research institutions, the semantic representation model (scientific research ontology) [10, 11] was designed, which included 13 core classes, 32 object attributes, and 74 data attributes. The scientific research knowledge map was constructed based on RDF-ETL tool, and about 1.16 billion RDF triples were generated [12]. In terms of knowledge mapping in agricultural thematic areas, rice thematic areas and pest thematic knowledge map were initially constructed [13] through entity extraction, association annotation, concept disambiguation and knowledge fusion of multi-source heterogeneous data related to rice genes and pests in data warehousing. Thus the new intelligent service functions such as knowledge map-driven semantic retrieval and visual navigation analysis were preliminarily developed, and other multi-scenario applications were still in practice. Entity extraction, association annotation, concept disambiguation, and knowledge fusion for rice genes, diseases and insect pests related data in data warehouse were conducted, and new intelligent service functions such as knowledge map driven semantic retrieval and visual navigation analysis were also preliminary developed. Other multi-scenario applications were still in practice.

3.4 Platform Service Functions

A multi-scene integrated service model of ubiquitous intelligent knowledge service and professional computing service was proposed to solve the adaptability and agility of system services, and an integrated service platform was built . Ubiquitous intelligent knowledge service included knowledge discovery service, data service, information service, knowledge application service, and open tool service. The professional computing service provided knowledge computing services for researchers in specific fields of agriculture.

3.4.1 Knowledge Discovery Services

Knowledge discovery is the key to achieve accurate knowledge search and acquisition. In order to meet the needs of users' knowledge acquisition, based on the self-built comprehensive high-quality big data warehouse for agricultural research, semantic knowledge base and multi-factor intelligent ranking model, the 'agricultural knowledge search' engine with universal domain and semantic intelligence was developed, which realized multi-angle and multi-dimensional retrieval methods such as unified retrieval, advanced retrieval and search word association, and provided users with cross-resource and cross-language knowledge retrieval and acquisition services covering information, literature, data and reports.

3.4.2 Data Services

This platform provided data open sharing, professional processing, operation hosting and other multi-form data services. Data types included scientific data and statistical data. Scientific data included agricultural resource zoning data, livestock and poultry feed and genetic parameters data and other types of scientific data. Statistical data covered the world crop production, production index, animal husbandry, meat production, animal husbandry processing, agricultural output value, crop processing products, population, import and export trade, etc.

In terms of data calculation, analysis, visualization and other services, the platform provided 10 kinds of data analysis tools, which could be used for statistical analysis of public or user-uploaded data and graphical display. The platform could provide users with a variety of prediction models to achieve online data analysis and prediction. In addition, the platform could also provide users with a series of report templates, which could generate and download charts and model analysis results obtained in the process of data analysis.

3.4.3 Information Services

To meet the needs of agricultural science and technology innovation, agricultural industry, agricultural production and operation, and government management decision-making, content analysis, scenario analysis, network analysis, model/simulation, bibliometric analysis and technology foresight were used to collect, track, mine, collate and analyze massive data resources. It could provide users with specialized, disciplined and personalized analysis reports based on agricultural scientific research big data, as well as analysis report on academic trends. At present, the system has integrated international industry reports, macro development reports, agricultural zoning reports, industrial analysis reports, competitiveness analysis reports, high-end forum reports, and scientific research situation reports. Users could read and share intelligence analysis reports on line. In addition, the platform supported users to customize reports online, and the on-demand customization of reports was realized through requirements confirmation, protocol signing and report delivery processes.

3.4.4 Knowledge Application Services

Knowledge application is to optimize and integrate data resources by using relevant algorithms and models according to specific needs of users and combined with demand scenarios, so that massive data has a systematic structure and system, which is convenient for users to understand and absorb knowledge. Based on the platform self-built resources, Internet resources and third-party resources, combined with agricultural professional knowledge and visual display methods, totally 10 knowledge applications were constructed, such as intelligent diagnosis of animal and plant diseases and insect pests, international trade analysis of agricultural products, and agricultural zoning analysis. The application can provide users with intelligent recognition and intelligent Q&A of pest images. The application showed the analysis of international agricultural products trade trends in multi-dimensional geographical space by embedding maps.

3.4.5 Open Tool Services

Open tool service refers to the batch acquisition of data resources in a discipline or a software system through API. Data-intensive agricultural research intelligent knowledge service platform implemented open access policy and provided OpenAPI open platform interface, by constructing an open platform of data resources based on OpenAPI and introducing a third-party authentication developer group, and API interface docking service of existing resources of the system can be realized. At present, the collaborative construction and management system of agricultural thesaurus can provide open interface API of agricultural thesaurus data, and also can call API interface of agricultural encyclopedia terminology system to exchange data.

3.4.6 Professional Computing Services

With the development of data-intensive scientific research environment, scientific research teams in the field of agriculture are facing great challenges in data processing and scientific research support. Combined with the practical difficulties existing in the scientific research team in the field of agriculture, the data analysis and data computing needs of field experts were excavated, and field data computing services were provided for professional research teams, such as crop computing breeding, plant pest monitoring and early warning, and association analysis between genomic big data and phenotype, so as to provide reference for scientific researchers and shorten the research cycle effectively. In terms of crop breeding, aiming at the problems of large amount of data in professional fields, lack of unstructured data management and one-stop data analysis, the data calculation models were studied and developed, data management tools and platforms were provided, which assisted scientific research teams to use big data technology to improve crop breeding efficiency and accelerate the process of breeding modernization.

4 Achievements in Platform Construction

According to the resource construction idea of 'overall resource scale and key resource refinement', the data intensive intelligent knowledge service platform for agricultural scientific research followed the principle of 'construction and use at the same time', and continuously improved various service functions and the quality of data resources in the process of use. At present, the platform construction achieved remarkable results, which can be mainly reflected in two aspects of data scale system and platform service.

In the construction of data scale system, the platform integrated more than 20 types of agricultural research big data resources, including scientific and technological literature, patents, regulations, information, projects, achievements, experts, institutions, reports, scientific data, statistical data, encyclopedias and agricultural basic knowledge bases. The total amount of resources was nearly 1 billion, forming eight data systems, including literature data, scientific data, industrial data, strategic intelligence data, agricultural intelligence data, Internet data, basic corpus data and user behavior big data, creating a high-quality agricultural research big data 'granary'.

In terms of platform services, the platform provided ubiquitous intelligent knowledge services such as 'one-stop' knowledge discovery, data services, knowledge applications, intelligence services, and model tools based on different application scenarios for different user groups such as agricultural managers, researchers, industrial personnel and college students, as well as knowledge computing services in the field of agriculture, and provided scientific and technological information support for major scientific research plans, high-end think tanks, agricultural production, government management and decision-making, and agricultural enterprises at home and abroad. The annual number of visits of the platform was more than 1.5 million. Users covered 161 countries and regions around the world, as well as 34 provinces (cities and districts) in China. The platform has served 60 academicians in the field of agriculture, 40,000 users of 875 scientific research universities, institutes and agricultural enterprises, and dozens of national major research groups. The platform gave full play to the existing resource advantages and service capabilities, and effectively supported the project application, achievement award and other related work of scientific researchers with comprehensive data and accurate services, which was highly appreciated by the majority of users.

5 Summary and Prospect

Emerging technologies such as big data are profoundly changing the paradigm of scientific research. Data intensive scientific research paradigm has become the mainstream paradigm of scientific and technological innovation and development. The massive acquisition of data, the sharing of data resources, cross domain cooperation and the co-construction and sharing of scientific research infrastructure have gradually become the key factors affecting the development of scientific research. The construction of data intensive agricultural scientific research intelligent knowledge service platform responded to the new requirements of scientific research paradigm change on scientific and technological information service and knowledge service, and provided strong scientific and technological information guarantee and knowledge service support for scientific and technological innovation in the agricultural field, so as to meet the new demand of data intensive agricultural scientific research innovation for domain knowledge service. The purpose of the platform's construction and service was to integrate multi-source scientific and technological information in the agricultural field, open up the data island, mine knowledge associations of various data resources, enhance the value of data resources, promote field knowledge discovery, and serve agricultural scientific and technological innovation and management, and become a high-end exchange platform for scientific and technological information to support data-intensive agricultural scientific research. Since the construction, this platform has made great progress in big data integration and governance, knowledge organization and association, providing intelligent knowledge services for users. It has also created a comprehensive scientific research informatization application demonstration oriented to the field of agricultural scientific and technological innovation. Looking forward to the future, the research team would continue to promote the construction of agricultural intelligent knowledge service platform from the aspects of big data construction of agricultural scientific research, development of key technologies of knowledge service, data application scenarios and knowledge discovery.

(1) Improving the construction of agricultural scientific research big data center. On the basis of continuously expanding the scale of data resources, it is strived to create high-quality data resources. Facing the major national strategic needs during the '14th Five-Year Plan' period and the overall planning of Chinese Academy of Agricultural Sciences and its subordinate research institutes, integrate and innovate the application of advanced technologies such as big data, cloud computing, artificial intelligence, 5G and robot, and systematically carry out technological innovation in the construction, collection and processing, aggregation and integration, governance and control of multi-source heterogeneous agricultural scientific research big data resources.
(2) Breaking through the research and development of key technologies of domain knowledge service and building an intelligent knowledge service system in the field of agriculture. In the field of agricultural knowledge graphs and other semantic knowledge base automatic construction technologies, semantic search

technologies, etc., there are still problems such as shallow knowledge representation and low accuracy of knowledge acquisition in practice. Therefore, it is necessary to further study the automatic construction technology of large-scale semantic knowledge based on high-quality agricultural science and technology big data. It is necessary to develop tools for automatic learning and construction, evolutionary updating and collaborative processing of semantic knowledge bases. The breakthroughs in multivariate data governance and associated computing methods should be looked for to achieve computability of semantic features. A big data-driven intelligent agricultural open knowledge service system should be built, and the independent research and development and integrated application of core key technologies should be strengthened, such as human–computer interaction and semantic intelligent search, and provide core algorithms and platform support for data-intensive agricultural intelligent knowledge services.

(3) Improving platform functions and expand data application scenarios. The improvement of high-tech and high-performance computing capabilities, as well as the exchange and sharing of related application technologies such as data analysis and knowledge mining, should be accelerated to form an integrated development trend of 'scientific research data' - 'mining analysis' - 'field application', and integrate domestic and foreign applications. The knowledge computing models and methods and tools in agriculture-related fields were integrated into the data-intensive agricultural scientific research intelligent knowledge service platform. At the same time, according to the specific needs of the agricultural professional field, such as intelligent computational breeding of crops based on epigenetic theory and technology, plant disease and pest monitoring and early warning, and intelligent prediction of agricultural green development, etc., the standardized and regularized process methods should be designed to improve the function of the platform which can provide convenience for scientists in the field and serve data-intensive scientific research.

(4) Developing a knowledge computing model to improve the efficiency of knowledge discovery. Guided by the optimization, integration and efficient utilization of scientific research big data resources, actively responding to the new needs and challenges of the fourth paradigm of data-intensive scientific research, the commanding heights of the world's scientific and technological frontier of biological computing breeding based on 'data', 'computing power' and 'algorithm' must be seized. The interaction and integration of big data and artificial intelligence technology and agricultural scientific research activities need to be deepened. The transformation and development of digital agriculture and digital scientific research should be accelerated to ensure the safety of national agricultural technology innovation and self-reliance, assisting in the high-quality development of agricultural and rural areas, and ensuring national food security.

Acknowledgements This work was partially supported by National Key R&D Program of China(2021ZD0113705) . Furthermore, we thank the anonymous experts for their constructive comments.

References

1. Hua Z, Ruixue Z, Huimin J, et al (2020) Construction and service of big data warehouse of agricultural science and technology. Digit Libr Forum (8):48–55
2. Hui C, Huiling R, Yi L (2021) Facing the development needs of medical science and technology, making the 14th five year plan of medical library. Digit Libr Forum (5):31–36
3. Yingying Z (2018) Exploration on the application of machine learning in library discover system-taking the discover tool yewno based on knowledge graph as example. J Libr Inf Sci Agric 30(07):47–50
4. Tan S, Zheng L, Yunpeng C, et al (2019) Analysis and design of a new generation of open knowledge service system integrating knowledge organization and cognitive computing. J Libr Sci China 45(3):38–48
5. Li Q, Jing X, Zhijun C et al (2019) Designing smart knowledge services with sci-tech big data. Data Anal Knowl Discov 3(01):4–14
6. Liqin Z, Hao F, Jianpeng P (2017) Research on the knowledge service framework. Res Libr Sci (21):53–59
7. Zhonghua D, Zhifang L (2015) Analysis of the needs of the fourth paradigm of science research based on the intelligence science perspective. Inf Sci 33(7):3–6, 20
8. Rongrong L, Donghui X, Tao L, et al (2015) Discussions on collaborative innovation model in agricultural scientific research. Manag Agric Sci Technol 34(5):15–18
9. Tan S, Yongwen H, Guojian X, et al (2021) Considerations for the development of agricultural informatization driven by a new generation of information technologies. J Libr Inf Sci Agric 33(3):4–15
10. Jiao L, Tan S, Yongwen H, et al (2021) Construction of scientific knowledge graph by integrating thematic knowledge and scientific literature. Digit Libr Forum (1):2–9
11. Jiao L (2021) Research on generation of scientific research review based on knowledge graph. Chinese Academy of Agricultural Sciences, Beijing
12. Li J, Xian GJ, Zhao RX et al (2021) RDFAdaptor: efficient ETL plugins for RDF data process. J Data Inf Sci 6(3):123–145
13. Yue L, Tan S, Guojian X et al (2021) Research on ontology construction of crop diseases and pests for deep fusion of multi-source data. Digit Libr Forum 02:2–10

Ruixue Zhao, deputy director of Agricultural Information Institute of Chinese Academy of Agricultural Sciences, has more than 20 years of experience in digital library, knowledge organization, knowledge service, and Scientific data Management, spearheading projects like National Key Technology R&D program, basic project of Ministry of Science and Technology, and special project of National Science and Technology Library (NSTL), etc. She also serves as Director of Library and Information Branch of Chinese Society of Agronomy, deputy director of Chinese National Data Center for Agricultural Sciences, committee member of Science and Technology Libraries (STL) in IFLA.

Application and Prospect of Blockchain Technology in the Energy Internet

Meicheng Li, Longze Wang, Yan Zhang, Xiaojun Lv, Delong Zhang, and Shucen Jiao

Abstract Blockchain is booming as a new generation of information technology. The basic features of blockchain technology, including decentralization, fairness, and expansibility, are consistent with the construction of an Energy Internet. According to the technical characteristics of blockchain and the development direction of the Energy Internet, this chapter analyzes the technical architecture of the energy blockchain. This chapter outlines the current situation of blockchain technology in the application of the Energy Internet using four typical application scenarios. Moreover, it identifies the key scientific problems to be solved in the application of energy blockchain technology and looks toward the development vision of an Energy Internet–based blockchain. Finally, this chapter provides guidance for promoting the application of blockchain technology in China's Energy Internet construction.

Keywords Blockchain technology · Energy Internet · Energy blockchain · Scientific problems · Prospect forecast

1 Introduction

At the general debate of the 75th Session of the United Nations General Assembly on 22 September 2020, President Xi Jinping solemnly announced that China would strive to achieve a peak in carbon dioxide emissions by 2030 and carbon neutrality by 2060. This means that China's economy will officially enter the path of low-carbon transformation. In the coming years, China will continue to adjust its economic growth, industrial structure, technological path, business model, and even the lifestyle of its residents [1]. The energy industry accounts for more than 80% of the country's total carbon dioxide emissions, which is a key area necessary to achieve the carbon

M. Li (✉) · L. Wang · X. Lv · D. Zhang
School of New Energy, North China Electric Power University, Beijing 102206, China
e-mail: mcli@ncepu.edu.cn

Y. Zhang · S. Jiao
School of Economic and Management, North China Electric Power University, Beijing 102206, China

China's e-Science Blue Book 2023, https://doi.org/10.1007/978-981-99-8270-7_10

peak and carbon–neutral goals [2]. Against the backdrop of low-carbon transformation of energy structure, the Energy Internet industry and technology have ushered in significant development opportunities.

The core of the Energy Internet is electricity, which can organically combine energy technology, information technology, and intelligent management technology. This effort has achieved the coupling and complementarity of various energy sources, including electricity, gas, oil, and coal, and thus meets the diversified energy demand of electricity, heating, and cooling [3]. Under the guidance of the carbon peak and carbon–neutral goals, the installed capacity of wind power and solar power has continued to rise. A large number of unstable power supplies connected to the Energy Internet has made the coordinated and optimal operation of energy an increasingly urgent task. In addition, the carbon peak and carbon–neutral goals have accelerated the large-scale development of the Energy Internet. This development requires an energy system with a complex structure, extensive equipment, and complex technology. This structure also has typical nonlinear random characteristics and multiscale dynamic characteristics. The traditional energy dispatching, trading, and management methods do not meet these new requirements for Energy Internet planning, construction, and operation management [4].

In his speech on promoting the innovation and development of blockchain technology and industry. President Xi Jinping stressed that "we should promote the combination of blockchain underlying technology services and new smart cities, and explore the application of information infrastructure, smart transportation, energy and power." The president's guidance necessitates the joint development of blockchain technology and industrial innovation and development, and promotes the need for accelerated integration and functional expansion of blockchain technology and an Energy Internet. In this chapter, we analyze the application mode and basic architecture of blockchain technology for the construction and development of an Energy Internet. We introduce the application status of blockchain technology according to four typical scenarios of energy supply, energy consumption, energy trading, and energy data management. Finally, we identify the key scientific problems facing energy blockchain and propose suggestions for the development of an Energy Internet based on blockchain technology.

2 Background Information

2.1 Energy Internet

Based on the existing energy supply system and distribution network, the Energy Internet has adopted advanced power electronic technology and information technology. The smart grid, thermal power network, natural gas network, hydrogen energy network, and other energy networks have been integrated to realize energy interaction and network sharing with a two-way flow of energy, information, and value [5]. In

2011, U.S. scholar Jeremy Rifkin in his publication *The Third Industrial Revolution* proposed the idea of an Energy Internet as a new energy utilization system characterized by an in-depth combination of energy technology and information technology. At the United Nations Development Summit in September 2015, President Xi Jinping proposed a China initiative to "Explore building a global Energy Internet to meet global electricity demand in a clean and green way." In 2016, the Chinese government noted in its "Guiding Opinions on Promoting the Development of Internet + Smart Energy" (FGAI Energy [2016] No. 392) that the Energy Internet is a new form of energy industry development with the deep integration of Internet access and energy production, transmission, storage, consumption, and energy market [6].

Against the backdrop of the low-carbon transformation of the energy industry and the development and progress of information technology, the development pace of Energy Internet technology is constantly accelerating. At present, the Energy Internet is based on extensive access to renewable energy generation, which covers various types of energy supply facilities in different regions. The Energy Internet connects micro energy networks consisting of distributed power generation, energy storage, intelligent power transformation, and energy saving equipment through Internet access and information technology to create multi-energy complementarity of energy sources [7]. As an intelligent energy system supporting clean and low-carbon transformation of energy and electricity, comprehensive utilization of energy, and flexible and convenient access of multiple subjects, the Energy Internet is an important extension of the power system in the energy field. In addition to providing traditional power transmission and distribution functions, the Energy Internet also has established an energy-sharing and service platform to better support energy nodes, such as low-capacity distributed power generation, electric vehicles, and smart home appliances, which can join the Energy Internet in a plug-and-play way [8]. In addition to the deepening of the concept of energy saving, low-carbon, and sustainable development, the coupling of energy interaction and carbon emissions has become an urgent issue affecting society. Understanding how to integrate carbon emission verification and carbon emission trading into the Energy Internet is an essential problem to be considered in the development of this power system [9].

2.2 Blockchain Technology

Blockchain is a decentralized distributed ledger database, which is a string of data blocks associated with cryptography. Within each data block, information that has been effectively confirmed by multiple businesses is arranged through time series and formed in a chain structure. According to the "White Paper on China's Blockchain Technology and Application Development" issued in 2016 by the Ministry of Industry and Information Technology, blockchain technology includes the use of the chain to verify the data structure and data storage, distributed node consensus algorithm to generate and update the data, cryptography to ensure the security of data transmission and access, and intelligent automation of script code contracts to program and operate data in a new kind of distributed infrastructure and computing paradigm. Decentralization, security, and reliability are typical characteristics of block chain technology.

According to different ways to participate, blockchain technology can be categorized in three forms: public chain, alliance chain, and private chain. Among these chains, the public chain is the earliest form of blockchain. It follows the principle of complete decentralization and complete peer of nodes. Any node can participate in record maintenance, but the transaction volume is limited. The alliance chain is used primarily for specific organizations, and the predetermined nodes participate in record maintenance. The consensus process is completed by a group of trusted nodes with a certain number of nodes, and the internal nodes are not completely equal. Private chains run only in an internal environment, and only a few internal users can record and access these chains. Internal rules can be easily modified, and restrictions on transaction volume and transaction speed are relatively low, while a-Nodes are quite limited [10].

The characteristics of blockchain technology, including decentralization, openness, transparency, and sharing, are consistent with the construction direction of holographic perception, open sharing, and integrated innovation of the Energy Internet. This technology can effectively solve the problems of data integration, network security, multiagent coordination, and other aspects in the construction of the Energy Internet system. Hence, this work analyzes the application of blockchain technology in the Energy Internet and discusses the application mode and future development direction of blockchain technology in various scenarios.

3 Blockchain Energy Internet Technology Architecture

The application of blockchain technology in the Energy Internet field is in its infancy. With the core idea of blockchain decentralization and distributed data structure, it will be helpful to build the information platform of an energy system. Through the effective integration of information flow and energy flow, the energy interaction and utilization efficiency of the Energy Internet can be improved. Carbon emissions of the entire energy system can be effectively reduced, which will further optimize the value stream flow within the energy system. The Energy Internet technology architecture based on blockchain technology is shown in Fig. 1.

With the development of the new energy industry and the progress of energy science and technology, the Energy Internet has gradually formed an operation mode of horizontal multi-energy complementarity and vertical source network load storage coordination. On the energy supply side, the Energy Internet gives priority to the development and utilization of renewable energy, such as wind power, solar power, and hydropower. It relies on fossil fuel to alleviate the pressure of large-scale unstable power on the operation and dispatching of the energy system and reasonably allocates energy storage facilities to ensure the stability of the energy supply. On the energy demand side, the diversified energy needs of different users for electricity, heating,

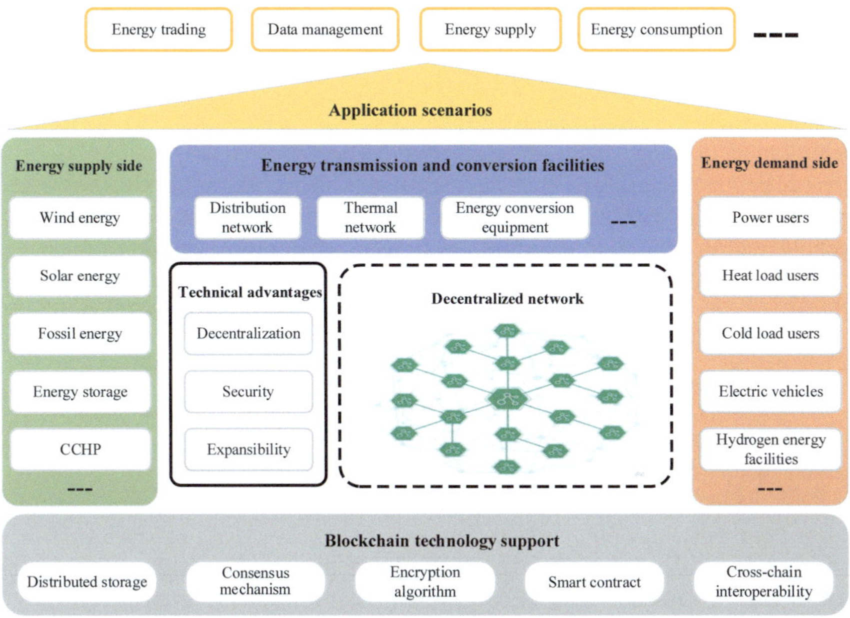

Fig. 1 Energy Internet technology architecture based on the blockchain

and cooling can be met through advanced energy scheduling and management means. Innovative application scenarios, such as charging and discharging of new energy vehicles and hydrogen energy production, have been actively expanded [11]. By connecting energy transmission and conversion, reliance on new technologies, such as an intelligent distribution network and thermal and cold power transmission, it is possible to establish efficient transmission channels at both ends of energy supply and demand. Combined with various energy conversion equipment, including heat pumps and phase change energy storage, the Energy Internet can promote the interaction and optimization of different energy flows [12]. In terms of energy data management, the Energy Internet relies on information technologies, such as big data analysis, machine learning, and prediction. It also integrates internal operational data, meteorological data, power grid data, and power market data of energy. In addition, big data analysis, load forecasting, and power generation forecasting can be carried out to realize the efficient recycling of data flow.

Blockchain technologies, such as distributed data storage, consensus mechanism, encryption algorithm, smart contract, and cross-chain interoperability, have played a key supporting role in the information interaction of the Energy Internet. Unlike traditional ways of storing all data of the energy system in one center, distributed storage allows all nodes in the system to store and manage data, and thus can form a distributed ledger maintained by multiple parties. The consensus mechanism is the key element needed to ensure that all participants in the energy system can form a unified account record. At present, the common consensus mechanisms include workload proof, rights and interests proof, agency rights and interests proof, and Byzantine fault-tolerant mechanisms. Hash algorithm and asymmetric encryption technology constitute the encryption algorithm technology of the blockchain network. With the help of modern cryptography, the trusted interaction of multiparty subject information can be guaranteed [13]. Smart contracts enable blockchain applications to expand from simple transfer transactions to other fields, such as energy and electricity. This technology can be used to deploy all terms and logical processes related to energy contracts. The automatic execution of the contract code is realized through event triggering, which avoids the low efficiency and potential safety hazards of human control. Cross-chain interoperability is an emerging technology in the blockchain. It can enable different blockchain networks to communicate or share information with each other. It is consistent with the characteristics of wide area interconnection and multisystem interaction of the Energy Internet. Cross-chain interoperability has become one of the key technologies of the future [14].

With the technical support of the blockchain, a decentralized network platform can be built. The energy supply side, demand side, transmission and conversion, data management, and other subjects can participate in the network platform as independent nodes. This ensures that the Energy Internet can offer technical advantages, such as decentralization, security, and expansibility. Decentralization is the most typical feature of an energy blockchain network. As the number of participating nodes in the energy system increases, the credibility of the system also increases [15] to form an energy interaction system of multilateral trust. The security of the energy blockchain is reflected in its ability to prevent forgery and tampering. The blockchain validates the transaction data occurring in a certain period of time and packages it into data blocks. Each block is linked in an orderly way with the previous block through cryptographic algorithm. The flexible expansibility of blockchain technology makes its integration with the Energy Internet possible. The distributed operation mode can be used to build a flat network platform and break the inefficient modes of centralized operation, control, and operation. Blockchain technology improves the interaction efficiency of energy flow, information flow, and value flow in all links of the Energy Internet. On the basis of the technical advantages of the energy blockchain, blockchain technology has been applied in many Energy Internet scenarios. The specific conditions are described in the next section.

4 Application of Blockchain Technology in the Energy Internet

The Energy Internet based on blockchain technology has adopted a midsize desktop interface design and has carried out relevant applications in energy supply, energy consumption, energy trading, and data management. This technology provides strong support for various energy businesses.

4.1 Energy Supply

The wide access to distributed renewable energy has promoted the point-to-point interaction between multilateral power trading and energy. The distributed energy management mode has some problems, however, such as high management cost, low operation efficiency, opaque data, and difficulty protecting privacy. At the same time, the energy and power industry has actively participated in carbon emission trading to realize low-carbon transformation and development. The carbon market still has problems in data collection, credit supervision, information circulation, and other aspects [16]. Blockchain technology is integrated into the energy supply side to build an equal, trustworthy, safe, and efficient network platform. The supply of information of the energy enterprises and equipment is linked for storing relevant product and service information through chain authentication. The time stamp recording and allocation mechanism can be used to address real-time queries for relevant information, such as an energy scheduling and management business. This improves the energy management efficiency of multiple energy coordination and complementarity. In addition, the blockchain can track and record the carbon emission data generated in the process of energy supply, which uses the blockchain data structure and distributed ledger technology. This carbon emission verification and trading information will be broadcast and stored throughout the network, which then provides support for the verification and supervision of carbon emission data. The structural diagram of an energy supply platform based on blockchain technology is shown in Fig. 2.

Many countries and regions have carried out energy supply management projects based on blockchain technology. Globally, the United States is one of the most active countries in the development and application of energy blockchain technology and developed the world's first energy blockchain project in 2016. This project collects

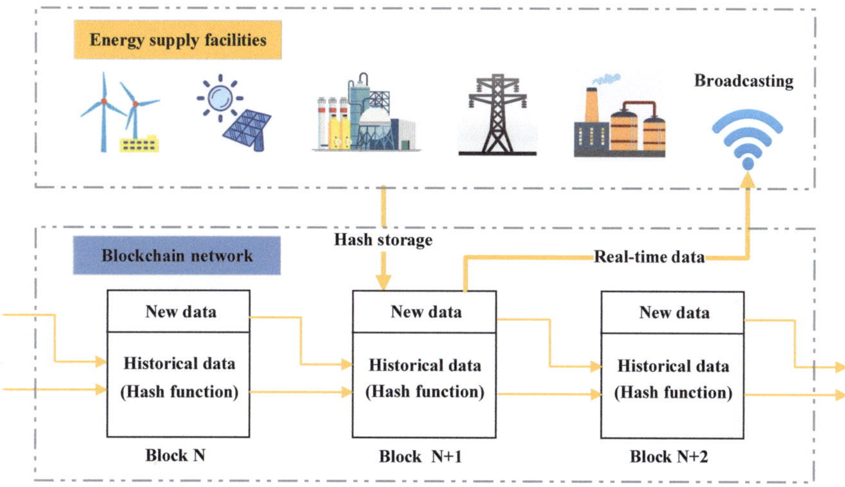

Fig. 2 Structural diagram of an energy supply platform based on blockchain technology

the power supply information of distributed photovoltaics in Brooklyn, New York, and uses blockchain technology to distribute surplus power to energy users in need. In 2017, WePower UAB in the United States established the WePower platform. This blockchain token technology virtually maps renewable energy electricity into energy tokens that can be circulated efficiently. It has improved scheduling efficiency and flexibility on the energy supply side [17]. The practice of an energy blockchain in China started a few years later. In July 2020, the Beijing Electric Power Trading Center launched the trial operation of a voucher trading system based on blockchain technology. The blockchain fully supports the implementation of a renewable energy consumption guarantee mechanism. The Haishu District in Ningbo officially launched a carbon housekeeper platform in July 2021. By relying on the digital economy portal, a data connection between the government and enterprise was realized. Full coverage of carbon emission data monitoring of industrial enterprises was achieved, and thus provided a data basis for the issuance of carbon quotas for industrial enterprises.

4.2 Energy Consumption

The demand-side response is an important means of Energy Internet supply and demand interaction. This response can optimize the resource allocation at both ends of energy supply and demand, alleviate demand for electricity from peak load, and improve the overall operation efficiency of the energy system [18]. Although China launched a series of demand-side response demonstration projects in 2012, the effectiveness of the project implementation was insufficient because of the cumbersome active response process at the energy consumption end, the low level of fine management, and the difficulty of information interconnection and sharing. As shown in Fig. 3, using blockchain technology, the supply and demand ends of the energy system have been connected directly to form an interconnected network platform. Energy consumers, service providers, government agencies, regulatory authorities, and other nodes are connected and share the information of the energy system in real time. This provides a basic information guarantee for active response and regulation on the energy consumption side [19]. In addition, energy consumers can find suitable energy suppliers by accessing the information of relevant blocks. The smart contract in the platform will match the appropriate supply and demand sides to ensure real-time confirmation, real-time execution, and automatic clearing of energy transactions to simplify the process of energy consumption.

To improve the convenience of energy consumption and guide energy users to actively consume clean and low-carbon renewable energy, Germany's Innogy Innovation Hub company launched the Share&Charge project. This project integrates the information of electric vehicle users and charging piles on the chain, and users can find a charging pile that is powered by renewable energy to charge the vehicle using a mobile phone program [20]. BittWatt of Singapore has established an intelligent

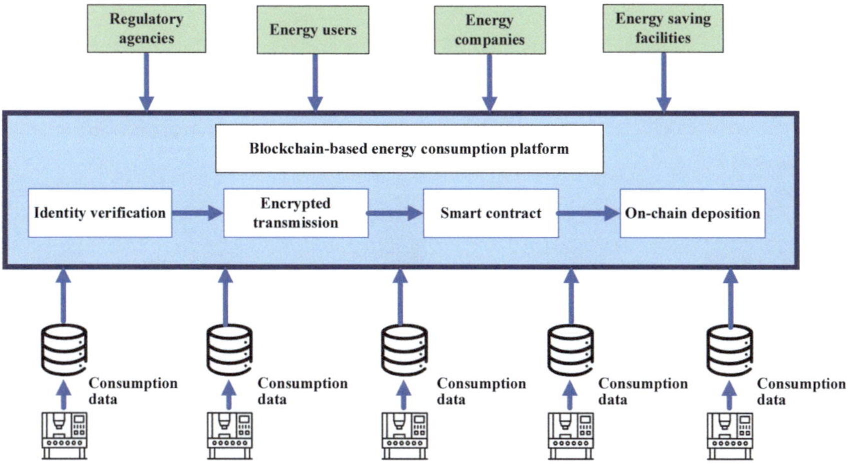

Fig. 3 Structural diagram of an energy consumption platform based on blockchain technology

energy consumption platform based on blockchain technology to solve the problems of delay and inefficiency in the interconnection of energy supply and demand. This program has achieved the real-time matching of energy supply and demand and the automatic transfer of digital assets. China's Shenzhen Zero Carbon Technology Co., Ltd. (Xarbon) has combined blockchain technology and carbon sink business. Different types of energy consumption data are intelligently analyzed to develop a "carbon sink chain" intelligent platform to encourage the public to consume green energy [21].

4.3 Energy Trading

The diversification of the main participants in the Energy Internet has driven the need for higher requirements for information transparency and transaction efficiency of energy transactions. The traditional electricity trading mode no longer meets the diversified needs of the energy market. As shown in Fig. 4, in the energy-trading platform based on blockchain technology, energy producers and sellers can encrypt and transmit the quotation to the blockchain system before the beginning of the trading cycle. Through a two-way auction, the supplier and the demander can reach the transaction intention, and the transaction contract will come into force after being signed digitally by the buyer and the seller and the power grid enterprise. Smart contracts will automatically perform energy scheduling, expense settlement, and other businesses, reduce personnel workload, and save labor costs [22]. At the same time, the blockchain system can set a reputation value list to exclude malicious nodes that have low value from energy transactions. This approach reduces the security risks

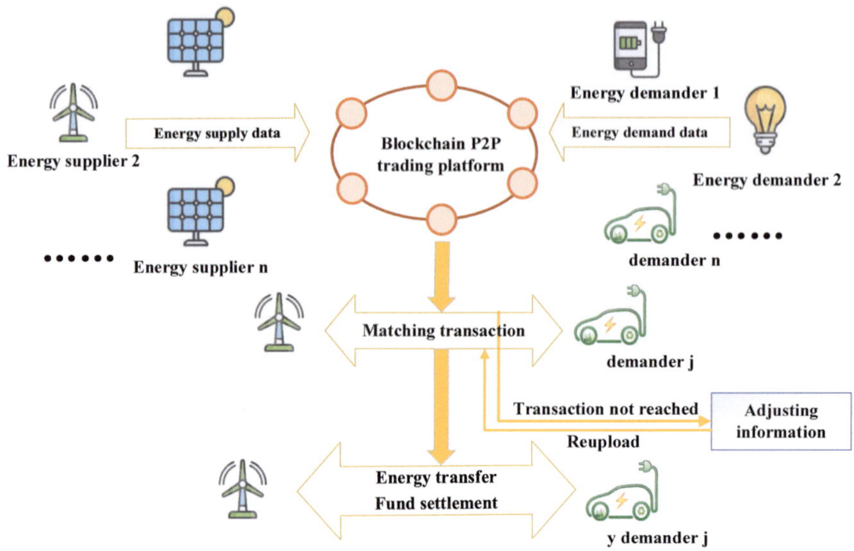

Fig. 4 Structural diagram of an energy-trading platform based on blockchain technology

of the Energy Internet and blockchain systems and ensures the normal progress of energy transactions.

The Piclo project in the United Kingdom is an earlier energy-trading blockchain platform, which provides direct energy interaction services by matching transactions according to the needs of users and suppliers. The Australian power ledger project has made full use of local lighting resources to build a P2P solar power-trading system based on blockchain technology. This project has solved the problem of mutual trust between energy supply and demand [23]. In China, the Shekou energy blockchain project launched by the China Merchants Charity Foundation will connect clean power located in Shekou Nanhaiyiku to participate in the virtual transaction of clean power. The smart contract in this project has been used to enable the supply and demand sides to complete energy transactions [24]. State Grid Shandong Electric Power Company has created an intelligent distributed energy-trading platform based on blockchain technology, which realizes the purchase and sale of electricity among photovoltaics, energy storage, and users in a microgrid to improve the transparency and operation efficiency of energy trading.

4.4 Energy Data Management

The open sharing of energy data can promote the revolution of energy production and consumption with digital assets. This ability to share big data plays an important role in improving the energy utilization efficiency and operation management efficiency of the Energy Internet. The disclosure of valuable information, however, poses many problems, such as the diversification of data structure, the risk of privacy disclosure, and the difficulty of ensuring information security [25]. The trusted mechanism of blockchain technology can build a highly secure data management platform for the Energy Internet. This technology supports the safe interaction and open sharing of diversified data at different levels. At the micro level, users can view their energy data anytime and anywhere and release energy demand information. In addition, energy suppliers can manage energy supply information through an integrated information platform and use various decision support systems to formulate energy supply plans. At the macro level, the distributed blockchain network stores energy supply and demand information that does not involve privacy, which can be extracted and utilized by all kinds of participants. Massive amounts of actual data are necessary for the research and development of renewable energy, energy conservation, and consumption reduction. Thus, building a data management platform based on blockchain technology can better provide a strong data foundation for Energy Internet science and technology research and technological innovation [26]. The structural diagram of a data management platform based on blockchain technology is shown in shown in Fig. 5.

To strengthen the data management of the energy industry, the Filament Company of the United States proposed Blocklet chip hardware and blockchain software solutions in 2018. The company applies these solutions to machines, instruments, or other edge equipment to realize the functions of safe transaction, real-time monitoring, asset tracking, and management of embedded energy equipment. PONTON of Germany launched a Gridchain for the real-time management of power grids. This system improves coordination among power generation, transmission, distribution, and dispatching and also provides a solution for power grid congestion management [27]. In China, the State Grid credit investigation company uses blockchain technology to integrate internal power data, e-commerce data, industry and commerce,

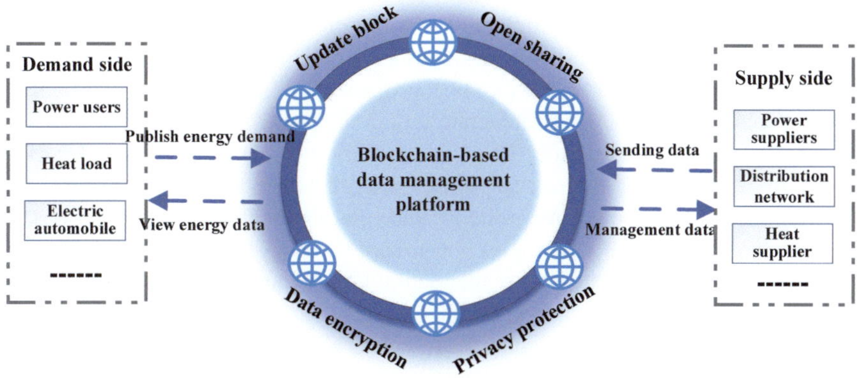

Fig. 5 Structural diagram of a data management platform based on blockchain technology

justice, and other external industry data of the State Grid Corporation of China. More than 10,000 pieces of black-and-white list data are stored and shared by encryption, which optimizes the management mode of credit data. China Southern Power Grid Corporation also invested in and built a green certificate trading platform. The blockchain distributed ledger function is used to manage user account information and transaction information in real time. This function is used to avoid violations such as artificially modifying parameters and tampering with transaction data.

5 Scientific Problems in the Application of Energy Blockchain Technology

The theoretical research of energy blockchain has achieved considerable research results, and relevant demonstration projects have been preliminarily put into practice. Because of the characteristics of blockchain technology and the physical characteristics of the Energy Internet, the application of blockchain technology in the Energy Internet still requires some adjustments. These key scientific problems are as follows:

1. Blockchain technology still faces a series of bottlenecks. First, the blockchain consensus mechanism poses the problem that it is difficult to balance the node size, operation efficiency, and fault tolerance [28]. The workload proof mechanism offers significant advantages in node size and fault tolerance, but it takes too long to reach a consensus and cannot meet the operational requirements of the energy system. Although Byzantine and proxy proof of rights and interest mechanisms can improve operational efficiency, some nodes are excluded from information consensus verification, resulting in reduced fault tolerance. Second, the current application of blockchain technology has focused primarily on business scenarios that do not require high real-time management and throughput. At present, the ability of blockchain technology to mine innovative business scenarios is relatively insufficient. Therefore, the maturity of blockchain technology needs to be improved, and continuous technical iteration is needed to meet the application needs of practical projects.
2. The application of blockchain technology in the field of energy and power also faces limitations. First, under the guidance of the goal of carbon peak and carbon neutralization, the installed scale of new energy power, such as wind and light, continues to rise, and the intermittence and volatility of new energy power generation have formed a more complex way to supply energy [29]. Second, the Energy Internet will open up the energy interaction throughout the source network load storage. The Energy Internet also can integrate the application of energy-saving and consumption-reducing equipment, such as absorption heat pumps and waste heat recovery. This integration could solve the diversified energy demand of electricity, heating, and cooling as well as the complex physical structure of the energy system, which may make it difficult to integrate various information in the

blockchain. In addition, the construction cost of the energy blockchain network platform is high. If the existing energy supply and demand equipment and transmission and conversion systems are transformed and updated comprehensively, significant investment costs will be required. This introduces further challenges to the large-scale application of blockchain technology in the field of energy.

3. The promotion and application of energy blockchain technology faces problems, such as difficult supervision and blame. The blockchain data structure can ensure that the information on the blockchain network can be traced but not be tampered with. When it comes to offline execution, real-time energy interaction, and other scenarios, however, online records cannot cover all business processes, and may pose problems with the data on the chain being inconsistent with energy information off the chain. Therefore, when trying to reduce human intervention, it remains difficult to ensure that the energy information outside the chain is truly and reliably digitized on the chain. The Energy Internet also is related to all aspects of the national economy and to people's livelihood. Especially in regard to the development trend of energy conservation and low carbon, the requirements of energy scheduling and trading have become increasingly strict. At this stage, the energy blockchain system has been dominated by an alliance blockchain that generally excludes regulators. This system cannot be used to guarantee the legality and compliance of blockchain technology application. Therefore, it is essential to establish an energy blockchain supervision mechanism that is suitable for future development [30].

6 Prospect of Energy Internet Based on Blockchain Technology

In light of the key scientific problems faced by the application of blockchain technology to the Energy Internet, the theoretical research and technological innovation of energy blockchain in the future may focus on the following three aspects: high operational efficiency, high scalability, and effective supervision.

6.1 High Operating Efficiency

The energy system has the characteristics of instantaneous response and real-time balance. It is necessary to solve the performance problem of the blockchain. Only by improving the operation efficiency of the energy blockchain network can this technology be applied to the construction of the Energy Internet on a large scale [31].

First, starting from the underlying architecture of blockchain, a consensus mechanism has been actively explored to meet the needs of the Energy Internet. A consensus verification method with both fault tolerance and response efficiency should be designed to meet the actual scenario of large-scale access of energy supply and demand nodes. Second, by continuously improving and optimizing the encryption algorithm of the blockchain and by using the latest research results of modern cryptography, we can reconstruct the formation mechanism of data blocks and improve the asymmetric encryption efficiency during energy information interaction. This mechanism would enable more efficient storage as well as the management of massive amounts of energy data. In addition, the effective integration of energy blockchain with artificial intelligence, 5G communication, Internet of things, and other information technologies has been accelerated. A low-cost and efficient communication system that is widely applied using 5G should be arranged at the terminal of energy facilities. This system would be able to effectively manage all kinds of data at both ends of the energy supply and demand and transmission system and thus would improve the circulation efficiency of energy information.

6.2 High Scalability

With the diversified and large-scale development of the Energy Internet, higher requirements will be needed for the scalability of energy blockchain technology. Therefore, operational characteristics should be the focus of the future development of the Energy Internet. By optimizing the platform architecture, improving the smart contract, and taking full advantages of cross-chain technology, the applicability and practicability of blockchain technology in the Energy Internet can be effectively improved.

For the energy blockchain network platform, the open-source structure and technical advantages of the public chain should be fully utilized. The underlying architecture to meet the multitype application scenarios of the Energy Internet has been studied. The modularization of new energy power generation facilities, energy-saving and consumption reduction equipment, and an intelligent regulation system have been embedded into the blockchain platform architecture to achieve the autonomous cooperation and cascade utilization of multi-energy networks. Regarding smart contracts for the energy blockchain, the operation mechanism of various energy equipment should be studied extensively. The energy flow and information flow of energy production, transmission, transformation, utilization, and other links must be effectively integrated. With the help of a high-performance, programmable development language and computing engine, a modular energy management intelligent contract can be constructed. This would shorten the development cycle of energy blockchain smart contracts and reduce the cost of contract deployment and transfer [32]. In terms of cross-chain technology utilization of the blockchain, cross-chain interoperability technologies, such as notarization mechanism, side chain relay, and hash lock, should be adopted to break the "data islands" of different energy systems and different energy attributes. The information interaction between heterogeneous energy subsystems has promoted energy interaction, which can be used to balance the energy-wide area of the Energy Internet.

6.3 Effective Supervision

The development of the energy industry is closely related to the national economy, energy conservation, and low carbon emissions. It is necessary to introduce a practical and effective regulatory mechanism to ensure the safety and credibility of energy information. Only in this way can we widely promote and apply energy blockchain technology and support the high-quality and sustainable development of the Energy Internet.

Furthermore, we should strengthen the supervision and accountability of all participants in the energy blockchain through technical means. In a decentralized, collaborative, and autonomous network platform, a high-security encryption algorithm should be introduced to improve data security and privacy protection of the system. With the help of blockchain cross-chain interoperability and other technologies, the key information that does not involve privacy can be packaged into blocks and sent to the supervision node regularly. While protecting the privacy of energy interaction subjects, additional supervision of energy trading, carbon emission trading, and other markets is needed. The entire management process of the Energy Internet should be realized by penetrating the supervision mechanism. Multiple subjects involved in energy dispatching and trading should be strictly reviewed, and entity mapping between real identity and virtual network should be established. Active verification, tracking, and authentication of the uplink data can prevent in-process stop loss and accountability for all kinds of Energy Internet businesses [33].

7 Summary

As a decentralized, open, transparent, and fair emerging technology, blockchain plays a fundamental and leading role in promoting data sharing, optimizing business processes, reducing operating costs, and improving coordination efficiency. The application of blockchain technology has been widely recognized and valued. It has become an emerging technology direction in the key areas of the strategic layout of many countries and has opened up a new track for national competition. The Energy Internet is a typical example of the deep integration of the energy industry and information technology. Its basic characteristics of openness, interconnection, and peer-to-peer sharing are highly consistent with the basic design of the blockchain, making the Energy Internet one of the more important fields in blockchain integrated application. The Energy Internet based on blockchain technology has revealed a unique technical advantage in energy supply, energy consumption, energy trading, data management, and other aspects, and as such, has provided new ideas for the development of Energy Internet technology and the energy industry.

Because the integrated application of blockchain in the energy field is still in the exploratory stage, the Energy Internet based on blockchain technology still faces problems, such as technical bottleneck, difficult expansion, and weak supervision. We need to accelerate theoretical research and technological innovation, strive to improve the operational efficiency of the energy blockchain, and enhance the applicability of blockchain technology in the energy system. In addition, independent technological innovation and mechanism reforms can improve the effective supervision of energy blockchain projects. In the future, blockchain technology should be combined with the energy industry to build an integrated Energy Internet system that is suitable for China's sustainable economic and social development. This development will greatly promote the transformation and upgrading of China's energy structure and contribute to the realization of China's strategic goal of reaching its carbon peak and carbon-neutralization goals.

References

1. Shuyin C, Yi L, Xiao C (2019) Research on model of blockchain-enabled power carbon emission trade considering credit scoring mechanism. Electr Power Constr 40(1):104–111
2. Song J, Zhang P, Alkubati M et al (2021, in Press) Research advances on blockchain-as-a-service: architectures, applications and challenges. Digit Commun Netw
3. Jogunola O, Adebisi B, Ikpehai A et al (2021) Consensus algorithms and deep reinforcement learning in energy market: a review. IEEE Internet Things J 8(6):4211–4227
4. Miglani A, Kumar N, Chamola V et al (2020) Blockchain for internet of energy management: review, solutions, and challenges. Comput Commun 151:395–418
5. Hasankhani A, Mehdi Hakimi S, Bisheh-Niasar M et al (2021) Blockchain technology in the future smart grids: a comprehensive review and frameworks. Int J Electr Power Energy Syst 129:106811
6. National Development and Reform Commission of the People' s Republic of China, Guidance on promoting the development of Internet plus smart energy, 24 Feb 2016. http://www.nea.gov.cn/2016-02/29/c_135141026.htm
7. Wu H, Li H, Gu X (2020) Optimal energy management for microgrids considering uncertainties in renewable energy generation and load demand. Processes 8(9):1086
8. Huang Y-F, Yang L-C, Lin J-Y (2012) An efficient energy data gathering based on grid-chain for wireless sensor networks. In: 4th international conference on awareness science and technology, pp 78–82
9. Rui H, Huan L, Yang H et al (2020) Research on secure transmission and storage of energy IoT information based on Blockchain. Peer-to-Peer Netw Appl 13(4):1225–1235
10. Tan S, Wang X, Jiang C (2019) Privacy-preserving energy scheduling for ESCOs based on energy blockchain network. Energies 12(8):1530
11. Jing R, Xie M-N, Wang F-X et al (2020) Fair P2P energy trading between residential and commercial multi-energy systems enabling integrated demand-side management. Appl Energy 262:114551
12. Zhao Y, Peng K, Xu B et al (2019) Status and prospect of pilot project of energy blockchain. Autom Electr Power Syst 43(07):14–22+58
13. Yao G (2020) Development progress and enlightenment of international energy blockchain. J Nanjing Univ Posts Telecommun (Natural Science Edition) 40(05):215–224
14. Lin J, Pipattanasomporn M, Rahman S (2019) Comparative analysis of auction mechanisms and bidding strategies for P2P solar transactive energy markets. Appl Energy 255:113687
15. Astriani Y, Shafiullah G, Shahnia F (2021) Incentive determination of a demand response program for microgrids. Appl Energy 292:116624
16. Zhang Y-J, Liang T, Jin Y-L et al (2020) The impact of carbon trading on economic output and carbon emissions reduction in China's industrial sectors. Appl Energy 260:114290

17. Aghajani G, Shayanfar H, Shayeghi H (2017) Demand side management in a smart micro-grid in the presence of renewable generation and demand response. Energy 126:622–637
18. Castro M, Colclough S, Machado B et al (2020) European legislation and incentives programmes for demand Side management. Sol Energy 200:114–124
19. Mengelkamp E, Gärttner J, Rock K et al (2018) Designing microgrid energy markets. Appl Energy 210:870–880
20. Foti M, Vavalis M (2021) What blockchain can do for power grids? Blockchain: Res Appl 2(1):100008
21. Wang Q, Jia Z, Wang T et al (2020) A highly parallelized PIM-based accelerator for transaction-based blockchain in IoT environment. IEEE Internet Things J 7(5):4072–4083
22. Zutshi A, Grilo A, Nodehi T (2021) The value proposition of blockchain technologies and its impact on Digital Platforms. Comput Ind Eng 155:107187
23. Sawa T (2019) Blockchain technology outline and its application to field of power and energy system. Electr Eng Jpn 206(2):11–15
24. Andoni M, Robu V, Flynn D et al (2019) Blockchain technology in the energy sector: a systematic review of challenges and opportunities. Renew Sustain Energy Rev 100:143–174
25. Wang Q, Li R, Zhan L (2021) Blockchain technology in the energy sector: from basic research to real world applications. Comput Sci Rev 39:100362
26. Ante L, Steinmetz F, Fiedler I (2021) Blockchain and energy: a bibliometric analysis and review. Renew Sustain Energy Rev 137:110597
27. Yang S, Zhang D, Li D (2019) A calculation model for CO_2 emission reduction of energy internet: a case study of Yanqing. Sustainability 11(9):2502
28. Sang Z-X, Huang J-Q, Du Z et al (2021) The power optimization on tie-line for the island energy internet based on interactive distribution network. Sustain Energy Technol Assess 45:101148
29. Chen G, Liang Z, Dong Y (2020) Analysis and reflection on the marketization construction of electric power with Chinese characteristics based on energy transformation. Proceedings CSEE 40(02):369–379
30. Zhou K, Yang S, Shao Z (2016) Energy internet: the business perspective. Appl Energy 178:212–222
31. Xue Y (2015) Energy internet or comprehensive energy network? J Mod Power Syst Clean Energy 3(3):297–301
32. Huang A, Crow M, Heydt G et al (2011) The future renewable electric energy delivery and management (FREEDM) system: the energy internet. Proc IEEE 99(1):133–148
33. Shen H, Ali S, Alharthi M et al (2021) Carbon-free energy and sustainable environment: the role of human capital and technological revolutions in attaining SDGs. Sustainability 13(5):2636

Meicheng Li is a professor/doctoral supervisor, the Dean of the school of new energy of North China Electric Power University and enjoys the special government subsidy of the State Council. He has been the Yangtze River scholar professor, leading talents in science and technology of national ten-thousand-person plan, leading talents of young and middle-aged in scientific and technological innovation of the Ministry of science and technology, member of science and Technology Commission of the Ministry of Education, member of the Education Instruction Committee of the Ministry of Education, 100 leading talents in science and technology in Beijing. He has served as the evaluation expert of the national science and technology award reviewer, vice chairman of IEEE PES energy development and power generation technology committee, chairman of energy storage materials and Devices Technology Subcommittee, and deputy director of the youth working committee of China Renewable Energy Society.

He is mainly engaged in research on new energy, energy storage technology and intelligent energy system. He has published nearly 300 papers in domestic and foreign journals such as Nature Energy, Joule, and has been grated 58 patents in China and the United States, 5 software copyrights, and compiled 8 publications in Chinese and English. As the first author, he won the first prize of Provincial Natural Science and 6 Science and Technology Awards such as Beijing Science and Technology Award. He won the "Power Technology Innovation Award" in 2019 and was selected into the 2020 list of the world's top 2% scientists released by Stanford University.

Open Source Ecosystem in New Era: Pattern and Trend

Liyun Yang, Hanghang Geng, Yuxia Zhang, and Minghui Zhou

Abstract At the end of the twentieth century, open source provided new ideas of user driven innovation, low cost and high quality for software development, and therefore achieved great success. An increasing number of companies and organizations have participated in open source, from hybrid commercial-open source projects, and driven the construction of various business models around open source software technology and platform. The intensive commercial participation in open source has promoted the cooperation and interest relationships among participants and formed an "open source software ecosystem." Around 1997, open source software began to develop in China. At that time, it mainly focused on the development of basic software such as operating system and made some progress, but the effect of open source software was not significant enough. In recent years, with the rapid development of the new generation of information technology, open source software has gradually been valued and developed rapidly in China. This paper studies the elements and commercialization of open source software ecosystem, and provides theoretical supports for the construction of open source software ecosystem in China by combining with the current development status in China.

Keywords Open source ecosystem · Open source commercialization · Open source elements

L. Yang (✉) · H. Geng
China Electronics Standardization Institute, Beijing, China
e-mail: yangly@cesi.cn

H. Geng
e-mail: genghh@cesi.cn

Y. Zhang
Beijing Institute of Technology, Beijing, China
e-mail: yuxiazh@bit.edu.cn

M. Zhou
Peking University, Beijing, China
e-mail: zhmh@pku.edu.cn

China's e-Science Blue Book 2023, https://doi.org/10.1007/978-981-99-8270-7_11

With idealism as its origin and the Internet as its carrier, the open source model has effectively gathered user innovation and collective intelligence, developed a considerable number of excellent open source software, and then, with the help of commercialization, created great opportunities for the whole industry and society. With the development of open source software, the continuous influx of social forces and the formation of supply chain relations make the ecologization of open source software a trend. The complexity in the nature of open source ecosystem is reflected in the universality of social groups participating in open source, the openness of the environment in which open source activities are located, and the uncertainty caused by the interaction and dependence between open source objects and subjects [1], which fills the success of open source with variability. Therefore, it is imperative to understand and clarify the elements and commercialization model of open source ecosystem.

In this paper, we first introduce the key elements of open source ecosystem to define the connotation and extension of open source. Then, we analyze the historical evolution and current development of open source commercialization to understand the key to the sustainable development of open source ecosystem. Finally, we put forward some suggestions on the development of local open source ecosystem construction in China based on the research.

1 Key Elements of Open Source Ecosystem

With the rapid development of open source software and the continuous influx of social forces, open source software ecosystem has become a core key word for the development of open source. The term ecosystem comes from biology and was first proposed by British ecologist Taylor in 1935. It refers to a unified whole composed of biological community and inorganic environment and is an open system [2]. In the year 2005, Messerschmitt and Szyperski put forward the concept of "software ecosystem" for the first time [3], pointing out that the boundaries between software suppliers and users are no longer clear, and those original competitors will cooperate in the same software ecosystem for mutual benefit. Accordingly, commercial organizations, non-profit institutions, and individual developers and users have established mutually beneficial and interdependent relationships around the development, application and market of core open source technologies (software), thus forming a unified and open whole, which is the open source software ecosystem. In exploring and promoting the development of open source software ecosystem, China Electronics Standardization Institute takes standardization as one of the approaches to conduct researches, organizes the entire industry to systematically analyze the core and key points for developing the open source ecosystem, and summarizes and proposes four major factors affecting its development, namely, open source rules, open source objects, participants and open source infrastructure [4].

1.1 Open Source Rules

Open source rules are the cornerstone of maintaining the operation of the open source world, including but not limited to open source licenses, governance rules of open source operations and application evaluation rules. The more well-known explicit open source rules are the open source licenses, which are essentially "non-standard" authorized contract texts, stipulating the license terms for enterprises or developers to jointly develop and use the source code of open source software based on the distribution, application and intellectual property rights of one or more types of codes that they jointly abide by. In case of disputes, they will not take effect until a court decision is made. Software copyright is one of the most frequently researched software intellectual property rights and is closely related to the open source licenses. The copyright of open source software is protected as that of private software. The main difference lies in their licensing model, i.e., the use of open source software requires the authorization of the open source licenses.

The Open Source Initiative (OSI, an international organization that sets standards for open source licenses) has established a set of standards for open source licenses, known as "Open Source Definition (OSD)," and registered a certification mark: the OSI Logo. this logo can be placed on the released software so that it can be easily identified as open source and the license used complies with the definition of open source. Licenses that are generally considered by the industry to meet the definition of open source are referred to as open source licenses. Among the many source licenses in the world, there is currently an OSI certified license in China, namely the Mulan PSL [5].

In addition to the open source license, the governance rules and application evaluation rules of the open source operation have long existed in the open source world, and are crucial implicit rules for the application and ecological development of software. The governance rules of open source operation are divided into 4 levels: open source foundation/community, open source enterprises, open source projects and open source contributors. They are a set of methodology followed by various entities such as open source community, open source foundation and open source enterprises when engaging in open source practice. For example, they explain how to organize an efficient and orderly open source governance structure, how to follow a set of compliant and safe processes to organize the research and development of open source products by technicians, how to measure the health of the growth and development of the open source community, and how to evaluate the incubation level of open source projects.

Taking Mulan Open Source Community cultivated by China's key R&D projects as an example, among the thirteen open source projects incubated by its independent operation, OpenDigger project constructs the activity index and collaboration influence index of open source projects based on Git log and software product library, and makes a quantitative analysis of the open source projects operated in Mulan Open Source Community from multiple dimensions and displays the results, as shown in Table 1.

Table 1 Activity of some projects in Mulan open source community

Ranking	Project	Activity
1	Ossrs/srs	4048
2	Kubeovn/kube-ovn	3346
3	X-lab2017/open-digger	1571
4	MonkSoil/Furion	370
5	Rcore-os/zCore	279
6	Oceanbase/obclient	40
7	Cas-bigdatalab/piflow	37

The gradual deepening and continuous improvement of the research on open source licenses and open source standards in China reflects the development of China's local open source ecosystem from the disorderly grass-roots stage to the systematic and large-scale direction, laying a good foundation and providing theoretical support for China's development and cultivation of open source industry in the next stage.

1.2 *Open Source Objects*

Open source objects are the core of intelligent innovation of open source group. The first choice of open source objects are open source codes and open source projects built on them, which is the most intuitive understanding in the industry. With the rapid development of open source and the deepening of open source research, the connotation of open source has been greatly expanded. From the technical perspective, open source objects have extended to hardware, data and other aspects, which can be summarized into three categories: open source software, open source hardware and open data sets. In addition, the extension of open source objects has also changed, diverging from the technical level to other levels, resulting in the emergence of open science, open source economy, modern service industry, etc.

Open science is a scientific practice in which people collaborate and contribute to research data, laboratory notes and research processes provided free of charge. It is a prime example of the effectiveness of science itself as an open source movement, employing practices such as public dissemination of information and peer review of scientific literature. Open source economy refers to the innovative cooperation model. As an innovative cooperation model, open source reshapes the social production model, which is in line with the two elements in economics (namely the production efficiency and distribution efficiency of commodities). From an economic perspective, the open source economy is mainly reflected in the "delivery efficiency on the supply side of commodities." More efficient production and delivery can be achieved through open source innovative collaboration models.

Currently, the research on open source objects is still ongoing, it will become clearer, more mature and richer with the continuous development of open source in China.

1.3 Participants

Open source participants are the driving force for the continuous development of open source software, including but not limited to contributors (individuals/enterprises), users (end users), operators (organizers/managers), and collaborators (laws/regulations). The open source ecosystem has gathered a large number of participants including enterprises, developers, open source foundations, open source communities, governments, scientific research institutions, etc., and continues to absorb contributions from all walks of life in the world. It has a rapid development speed and great innovation potential [6].

From the perspective of developers, according to the statistics of GitHub, the world's largest open source project hosting platform, there were 16 million users in 2020 compared with the previous year, and the number of users is expected to reach 100 million in 2025. CSDN has gathered the largest number of developers in China. According to the "2020–2021 Survey Report on Developers in China" released by CSDN at the beginning of 2021, the number of CSDN de-duplicated users has reached 32 million, and maintained a rapid growth trend, of which 81% are developers under 30 years old and 13% are students, laying a foundation for the group of contributors to the open source ecosystem, as shown in Fig. 1.

As a third party neutral, open source foundation gathers the capabilities of developers and open source enterprises, incubates and cultivates open-source projects, provides intellectual property rights and legal support, and promotes the formation of an open source business and application ecosystem, thus playing an indispensable role in the development of the open-source ecosystem [7]. As the first open source foundation in China, the OpenAtom Foundation has been established. As for all those well-known open source foundations with their own characteristics around the world, since there are already a large number of related research materials, we will not elaborate on them here.

The development of open source ecosystem is also closely related to the trend of national policies. In the year 2021, China included "open source" in the outline of the "14th Five Year Plan", injecting strong impetus into the industrial development in China's open source ecosystem. As for other countries and governments, they have already participated in the construction of open source ecosystem from the aspects of government procurement and the formulation of policies and regulations [8]. For example, the U.S. introduced the federal source code policy in 2016, stipulating that 20% of the software purchased by the U.S. authorities every year should be open source; The 12th article of the latest version of the *Digital Services Standard* issued by the UK in 2019 requires the government to open up new source code and select the appropriate license for open source; The UK, France, Germany, Belgium, Brazil

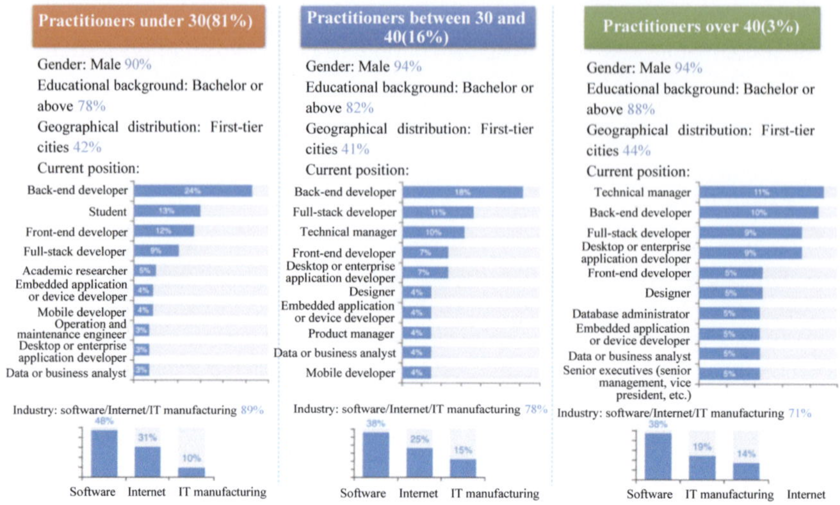

CSDN *2020-2021 Survey Report on Chinese Developers*

Fig. 1 Employment of Chinese developers from 2020 to 2021

and other countries have passed proposals requiring the government to compare commercial software and open source software before procurement, and give priority to open source software.

To sum up, China has a large stock of open source participants, which has continued to grow in recent years. In the future, China will have prominent late development advantages in leading technology, leading innovation and cultivating ecosystem.

1.4 Open Source Infrastructure

Open source infrastructure is the carrier of activities, which refers to the software and hardware platforms, tools and services provided for the purpose of serving open source projects, open source communities and open source ecosystem. The most common open source infrastructure is the open source code hosting platform, which is mainly used for code development, management, communication, modification, as well as proposing new development requirements, covering the full life cycle of open source software production. GitHub, GitLab and Gitee are common open source code hosting platforms in the industry. The academic community has also been studying the open source intelligence mechanism for a long time and has built Trustie open source collaboration platform [9]. In 2021, the Open Source Development Committee of China Computer Federation initiated GitLink based on Trustie, and jointly promoted the exploration of new open innovation service platform with enterprises, research institutes and higher education institutions. Open source infrastructure also includes mailing lists, tool sets and issue tracking systems for technology development, community communication, developer convergence, promotion, etc.

Open source infrastructure is the hub of convergence and carrying of open source elements. The level of open source infrastructure construction will, to a certain extent, affect the development speed of open source objects, the stickiness of the participating subjects and the convergence of open-source ecosystem construction. It is also an important factor in the current development of open source in China.

2 Open Source Commercialization Model

The commercialization of open source is an important driving force for the vigorous and sustainable development of open source ecosystem. While in the past, developers regarded open source as a means of product iteration, with the arrival of the cloud service era, open source has become a means of sales market, that is, to make more people believe in their technical capabilities through open source, even if users will not participate in contributing a line of code. The success of open source commercialization depends on many factors, but the core lies in the product form of the open source project itself. Open source can effectively help enterprises reduce the cost of sales and free dissemination.

Through the continuous verification of the market, enterprise service open source has been fully proved to have great commercial potential. With the promotion of digital transformation in China in recent years, many new enterprises have been based on cloud and technology since their birth, and many traditional enterprises have turned to open source cloud technology. Meanwhile, China has quite a number of excellent programmers and technology entrepreneurs, as well as world leading Internet scenarios. Our products can serve not only China, but also the world. The following will introduce the history and current situation of the open source commercialization model, summarize the successful experience and potential challenges.

2.1 Historical Evolution of Open Source Commercialization Model

The historical evolution of the open source commercialization model can be divided into three distinct periods, as illustrated in Fig. 2:

(1) Freeware (1970s–1990s): The open source movement started in the mid-1970's, when software was developed by academics and enthusiasts, and the overall concept was to "make software free." At that time, there was no concept of a

Fig. 2 Historical evolution of open source commercialization model

clear business model, and most of the funding behind the development of "free software" came from the R&D funds of universities or enterprises.

(2) Support service (1990s–2000s): Since 1991, with the emergence of Linux, open source has been proved to be a better and faster way of software development, and its importance has become prominent for enterprises. Therefore, the open source community and enterprises have begun commercialization attempts. In 1998, OSI coined the term "Open Source" During the same period, the first business model of open source software, represented by Red Hat and MySQL, emerged as the times require, that is, to provide paid support and services on the basis of freeware.

(3) SaaS and Open Core (Since entering the twenty-first century): Mid-twenty-first century, the valuation of open source companies began to change, and with the increasing popularity of cloud computing as an infrastructure, enterprises started a SaaS model on the basis of open source. When an open source service runs in the cloud, users do not need to care whether the underlying code of the cloud service is open source or closed source, and it mainly provides SaaS cloud services in a software out-of-the-box manner. Typical representatives are GitHub, Elastic, MongoDB, Databricks, etc.

The open source software research team of the School of Computer Science of Peking University found that even in the same open source software ecosystem, there will be a variety of open source commercialization models, such as support services and SaaS in OpenStack [10]. The business model for a specific open source project is not only related to its period, but also to factors such as the type of open source project and the field of the company [11]. In addition, the cooperation between different companies and the relationship between companies and volunteers in open source commercial projects have been found to be key issues that must be faced in the sustainable development of open source projects [9, 11].

2.2 *Status and Development of Open Source Commercialization*

Open source commercialization has always been the focus of open source practitioners, research institutions and community media at home and abroad. We have studied and sorted out the development process and models of open source commercialization from different dimensions and perspectives. It is generally agreed that although open source still faces various problems such as "borrowlism" and "security and compliance" in its development process, open source software has become a part of each software company after nearly 50 years of evolution and development. As the open source development model is gradually widely recognized by the participants, the commercialization path and model of open source software are gradually clear. With the expansion of market share and user group of open source software, its commercial value has gradually caught up with or even exceeded that of closed source software, and more and more favored by the capital sector, thus becoming the mainstream trend of software development and commercial development.

2.2.1 The Commercial Value of Open Source Software Has Begun to Exceed that of Closed Source Software

Compared with closed source software, open source is more efficient, robust, and flexible because of its asynchronous collaboration, open access, loose management and other practices. According to the *Open Source Ecosystem White Paper (2020)* issued by China Academy of Information and Communications Technology, in 2019, the proportion of enterprises that have used open source technology in Chinese enterprises was 87.4%, with a year-on-year increase of 10.7%; Enterprises use the most open source software in database, accounting for 58.7%. Enterprises also use more than 50% of the open source software for big data and storage, accounting for 52.4% and 51.2% respectively, indicating that the use of open source technology by Chinese enterprises has gradually become the mainstream. China's enterprise open source software application areas are shown in Fig. 3.

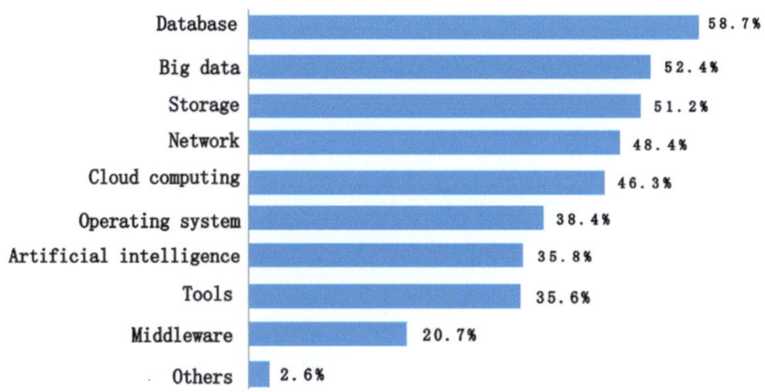

Fig. 3 Application fields of enterprise open source software

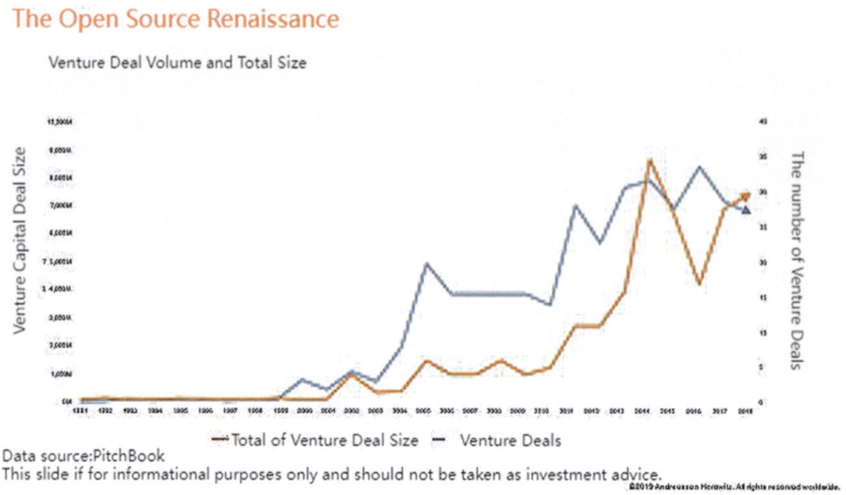

Fig. 4 Global open source funding trends (*Source* a16z)

2.2.2 The Number of Initial Public Offerings (IPOs) and Venture Capital (VC) Investments of Commercial Open Source Software is on the Rise

According to a16z data (Fig. 4), from the perspective of the number and scale of financing in the open source field from 1991 to 2018, three-quarters of open source companies and 80% of the financing amount were generated after 2005, and showed a sharp growth trend over time.

There are quite a lot of open source commercialization investment and financing cases around the world, According to CSDN data statistics, Some of the more typical cases are: In 2008, MySQL was acquired by Sun (later acquired by Oracle) for $1 billion; in 2017, MongoDB submitted an IPO listing, with a market value of $18.5 billion; in 2018, Microsoft acquired GitHub for $7.5 billion; in 2019, Red Hat was acquired by IBM for $34 billion; in 2020, SUSE acquired Rancher Labs, the creator of the Kubernetes management platform; in 2021, Databricks, the open source company behind Spark, completed $1 billion in financing, with a current valuation of $28 billion by 2021, etc. All these cases show that open source software has successfully created commercial value and economic value, and the commercialization of open source is also in constant innovation and development.

Meanwhile, the commercialization of open source has developed rapidly in the Chinese market, with a large influx of capital. Judging from the financing of open source commercialization in China from 2021, 80% of the capital, represented by PingCAP, Taosi Data, etc., focuses on basic open source software, such as operating systems, databases, cloud computing and other fields. Table 2 shows some of open souce investment and financing in China calculated by CSDN.

Table 2 Status of investment and financing of open source in China (CSDN in continuous update)

Time of establishment	Organization (company/ foundation/ individual)	Core open source projects	Owner/ founder	Latest financing round	Financing disclosure time	Transaction amount
2015/4/17	PingCAP	TiDB	Liu Qi, Huang Dongxu	D Round	17/11/ 2020	USD270 million
2012/4/19	DCloud	Uni-app	Wang An	Equity financing	27/8/2020	Undisclosed
2012/9/10	Yunshen Internet (Beijing) Technology Co., Ltd.	Amaze UI	Chen Benfeng	C Round	15/8/2018	CNY 250 million
2014/10/21	Hangzhou FIT2CLOUD Information Technology Co., Ltd.	JumpServer	Guang Hongwei	C+ Round	16/4/2020	Undisclosed
2011/8/30	Wuhan Deepin Technology Co., Ltd.	Deepin	Wang Yong, Liu Wenhuan	B Round	3/4/2015	CNY Tens of millions
2011/4/15	Shanghai RT-Thread Electronic Technology Co., Ltd.	RT-Thread	Xiong Puxiang	Equity financing	7/1/2020	Undisclosed
2012/10/11	SequoiaDB	SequoiaDB	Wang Tao	D Round	13/10/ 2020	CNY Hundreds of millions
2017/5/16	Taosi Data	TDengine	Tao Jianhui	Pre-A Round	24/4/2020	Over USD 10 million
2013/9/17	Beijing EasyStack Technology Development Co., Ltd.	EasyStack	Chen Xilun	D Round	18/11/ 2019	CNY Hundreds of millions
2016/1/28	Shanghai Kyligence Information Technology Co., Ltd.	Kyligence	Han Qin	C Round	28/3/2019	USD25 million

Although there is a certain development in China's open source commercialization, it is still in the early stage as a whole. Due to the differences in the deployment mode of the underlying infrastructure, social structure and cultural differences, the way to form a benign business closed-loop model is still under further exploration. As for whether China's open source commercialization can fully replicate the existing foreign open source commercialization model, further proof of practice is needed. In general, optimizing open source products, building a good open source community and enhancing the stickiness of ecological partners are the key to the commercialization of open source; On the other hand, with the rapid development of cloud computing technology and services in China and the strong demand for digital transformation, it is expected that China's open source commercialization will stimulate more possibilities and expansion space.

3 Opportunities and Challenges Faced by the Construction of Open Source Ecosystem in China

In general, China has gradually changed from the original open source users to open source contributors, and the construction of open source ecosystem has developed from grassroots and fragmented to large-scale and systematic. The overall development situation is relatively optimistic, and there are institutional advantages and development opportunities that can focus everyone's strength on important tasks. However, there are also challenges such as an imperfect open source governance rules system, a lack of volunteer contribution culture, and open source applications that need further encouragement.

3.1 The Trend of Open Source Marketization is Improving, but the Ecological Environment of Domestic Open Source Commercialization Needs to be Further Created

According to the research data of China Academy of Information and Communications Technology's "Open Source Ecological White Paper 2020", among the open source user enterprises in China, 50.8% believe that the use of open source technology can reduce costs, 45.1% believe that the use of open source technology can greatly shorten the application deployment time, and 30.3% and 20.7% respectively believe that the use of open source technology can improve controllability and reduce the risk of trial-and-error. More than half of domestic enterprises have applied open source software in the field of databases, cloud computing has generally adopted open source cloud computing technology, more than 70% of enterprises have applied open source container technology, and more than 60% of enterprises have applied or are testing microservice frameworks. It can be seen that Chinese enterprises have a high degree of acceptance of open source technology, and the use of open source technology has become the mainstream.

From the perspective of industry applications in China, various industries are actively embracing open source. The Internet, finance, software and information technology service industries have become the main open source commercial enterprise service targets. Open source commercial enterprises account for the highest proportion of Internet services support at 40.8%, 32.6% of service support software and information technology service industries, and 31.6% of financial services support. Meanwhile, the formation of open source ecosystem in various industries in China has been accelerating. In 2021, the financial industry has just released the *Opinions on Regulating the Application and Development of Open Source Technology in the Financial Industry*, encouraging financial institutions to follow the four basic principles of "safe and controllable, compliant use, problem oriented and open innovation" and include the application of open source technology into their own information development planning. They are expected to establish and improve the application and management system of open source technology and actively participate in the construction of open source ecosystem. In addition, although the path of open source commercialization has been relatively clear, the mode of commercial transformation of open source software products in China is still in the exploratory period. Most open source start-ups are still unable to make profits, most domestic traditional enterprises do not support the open source commercialization model enough, and open source commercialization enterprises are often deployed in overseas markets.

3.2 The Advantages of Local Accumulation and Reserve of Talents and Technology

In terms of open source talents, Chinese contributors have been active in the open source community. According to the GitHub 2020 Digital Insight Report released in 2020, in the year of 2018, the number of contributors on GitHub was 1.6 times that of the 2017, and the number of newly registered users in China ranked second only to the U.S.. By 2020, the number of developers on GitHub in China has grown by 37% annually, ranking first in the world and contributing more than 5.5 million projects. By 2021, the overall number of open source developers in China has leapt to the top 3 in the world with a large scale.

China's contribution to the international open source community has increased year by year, among which, Chinese contribution to the Linux core ranked first in the list of nationalities in the latest version 4.2, and has become more active in the communities of OpenStack, Hadoop, and MySQL projects [12]. In order to more comprehensively show the growth process of China in the open source field in the past period, Combined with the survey data of the "2021 China Open Source Blue Book" by the China OSS Promotion Union, we present it through the patch contribution data of Chinese developers and Chinese technology companies to the

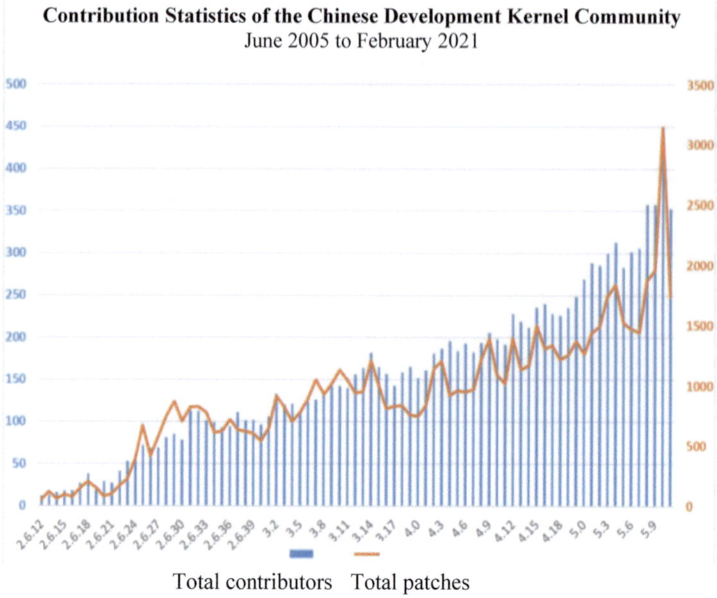

Fig. 5 Contribution statistics of the Chinese development kernel community

kernel.org community. kernel.org is the world's largest Linux Kernel open source project community. Linux Kernel project versions are usually released every 2–3 months, and we have compared the contributions of Chinese developers to Linux Kernel versions 2.6.12–5.11 in the kernel.org community. This data (see Fig. 5) provides a comprehensive picture of the rapid growth of Chinese developers.

In terms of open-source projects, China's independent open-source projects cover the whole technology stack in the IT field, including the low-level operating system, Internet of things operating system and compilers, edge computing, containers, middleware, micro services, database and big data, the front-end development of the upper level, mobile development and UI framework, as well as the artificial intelligence, operation and maintenance and other technical fields at present. Several excellent self-research open source projects have emerged in China in the fields of operating systems, databases, artificial intelligence, etc. including openEuler (HUAWEI open source server operating system), OpenCloudOS (Tencent open source server operating system), OpenAnolis (Alibaba open source server operating system), OceanBase (Anthem open source financial-grade distributed relational database), PaddlePaddle (Baidu open source deep learning framework), etc. At present, the open source foundation has carried out self-incubation work in the field of operating system. According to the *GitHub 2020 Digital Insight Report* released in 2020, Chinese enterprises such as Alibaba, Baidu and HUAWEI are gradually becoming important contributors to open source, as shown in Table 3. In addition to the massive influx of developers into open source projects to make contributions, it is also necessary to strengthen the cultivation of core code reviewers, so as to reduce the untimely response to the contribution to the development of open source projects caused by the rapid growth of ordinary developers [13].

3.3 Concerns About the Security of Open Source Hinder the Application of Technology, and the Governance of Open Source is Difficult

In addition to security loopholes, there are so many security issues in open source software, such as the large number of open source software, difficulties in overall management, fast technology update and iteration, high operation and maintenance costs, complex open source license and intellectual property issues have become the main reasons for affecting the application of open source software in Chinese enterprises, it is also the difficulty of open source governance. Open source supply chain risk will be an important challenge to the open source ecosystem for a long time [6], and a typical example is the recent Log4j in 2021. According to the survey data of China Academy of information and communication technology, in 2019, the proportion of enterprises that did not use open source technology for security reasons in China reached 43.8%, an increase of 8.6% over the previous year. Since 2020, there have been some cases of infringement of open source license in domestic courts. On the one hand, it shows that the awareness of domestic enterprises on the compliance and safe use of open source technology has been improved. On the other hand, it shows that the application of open source technology has gradually increased, and the demand of enterprises for open source governance is more urgent. Meanwhile, In terms of the open source governance rules that affect security, take the open source license as an example: the OSI mentioned in the article is established in the U.S., and it has approved the release of more than 80 international general open source licenses, most of which come from the U.S. and are released based on the legal context of the U.S. At present, in terms of international general open source license, there is only the Mulan PSL in China. The open source rules have not yet become a system, and the judgment experience of open source related legal cases is insufficient.

In summary, China has a late development advantage in ecological construction, especially in terms of talents, technical strength, application level open source projects, etc., but we still need to focus on breaking through the core foundation of open source project incubation, combing open source intellectual property risks and security loophole risks, and building and improving open source rules and systems.

Table 3 GitHub China enterprise open source project activity ranking

#	公司	活跃度	项目数量	Issue comment	Open issue	Open pull	Pull review comment	Merge pull	Star	Fork
1	Alibaba	1,571.10	1,496	1,30,558	33,947	29,097	22,615	17,471.6	2,16,980	68,864
2	PingCAP	778.4	151	1,39,255	8,138	25,401	61,538	18,880.4	18,008	5,058
3	Baidu	671.2	540	55,265	12,592	20,720	23,380	13,475.9	70,960	22,148
4	Tencent	432.3	388	21,446	8,599	10,264	2,870	7088.8	69,198	19,348
5	JD	153	74	20,126	4,504	4,483	2,043	3214.2	13,119	3,316
6	Huawei	101.8	200	10,322	1,709	2,930	3,867	2005.2	8,758	3,168
7	DiDi	89.4	63	3,114	1,290	827	207	508.7	20,489	3,907
8	Youzan	88.6	58	7,259	2,760	1,409	634	1068.1	9,509	5,820
9	Bytedance	59.2	85	1,973	645	785	659	514.9	14,034	1,671
10	WeBank	57.9	59	2,197	718	3,501	596	2411.9	5,225	1,902
11	Xiaomi	50.4	98	1,767	1,604	1,007	3,001	691.6	5,823	1,760
12	Meituan	46.9	68	1,356	564	305	17	147	10,879	2,573
13	Bilibili	42.7	51	1,306	446	132	52	66.4	10,278	2,295
14	360	39.8	147	1,769	810	441	40	231.7	8,105	1,914
15	Juejin	39.5	26	3,866	578	661	3,624	546.9	4,208	810
16	Ctrip	36.9	25	2,346	537	216	276	130.5	6,196	2,562
17	Linux China	34.1	16	226	10	3,862	11	3123.3	482	302
18	Netease	25	119	1,603	777	313	32	149.6	3,880	1,445
19	Deepin	18.6	267	2,555	931	326	21	132.7	1,339	821
20	Qunar	7.1	43	113	54	56	10	9.9	1,653	478
21	Vipshop	7.1	14	112	127	66	0	14	1,604	421
22	Douban	3.7	41	98	43	158	58	1238.6	508	99

4 Subsequent Development Suggestions

The "the 14th Five-Year Plan" and other national policies have clearly supported open source software development, and China's open source environment has also been significantly improved. In 2020, China established the first open-source foundation, namely the OpenAtom Foundation, and formulated the first OSI certified bilingual license in Chinese and English, namely the Mulan PSL, established the Mulan open-source community based on the key national R&D program projects, and developed a number of open source project hosting platforms. At the same time, many technology companies in China have taken the initiative to open up private software projects. The diversity of application fields and the depth of the market in China are expediting future technological changes and preparing to lead, such as the huge potential in the new generation of artificial intelligence, 5G, Internet of vehicles, industrial Internet, ubiquitous operation system, etc. In general, the development momentum of China's open source software ecosystem is improving, and all open source elements of the development of open source ecosystem are advancing side by side, accelerating forward.

In order to further develop the open source ecosystem, we put forward the following suggestions:

(1) Implement the *14th Five Year Plan for National Economic and Social Development and the Outline of Long-term Objectives for 2035 of the People's Republic of China*, and provide a more friendly development environment for the development of open source ecosystem.

    Specific measures: Introduce corresponding policies to guide the government, enterprises and institutions to support open source software, such as supporting the government's procurement policies to favor open source software products, liberalizing the restrictions on the establishment of open source foundations, encouraging the participation of social capital, and setting corresponding local tax policies; Strengthen the support for innovative enterprises exploring the commercialization of open source; Promote investment institutions to pay attention to and understand open source enterprises, improve the evaluation ability of open source technologies and open source business models, and encourage early entry in the development of enterprises and projects; Guide the combination of capital and open source software, cultivate long-term strategic investment values, and promote the commercialization of open source.

(2) Promote the construction of China's open source developer community, improve open source ecosystem of enterprises, research institutes and higher education institutions, and provide basic driving force for the development of open source ecosystem.

    Specific measures: encourage the development of developer community, hold related open source activities, promote the growth of Chinese open source developers, the improvement of open source application and creation level and open source culture; Continue to strengthen the construction of local open source communities and open source code hosting platforms, and encourage open source developers to discover, communicate, share, innovate applications and promote open source projects; Integrate open source technology and open source training into the existing education system and build an open source innovative science and education platform; Joint schools, enterprises and society to set up open source practice courses, increase the scale of domestic open source developers and improve the construction of open source human resources database through open source universities and vocational training, employment guidance and comprehensive evaluation of open source talents; Promote the standardization of open source, establish a perfect open source rule system, improve the risk prevention awareness of domestic open source industry, and establish a comprehensive open source risk prevention system across departments and fields. Industry, universities and research institutes united for cross-empowerment, give full play to their respective advantages and promote the all-round development of open source ecology in China.

(3) Establish an open source development fund to help and support the development of excellent open source projects and industries, and provide sustainable financial support for the development of open source ecosystem.

Specific measures: establish a special fund for open source, focusing on promoting the establishment of China's open source ecology and the improvement of open source supply chain; Improve the governance and operation capacity of open source communities and open source projects, provide targeted funding for open source community governance experts and core maintenance personnel of open source projects, and let competent and experienced professionals continue to focus on the development of open source communities and projects; Provide investment and financing services for excellent open source projects with rapid development; Provide infrastructure services and incubation for excellent start-up open source projects; Encourage science and technology enterprises to carry out open source transformation and provide guidance through the open source incubation platform.

(4) Take a global perspective, promote international cooperation in open source, and open up more space for the development of open source ecosystem.

Specific measures: attract well-known foreign open source projects to establish Chinese open spouse communities or joint ventures in China; Attract the global open source community foundation to set up branches in China; Encourage domestic enterprises, institutions and universities to undertake and host more international open source technology exchange activities, encourage global open source masters and elites to work in China, and reward developers to participate in open source communities and contribute open source projects; Help and support China's open source projects to go overseas and contribute Chinese wisdom to the world.

(5) Starting from China's national conditions and open source ecosystem, study the new swarm intelligence software ecological model, so as to better integrate the

production resources of the software industry and maximize the enthusiasm of developers.

Specific measures: break through the existing open source model framework, research and design a new model in line with China's national conditions and the law of software development and production, so as to overcome the problems of insufficient development incentive in the existing open source model and insufficient technology opening in the business model. Encourage competent and experienced professionals to systematically design new models in line with ecological laws; Provide research and demonstration conditions for the new model of start-up; Encourage the platform development and practice pilot of new models, and continue to focus on the better solution of possible models, so as to realize the leap forward progress of China's software ecology and surpass the existing ecosystem of western countries.

5 Conclusion

This paper analyzes the key elements of open source software ecosystem to support the understanding of the connotation and extension of open source, reveals the key to the sustainable development of open source ecosystem by introducing the evolution and current situation of open source commercialization, and puts forward some suggestions by analyzing the opportunities and challenges of open source ecosystem construction in China.

During the "14th five year plan" period, under the new development pattern of mutual promotion of domestic and international dual cycles with domestic large cycle as the main body, relying on the open source model to upgrade China's scientific and technological innovation governance system has become an important way to break through the key and difficult points of information technology, and has also brought new opportunities and new requirements to China's key technological innovation and the development of software industry. All sectors of industry, education, research and application in China should strengthen coordination, to further strengthen the synergy of the key elements of the industrial supply chain and the open source ecosystem, improve the mechanism, integrate resources, explore effective paths, and support the commercial development of the open source, so as to accelerate the pace of the construction of the open source ecosystem in China.

References

1. National Natural Science Foundation of China, Chinese Academy of Sciences (2021) Software science and engineering: software ecosystem. Science Press, Beijing
2. Zhang DG, Li B, He P et al (2015) Characteristic study of open-source community based on software ecosystem. Comput Eng 41(11):106–113

3. Messerschmitt DG, Szyperski C (2003) Software ecosystem: understanding an indispensable technology and industry. MIT Press, Cambridge
4. China Electronics Standardization Institute (2021) Information technology-open source-overview and terminology of open source: CESA-2021-3-021. China Electronics Industry Standardization Technology Association, Beijing
5. Wu X, Wu JY, Zhou MH et al (2022) Selection of open source licenses: challenges and impact factors. J Softw 33(1):1–25
6. Zhou MH, Zhang YX, Tan X (2019) Software digital sociology. Chin Sci: Inf Sci 11:1399–1411
7. Overney C, Meinicke J, Kästner C et al (2020) How to not get rich: an empirical study of donations in open source. In: Proceedings of the ACM/IEEE 42nd international conference on software engineering. IEEE, New York, pp 1209–1221
8. Mei H (31 May 2016) Suggestions on the development strategy of open source software technology in China. http://casad.cas.cn/zkjs/jczx/zxbg_124305/2016n/201912/t20191223_4728861.html
9. Wang T, Yin G, Yu Y et al (2020) Group method and practice of software development based on group intelligence. Chin Sci: Inf Sci 50(03):318–334
10. Zhang Y, Zhou M, Mockus A et al (2019) Companies' participation in OSS development–an empirical study of OpenStack. IEEE Trans Softw Eng 47(10):2242–2259
11. Zhou M, Mockus A, Ma X et al (2016) Inflow and retention in OSS communities with commercial involvement: a case study of three hybrid projects. ACM Trans Softw Eng Methodol 25(2):1–29
12. Ou JS (2016) Open source from the perspective of enterprises: 5 years of practical experience of HUAWEI open source. Commun Chin Comput Soc 12(2):40–43
13. Tan X, Zhou M, Fitzgerald B (2020) Scaling open source communities: an empirical study of the Linux kernel. In: 2020 IEEE/ACM 42nd international conference on software engineering (ICSE). IEEE, New York, pp 1222–1234
14. Zhang Y, Zhou M, Stol KJ et al (2020) How do companies collaborate in open source ecosystems? An empirical study of OpenStack. In: 2020 IEEE/ACM 42nd international conference on software engineering (ICSE). IEEE, New York, pp 1196–1208

Yang Liyun director of Cloud Computing Research Office, Software Center, China Electronics Standardization Association, executive deputy secretary general of China Open Source Cloud League, mainland convener of cross strait cloud computing sub forum of the Ministry of Industry and Information Technology, and head of the Secretariat of Cloud Computing Standards Working Group of China National Information Technology Standardization Network. She has been engaged in the standardization of information technology for a long time, and has led and participated in more than 20 national standards and cloud computing open source technology group standards in the fields of cloud computing, middleware and programming languages. She has engaged in the construction and operation of the open source community, standardization and technical research for a long time, she is the core sponsor and leader of the Mulan open source community and the Mulan open source license.

The Construction and Application of Agricultural Information System for Black Soil Conservation Strategy

Yucheng Zhang, Zichen Liu, Zhaomin Hu, Xiaowei Cao, and Lei Li

Abstract Facing the national strategy of black soil conservation, this paper designs a black soil agricultural information system based on information technology means, integrated application of Internet of Things technology, artificial intelligence technology, etc., for the purpose of black soil conservation and utilization, combined with agricultural production management process, based on business flow and production management driven approach, to provide specific business applications for different users to realize black soil conservation in planting and production. The information system has built a four-level hierarchical management system for plots, with farm regional grid units developing planting and rotation planning based on system recommendations; production team regional grid units specifically implementing farm planning; natural plot units being units of operational programs, with production teams carrying out operations as such; and minimal data units being units of data collection, analysis and decision making, and implementation of precise operations. At present, the system has pushed more than 100 suggestions with an accuracy rate of over 80%, and we will further improve the functions. Finally, this paper outlooks the application prospect of the business flow-driven black soil agricultural information system based on the business flow, with a view to providing reference for exploring a new model of black soil conservation and utilization in China.

Keywords Black soil · Business flow · Planting management · Plot classification

1 Introduction

The annual commercial grain production in northeast China, the main distribution area of China's black soil, amounts for one-fifth of national grain production, and it serves as a "stabilizer" and "ballast" for China's food security. Long-term inappropriate cultivation and high intensity usage, on the other hand, have resulted in the problem of "quantity reduction and quality degradation of China's black soild", which

Y. Zhang (✉) · Z. Liu · Z. Hu · X. Cao · L. Li
Institute of Computing Technology, Chinese Academy of Sciences (CAS), Beijing, China
e-mail: zhangyucheng@ict.ac.cn

China's e-Science Blue Book 2023, https://doi.org/10.1007/978-981-99-8270-7_12

significantly affects national food security and regional ecological security [1]. The major goal of China's black soil conservation plan is to strike a balance between "using well" and "nurturing well". "Nurturing well" emphasizing soil, microorganisms, organic matter, and other basic species, while "using well" emphasizing customized and large-scale conservation tillage through scientific means.

The key to achieve of "using well" is the scientific decision-making process of agricultural output. And the core solution to the scientific organization and effective decision-making of agricultural production is how to collect, store and process information effectively. On the basis of special smart agricultural equipment, information technology and intelligent means, scientific evaluation of conservation tillage, and implementable recommendations are developed [2]. As a result, one of the most successful ways to using and nurturing the black soil well is to build agricultural field models to make decision for agricultural output based on effective information.

The research on these fields has been conducted for more than 40 years overseas, and a complete research system and structure has been formed, resulting in a series of agricultural field model systems, including DSSAT, APSIM, etc. [3–6]. Meanwhile, supported by agricultural simulation model systems and big data technologies, a series of production and farm management-decision support systems such as Climate FieldView and AgWorld have emerged and promoted worldwide in recent years [7, 8]. In combination with the development of agricultural machinery technology and basic mechanism research on soil and microorganisms, black soil conservation in the United States has achieved better results in recent years.

Incontrast, the study of agricultural informatization in China started late, but is developing fast, and has gained some breakthroughs in single-point technologies, such as agricultural field models and data collection. However, one of the biggest problem is the disconnection between the informatization system and the agricultural production process, which making it difficult to support large-scale agricultural production in a systematic way. Infact, based on business applications, the combination of information technology and the agricultural production process can both solve the problems in the agricultural data collection process and provide a framework for application support for agricultural models and algorithms.

To address the problem, the Chinese Academy of Sciences (CAS) organized the Strategic Priority Research Program of Chinese Academy of Sciences "Science and Technology Battle of Black Soil Granary" (STBBSG) in 2021. The Institute of Computing Technology of the Chinese Academy of Sciences (ICT) leads the construction of Dahewan Demonstration Area of STBBSG, aiming to explore the mode of "data decision and automatic execution". Based on business flow of Dahewan Farm, the ICT team has constructed a management system oriented to large-scale planting and achieved some results.

2 System Framework Design

The core of the business flow and management driven approach is to create a business system of information management and design a data organization and management model.

2.1 Design of Business Process and Application System

The core of large-scale production activities is the planning and task management of large area planting plots. The core of the Black Soil Agricultural Information System in Dahewan Demonstration Area includes planting planning business process and task management process, which are shown respectively in Figs. 1 and 2.

Planting planning is the core of scientific decision making in the process of black soil conservation, and make plans of crop species, fertilizer types, etc., for all plots of land. The planting planning subsystem invokes data analysis processes based on agricultural field models or artificial intelligence to analyze historical plot data and provide optimized planting recommendations. The technology department of Dahewan Farm makes plans according to the recommendations; and can plan in advance intra-field or inter-field rotation scheme to achieve fine and flexible planting scheduling. The function calling process of the planting planning subsystem is transparent to the staff of the farm technology department, which not only solves the problem of model application, but also reduces the manual intervention of production managers. This design process not only which ensures the penetration rate of

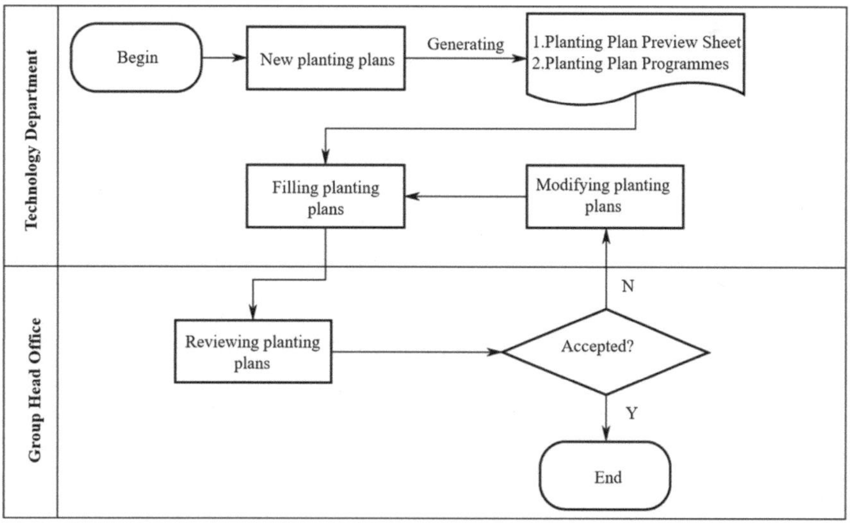

Fig. 1 Business process of planting planning

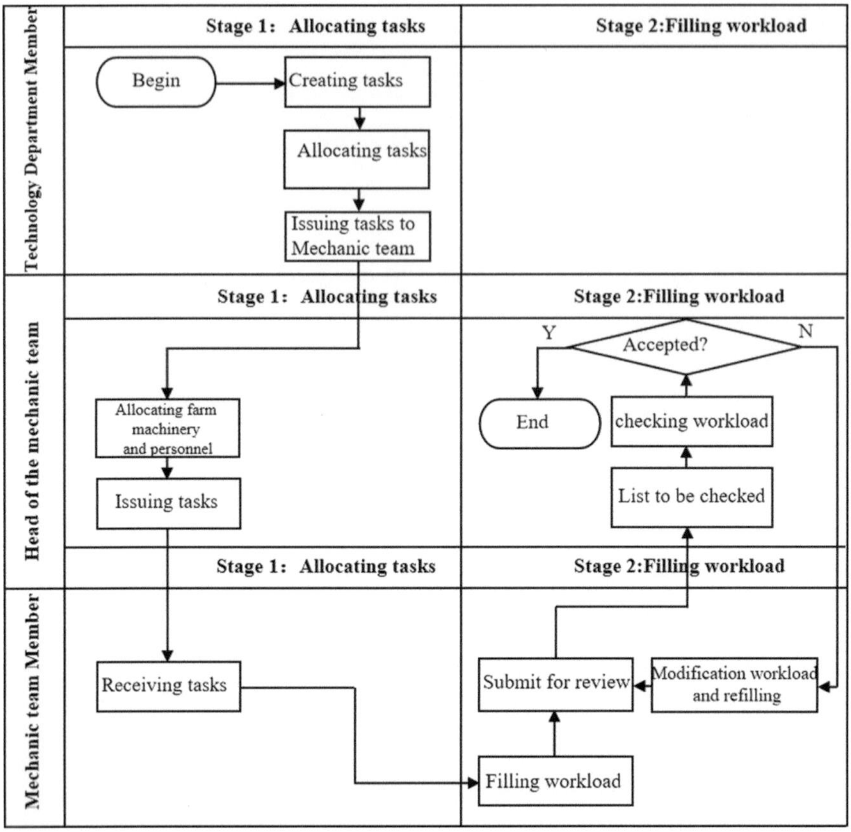

Fig. 2 Task management process

the system and the effectiveness of scientific planting decisions in the production process.

The task management process can play two major roles in the process of black soil conservation: first, to achieve standardized management of different business processes of the plots, to ensure that the operation process of each plot meets the requirements of the operation process management; second, to monitor the quality of operations based on information technology of agricultural machinery. The design of the task management process in the Dahewan Demonstration Area focuses on two aspects: first, the task allocation process, based on the planting plan, the technology department unified each plot of land to distribute operating specifications to different operators through the App; second, the operation execution and acceptance process, operators carry out operations after receiving task, and operation quality data will be automatically reported to the farm in real time. When the operation is completed, the farm manager will confirm the operation quality to ensure the operation process meets the operation specification. In the task allocation process, agricultural field models

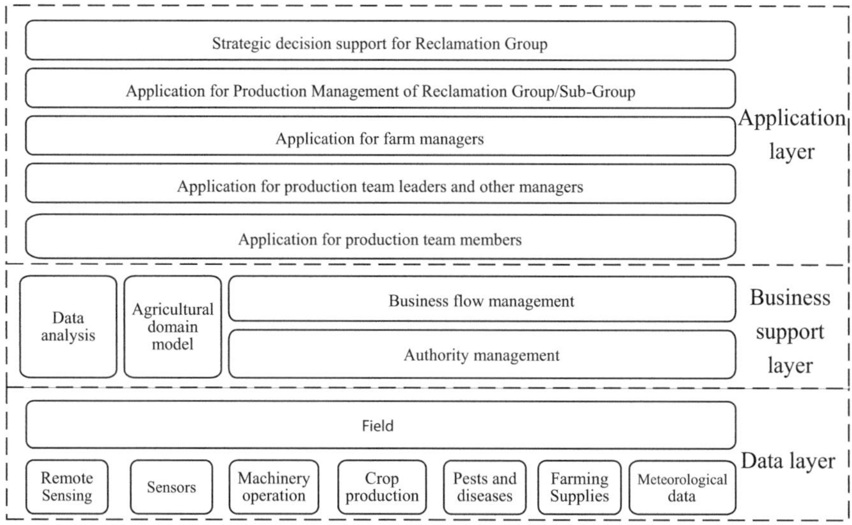

Fig. 3 Design of the application framework system in Dahewan demonstration area

and artificial intelligence technologies will also automatically provide optimization suggestions to improve the efficiency of task allocation management.

Based on the above process and combined with the data management requirements, the application framework system of Dahewan Demonstration Area is designed as shown in Fig. 3, which is divided into three major layers. Firstly, data layer is responsible for the organization of basic plot information such as remote sensing, farming operation and yield, etc. Secondly, Business support layer provides support for business applications based on the authority and business flow management, combined with data analysis engine and agricultural domain model. Thirdly, application layer develops specific business applications for different users of the Reclamation Group.

2.2 *Data Organization Method for Hierarchical Plot Grids*

Based on the above framework and the business need of the farm, data organization model adopts the idea of block-based plot hierarchical grid management, dividing the entire plot hierarchical grid into four levels to achieve organic integration of data at different resolutions.

1. Farm area grid unit. The farm area grid unit is the basic resource allocation and management grids for agricultural production, allocating agricultural machinery, personnel, sensor equipment, seeds, pesticides, fertilizers and other agricultural production factors. In actual production, when the operation pressure is high in a certain area, the system schedules operation resources among different grid

units to realize the dynamic allocation of resources in the agricultural production process.

2. Production team area grid unit. Specific production tasks, collection of production data, job scheduling and operational decisions are all performing in production team area grid units.
3. Natural plot unit. The natural plot unit is the basic plot of the planting operation, which corresponds to the concept of a plot in traditional agricultural production. The Black Soil Agricultural Information System outputs the operation plan with the natural plot unit as the smallest operation plan unit and gives the suggestion of rotation planning among the plots.
4. Minimum data unit. The minimum data unit is 10 m by 10 m, which is the smallest data collection, analysis and decision. Data about soil, water, fertilizer, crop phenotype, and yield are collected through information technology to form normalized evaluation indexes. Based on data analysis and agricultural field model, the Black Soil Agricultural Information System outputs recommendations for precision sowing and fertilization of the smallest data unit, and gives refined recommendations for regional rotation within the plot.

In the above grid system, farm area grid unit and production team area grid unit are macro management units based on production process, which provide macro level management and decision. The natural plot unit and minimum data unit are micro foundation unit for data organization and data analysis applications, designed from data management and technology perspective. The mapping between macro management units and micro foundation units are showed in Fig. 3. Through this division, the multi-level production management system realizes business management and solves the problem of data sources; the business management process unifies data users and data generators, which provides entire process support for agricultural production management and realizes effective data governance.

3 Technical Architecture Design and Implementation

The Black Soil Agricultural Information System is designed with business flow and production management as the core, data organization and analysis as the technical support, and micro service architecture as the basis. The overall technical architecture uses hierarchical modular design, which is shown in Fig. 4.

3.1 Network Access Management Layer

The network access management layer is responsible for the remote access of various agricultural monitoring terminals to realize the real-time data collection and analysis about weather, soil, crop, and operation. The network access layer adopts a variety of

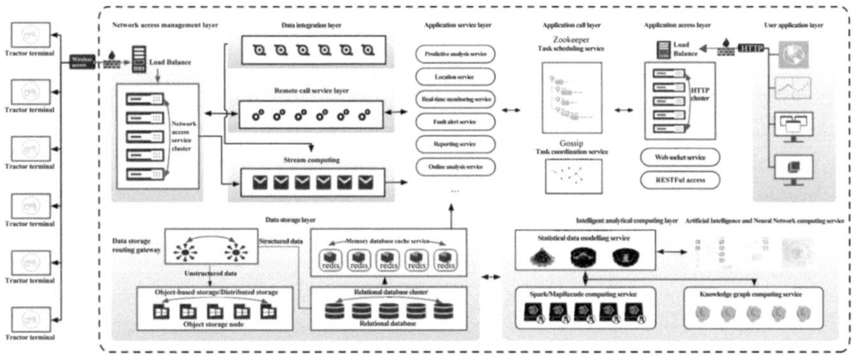

Fig. 4 Technical architecture design of black soil agricultural information system

transmission protocol modes. First, it adopts the self-defined TCP data transmission mode to realize lightweight dual-direction data transmission. Second, it uses a combination of CoMP and MQTT data transmission method to achieve the different QoS guarantee while improving the scalability of data transmission. Each network access management node adopts an event-driven asynchronous model to ensure standalone processing capability, and supports horizontal scaling in a load-balanced manner to receive massive data services and provide expansion support for accessing more IoT data. In terms of data security, TLS standard transmission protocol is adopted to ensure reliable link transmission. In terms of data fields, XXTEA, AES-256, chacha20-ietf and national encryption algorithms are used for secondary encryption.

3.2 Data Storage Layer

The data storage layer implements the storage management of the platform system and adopts a tiered storage architecture. The top layer is cache service composed of memory databases such as Redis/Memcache to ensure data reading speed. The cache data layer can be extended to Ehcache, MongoDB and other NoSQL database systems according to demand. The second layer is the relation database for storing structured data. Based on MySQL clusters, the relation database adopts the tiered storage model of NewSQL and big data storage systems to achieve the separation of hot and cold data and provide multi-level horizontal scaling support. The data storage layer also supports data reading and writing of distributed storage systems such as Object Storage and FastDFS, and provides storage for unstructured data including remote sensing images, images and video.

3.3 Data Integration Layer

The data integration layer provides support for the integration of different external systems and improves the scalability of the system, with the ability to call Kafka, HTTP Web Service, WebSocket, Federation and other interfaces. The data integration layer realizes the data interaction between the Black Soil Agricultural Information System and the existing information system of Reclamation Group, which ensures data synchronization and provides comprehensive data analysis service.

3.4 Remote Call Service Layer

As the core of the overall system scalability, the remote call layer realizes the control of specific business applications on remote terminals and supports micro-service architecture.The remote call service layer is implemented based on the gRPC framework, which currently provides a Java development call interface and can support C++/Python/JavaScript extensions according to the call requirements of specific business application.

3.5 Stream Computing Layer

Streaming computing provides real-time data processing for massive access modules. The stream computing layer takes Kafka as the core structure, transforming massive access to concurrent agricultural data into streaming data and achieving the loose coupling chain data process to improve the overall system throughput and provide better data processing capability in real time.

3.6 Application Support Module

The application support module includes three layers: application service layer, application scheduling layer, and application access layer. The application service layer implements business logic support for specific applications and provides a unified service interface for web applications and mobile applications based on the design idea of front-end and back-end separation. The application scheduling layer is implemented based on Zookeeper, and provides inter-cluster task coordination for business applications. The application access layer supports Web terminal services and mobile services through HTTP cluster that can provide three types of services including WebSocket, SockJS and RESTFul.

Based on the above basic framework, Black Soil Agricultural Information System has designed an intelligent analysis and calculation layer, which provides model and algorithm support for business processes based on a complete data organization and mainly provides three core functions. Firstly, The data statistical calculation service provides basic computing functions for cleaning and analyzing different data, and provides custom development functions based on BI system and scientific computing framework. Secondly, the knowledge graph computing service provides the storage and query functions of basic agronomic knowledge based on graph database. Thirdly, based on the integration of PyTorch framework, the artificial intelligence and neural network computing service supports the optimization of the operation and planning strategy of black soil conservation process.

4 Application Demonstration

The concept of planting planning and planting operation management has been promoted in the whole Reclamation Group. Based on the Black Soil Agricultural Information System of the Dahewan Demonstration Area, the planting management system with GIS is designed to provide the comprehensive process management of farming production in Reclamation Group, as shown in Table 1. The planting management system provides detailed records and data about the whole operational process, and has built complete operation data files which are combined to provide an additional source of data for the joint analysis of planting management and black soil evolution.

Table 1 Main functions of the planting management system

The planting management system	
Main modules	Main functions
Meteorological information	Collect data such as temperature, humidity, light, poise, rainfall and soil nutrients, etc. Support wireless transmission and solar power
Soil information	Collect data in the 50 cm depth soil such as soil moisture, organic matter, nitrogen, phosphorus, potassium, etc Support wireless transmission and solar power Support four-level plot information management
Planting planning	Upload and distribute plating plans
Field operation management	Manage each agricultural operation such as plowing, planting, management and harvesting
Agricultural basic data maintenance	Maintain basic information for farms and production teams
Data analysis	Analyze the data in different dimensions such yield, revenue, etc
Agricultural equipment management	Manage the agricultural equipment file information

The planting management system provides agronomic guidance strategies for the pre-planting, in-planting and harvesting stages according to the agricultural production chain. In the pre-planting stage, the system provides recommendations on additional fertilizer, cultivation methods, planting patterns and sowing period prediction by analyzing soil fertility, weather history and previous crops. In the in-planting stage, according to the crops information and remote sensing, the system provides the recommendations about precise seedling replenishment and control methods of pest and disease. In the harvesting stage, the system predicts the crop yield, and provide harvest date prediction and harvest method suggestions. The crop type, germination rate, fertilizer and chemical application management, and chemical formulation and dosage of each plot can be filled in and uploaded through the cell phone App of the agricultural technician. The planting management system provides automatic statistics on spring sowing plans, spring sowing progress and autumn harvesting progress, and keeps track of all production teams' work tasks and execution progress in real time.

In terms of data resource management, based on the four-level plot management system and combined with GIS technology, a variety of data in Dahewan Farm including soil nutrients (nitrogen, phosphorus, potassium, organic matter), black soil thickness, texture, weight, porosity, moisture, area, crop yield estimation, etc., have been digitally managed. The multi-level data management module not only provides business applications for different users at different levels, but also enables the classification and categorization of various types of data on the farm, effectively improving production management efficiency.

Based on the business application platform and data organization, preliminary agricultural model integration has been established, which contains a library of algorithms for crop phenotype data analysis, crop fertility analysis, growth model, meteorological analysis, sowing and harvest prediction, soil moisture, nutrient inversion, seedling rate, crop growth, crop identification, drought index, pest identification, straw cover, yield estimation, etc. Among them, the accuracy of 47 diseases based on six major crops including wheat, soybean, corn, tomato, alfalfa and sweet sorghum, is over 80%, which can effectively guide the prevention of pest and disease disasters.

According to the agronomic system of conservation tillage actually adopted in each plot, combined with the crop growth state sensed in real time in the data collection system, a dynamic simulation between crop growth state and environment is established to open up the information channel during pre-planting, in-planting and harvesting. The expert online decision and diagnosis push system established in 2022 has developed more than 100 recommendations with an accuracy rate higher than 80%. The current system, however, is mostly reliant on domain knowledge and expertise of specialists, and an intelligent system based on big data and autonomous decision-making across the entire process is being developed.

5 Conclusions and Outlook

Black Soil Agricultural Information System based on a business flow driven approach not only helps Dahewan Farm to identify basis resources and investigate the promotion of scientific decision-making approach for black soil conservation while efficiently lowering management costs. In the next step, the improvement and upgrading of system data and functions will be gradually promoted as followed.

1. Improve the degree of automated data gathering for agricultural production, which will significantly minimize human hours and narrow the area of manual input.
2. Increase the variety of agricultural production data, improve data flow and integration in each production stage, and thoroughly investigate the value of information systems in production decision-making to obtain more accurate support services.
3. Provide intelligent production decision support based on business flow management system, increasing the overall benefit more than 10%.

The Black Soil Agricultural Information System provides data-based, site-specific decision-making methods for black soil conservation and utilization, including agronomic modes such as straw mulching, rotation plans and planting plans to increase crop yield or improve soil quality. This system can also monitor the soil quality in a dynamic and real-time manner, providing complete data support for national and local departments to formulate relevant policies and measures.

Acknowledgements This study was supported by the Strategic Priority Research Program of Chinese Academy of Sciences(Grant No. XDA28120400).

References

1. Shen H (2021) Solving the problem of black soil conservation and utilization in the Northeast. Economic Daily 2021–07–26 (9).
2. Sun N, Zhang Y et al (2021) Agricultural simulator: using intelligent technology to get data flow for black land protection. Bull Chin Acad Sci 36(10):1165–1174
3. Jones JW, Hoogenboom G, Porter CH et al (2003) The DSSAT cropping system model. Eur J Agron 18(3–4):235–265
4. Dzotsi KA, Jones JW, Adiku SGK, et al (2010) Modeling soil and plant phosphorus within DSSAT. Ecol Model 221(23):2839–2849
5. Keating BA, Carberry PS, Hammer GL et al (2003) An overview of APSIM, a model designed for farming systems simulation. Eur J Agron 18(3–4):267–288
6. Robertson MJ, Lilley JM (2016) Simulation of growth, development and yield of canola (*Brassica napus*) in APSIM. Crop Pasture Sci 67(4):332–344
7. Kadoic N, Tomicic-Pupek K, Vrcek N (2020) Decision making on digital platforms in agriculture. In: 2020 43rd international convention on information, communication and electronic technology (MIPRO)

8. Zhao C, Yang X et al (2018) Review and outlook of agricultural information technology development in China. China Agric Digest Agric Eng 30(4):3–7

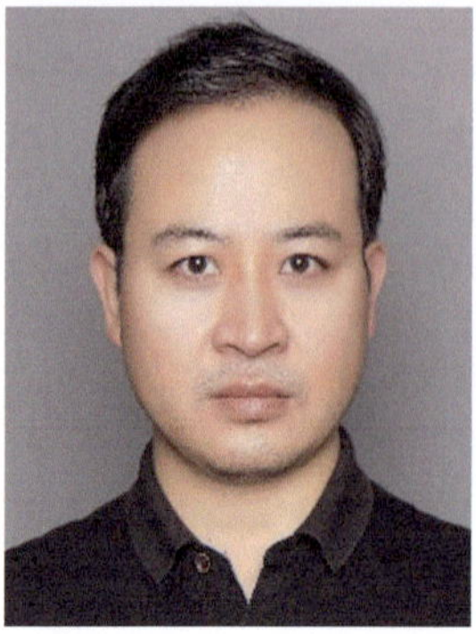

Yucheng Zhang Ph.D, Senior Engineer of Institute of Computing Technology, Chinese Academy of Sciences (CAS), Deputy Director of CAS Engineering Laboratory of Intelligent Agricultural Machinery and Equipment. He is mainly engaged in the research of intelligent agricultural machinery, control theory and method of complex agricultural system. He proposed to build the third generation of agricultural machinery technology system with information technology as the core, achieve parallel breakthroughs in "core technology, major equipment and industrial demonstration", and explore a new agricultural production model with the third generation of agricultural machinery as the core and data as the driver.

Establishment and Application of Regional Agricultural and Rural Data Asset-Based Management Mode

Juan Liu, Yunpeng Cui, Mengjia Huo, and Shumei Zhou

Abstract With the widespread popularization of big data concepts and technologies, data resources have become the core elements of digital agricultural rural construction, and data-intensive decision-making management has become the only way for regional digital economic development. Based on the theoretical analysis of data morphological evolution, data asset attributes and evidence-based data decision-making, this paper defines the business orientation of agricultural and rural data asset management, clarifies the function and architecture of regional agricultural and rural data platform construction, and constructs the business model of data asset management. Based on the core elements of problem, data and technology, this paper constructs four types of digital asset management models, namely, business digitalization, technology iteration, business panorama and decision-making intelligence. The study finally proposes that revitalizing agricultural and rural data assets requires cultivating big data decision-making consciousness, establishing demand-oriented data asset management ecology and methodology, and stimulating the effectiveness and vitality of regional agricultural and rural data "new infrastructure" project.

Keywords Agriculture and countryside · Data assets · Management model

1 Introduction

In April 2020, the document "Opinions on Building a More Perfect Market-oriented Allocation System and Mechanism of Factors" issued by the CPC Central Committee and The State Council listed data as a new factor of production [1]. In November 2021, the Ministry of Industry and Information Technology issued the "Big Data Industry Development Plan during the 14th Five-year Plan period", determining that data is an important factor of production in the new era and a basic strategic resource of the country [2]. How to turn data into assets has become an important issue faced by various sectors.

J. Liu (✉) · Y. Cui · M. Huo · S. Zhou
Agricultural Information Institute of Chinese Academy of Agricultural Sciences, Beijing, China
e-mail: diccaas@caas.cn

China's e-Science Blue Book 2023, https://doi.org/10.1007/978-981-99-8270-7_13

The research on data capitalization can be traced back to 1998. Weir [3] proposed that data mining can discover enterprise assets and realize the value of data. Since the organizing committee of the China Data Asset Management Summit (DAMS) first proposed "Data Asset Management" in 2015, the problem research has gradually increased. Zhang [4] proposed that asset management can promote the transformation of data from resources to assets and build the transaction mode of government data assets based on system, platform and manpower. Zhong [5] proposed to build a provincial data asset operation platform to mine data asset value after analyzing the bottleneck problem of government data assetization. Mu [6] conducted analysis and empirical research on key issues such as asset attributes of data resources, ownership definition, value evaluation and data trading, and clarified the asset management method of data resources.

The massive data resources created in the era of big data have great data mining and analysis application value, which has become a social consensus. However, data management in the field of agricultural and rural areas is affected by multiple factors such as weak basic conditions of informatization, low degree of business digitalization, and imperfect data system. The assetization of agricultural and rural data is still in its infancy. In recent years, the agricultural and rural management departments in various regions have carried out useful explorations in the construction of data platforms to support data asset management. However, there has been a lack of research and discussion on how data become assets in the field of agriculture and rural areas and what business models are included in data asset management. This paper combines with the practice of field tracking and regional agricultural and rural big data platform construction to discuss and analyse the related problems in order to provide reference and enlightenment for regional decision-making management departments to establish scientific big data platform construction methodology and operation and maintenance concept.

2 The Definition of Regional Agricultural and Rural Data Asset Management Business

Agricultural and rural areas cover many aspects of rural production, farmers' development and rural governance. Agricultural and rural data cover the original observation data, detection data, experimental data, survey data and statistical data generated in the development of urban and rural areas, agricultural and rural industrial development and rural social development. There are many types of data, including industrial data such as planting, livestock, fishery, and international data resources such as agriculture, global remote sensing, quality and safety, science and technology education, facilities and equipment, agricultural factors, resources and environment, disaster prevention and mitigation, disease prevention and control.

Regional agricultural and rural data refers to the collection of data related to agriculture, rural areas and farmers in a certain spatial range. In this study, regional agricultural and rural data asset management refers to the data asset management process of agricultural and rural management departments within the scope of administrative divisions such as provinces, cities and counties. It is also a set of business functions for decision-making management departments to plan, control and provide data and information assets to ensure data value discovery, including the development, implementation and supervision of agricultural and rural data plans, policies, programs, projects, processes, methods and procedures.

3 Regional Agricultural and Rural Data Asset Management Platform

Regional agricultural and rural management departments excavate data value to strengthen data asset management. They mostly take technology stack as the core to build agricultural and rural big data platforms, integrate data resources in order to improve data resource management and control ability, and support one-stop data aggregation, governance, analysis, mining and service application with the instrumental product.

3.1 Construction of Regional Agricultural and Rural Big Data Platform

Platform construction solves the problem of data interconnection and exchange and sharing among multi-sectors and multi-systems to realize the standardized processing of decentralized and heterogeneous data through the development and application of data standardization aggregation governance method technology. Through the research and application of professional model algorithms in agricultural and rural areas, the platform will solve the problems of insufficient data mining and insufficient application of deep data analysis, and realize the value discovery and deep application of cross-domain data. Through the exploratory data analysis and visualization technology, the platform will solve the problems of the data analysis of static display, light application service problems, and realize the different levels of data real-time dynamic calculation and interactive display.

3.2 Construction Framework of Regional Agricultural and Rural Big Data Platform

Platform construction involves five levels, including: data layer acquires and integrates the business system data, acquires the real-time flow data and collects the

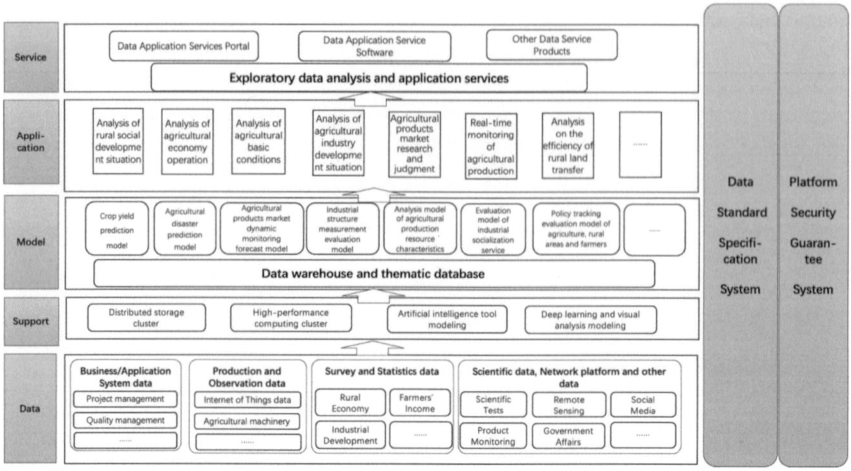

Fig. 1 Architecture of regional agricultural and rural big data platform

business survey data, supports the construction of data center (Fig. 1); according to the actual needs of data capacity and data type storage, the support layer involves distributed storage system, distributed computing system, deployment AI tool modeling and deep learning and visual analysis modeling tools, which provide basic environmental support for platform operation; the model layer carries out the construction of thematic database based on problem orientation, and carries out data modeling, supporting the loading and calling of various model algorithms based on component configuration and embedded technology; the application layer meets the needs of problem-oriented data analysis, carries out data mining analysis, and realizes the transformation of data to information and knowledge; the service layer realizes the release and service of data and data products through visualization, data report, data service terminal and other technologies and carriers.

4 Asset Management Mode of Regional Agricultural and Rural Data

Based on the construction and operation experience of regional agricultural and rural big data platform, the closed-loop business flow of problem/demand—data positioning—database construction—data analysis and mining—system and service application—service feedback and problem discovery is the business foundation for the continuous operation of the platform and the realization of data capitalization. However, the core of the business is to build a database based on the problem and demand and clarify the data analysis and mining method to support data-driven business. The establishment of regional agricultural and rural data asset management model based on the relationship combination of three core elements of problem-data-technology can form the implementation structure and specification of data asset business, and support the diagnosis and optimization design of data platform operation business.

The business model of regional agricultural and rural data asset management is located from two dimensions of data asset abundance and technical complexity. Data asset abundance and technical complexity are relative concepts, and data abundance represents whether the data type, structural degree and dimension are complex. Technical complexity represents whether the technology or model algorithm is generally moderate and the difficulty of mastering is complex.

In general, the more complex the problems need to be solved, the more decision-making elements involved, the greater the data richness and data dimension needed to carry out evidence-based decision management, and the more multidimensional the interaction mode, frequency and scope of data users and business scene quality inspection in business problem solving. On the other hand, the complexity of technology development and technology application determines the differences in the factors involved in problem solving and the use of management methods. And usually complex problems require more complex technical means, which is also consistent with the application of deep learning, neural network and other big data analysis technologies in big data environment to provide important tools for evidence-based decision-making management in complex scenarios.

Based on the analysis, this paper establishes a regional agricultural and rural data asset management framework based on two dimensions of big data asset abundance and technical application complexity, and forms a 2 * 2 regional agricultural and rural data asset management business model, which is divided into four types: business digital application, technology iterative application, business panoramic application and decision-making intelligent application. The characteristics and examples of each type are discussed respectively (Fig. 2).

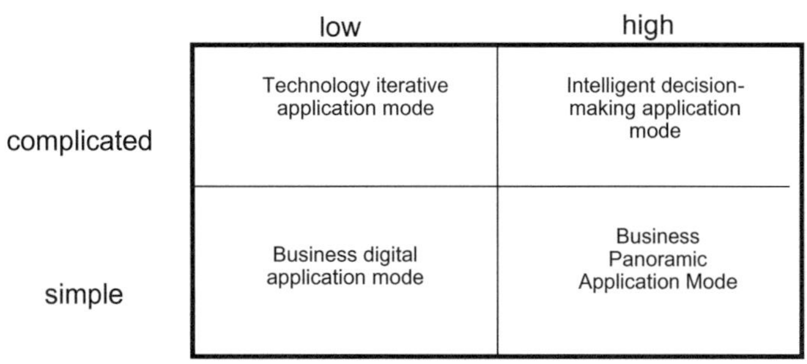

Fig. 2 Four business models of regional agricultural and rural data asset management

4.1 Business Digital Application Mode

This type of business scenario has low asset abundance and low technical complexity. In this type, the content, structure degree or data field properties of the data needed to solve the problem are relatively clear, but the data accumulation in the business field is less, or the degree of digitization of the accumulated data is low, and the technical means to solve the problem with data are mature and easy to use. In general, the users of this model have a relatively clear demand for digital business management and a clear business boundary, which belongs to the empty zone of agricultural and rural informatization construction. They can improve business processes through business system construction and accelerate the construction of digital resources in the field and improve business management efficiency by digitalization.

4.2 Technology Iterative Application Mode

This type of business scenarios has low data asset abundance and complex relatively technical. In this type, the typical feature of business scenario problem is that the content, structural degree or data field attribute of the required data are relatively clear, but the data analysis model needed to solve specific business problems is weak in adaptability and high in complexity to specific business scenarios. Evidence-based decision-making and management needs are relatively clear, and the data accumulation is not high. It is necessary to develop and optimize the model algorithm through data analysis, improve the ability of solving and explaining specific problems, and improve the accuracy of model algorithm and the adaptability of application scenarios.

4.3 Business Panoramic Application Mode

This type of business scenarios has high data asset abundance and simple technology. In this type of business problems, the data content is relatively rich, the degree of data structure is high, and the technical required to solve specific business problems is simple. It is more inclined to panoramic interpretation and basic situation judgment of business scenarios based on simple statistical analysis and correlation analysis. In this type of business, users' demand for business panoramic analysis is clear, but the data span is large and the space–time range is high. It is necessary to ensure data collection and progressive enrichment, and gradually improve the accuracy of data-driven business scene interpretation.

4.4 *Intelligent Decision-Making Application Mode*

This type corresponds to the situation where the data asset abundance is high and the required technology is relatively complex in business scenario problems. In this type, the typical characteristics of business scenario problems are that the data content requirements for solving decision-making management problems are rich, the degree of data structure is not high, and the technologies needed to solve decision-making problems are relatively complex, which are mostly in-depth analysis and application of complex business scenarios. In this type of business, it is difficult to obtain data and the complexity of data mining algorithm model is high. It is an important part of digital and intelligent decision-making management, and also the core link of big data technology application in agricultural and rural regional management.

5 Research Conclusions and Prospects

Based on the theoretical framework of data assets, this paper clarifies that regional agricultural and rural data asset management is a process of data analysis products that provide high accuracy reference and predictability for agricultural and rural management decision-making business scenarios. This paper analyzes the problems existing in the platform management of regional agricultural and rural data assets, constructs a five-element model of scene-data-technology-talent-system, and provides diagnosis and optimization basis for the operation and management of large-scale agricultural and rural data platforms. From the two dimensions of data abundance and technical complexity, this paper forms four application management modes of agricultural and rural data resource management, including digital business, iterative technology, panoramic business and intelligent decision-making. The regional construction of data asset management system and management methodology oriented to business problems and decision-making management value can guide asset management practices such as business-based data aggregation, data communication and data application.

References

1. Opinions of the CPC Central Committee and The State Council on building a more perfect market-oriented system and mechanism for the allocation of factors of production. People's Daily 2020-04-10 (1)
2. Circular of the ministry of industry and information technology on printing and distributing the "14th Five-Year" big data industry development plan. 15 Nov 2021
3. Weir J (1998) Data mining: exploring the corporate asset. J Inf Syst Manag 15(4):68–71
4. Lili Z, Tie L (2018) Analyze the realization path of big data asset management of government affairs. Econ Trade 23:151

5. Jun Z (2017)Research on the capitalization and transaction structure of government data. Fujian Comput 33(10):87–88. https://doi.org/10.16707/j.cnki.fjpc.2017.10.045
6. Yong M, Wei W, Ying Z, Yixing S (2017)Research on current situation, problems and countermeasures of asset management of data resources in China. E-Government 02:66–
7. Haijian L, Li Z (2021) Data becomes a factor of production: characteristics, mechanisms, and the evolution of value form. Shanghai Econ Rev 8:48–59. https://doi.org/10.3969/j.issn.1005-1309.2021.08.006
8. Yangyong Z, Yazhen Y (2018) Defining data assets based on the attributes of data. Big Data Res 4(6):65–76. https://doi.org/10.11959/j.issn.2096-0271.2018062
9. Provost F, Fawcett T (2013) Data science and its relationship to big data and data-driven decision making. Big Data 1(1):51–59
10. Yu J, Yao Q (2020) From data to evidence: research on the construction of evidence-based policy-making mechanism in the era of big data. Chin Public Admin 04:81–87. https://doi.org/10.19735/j.issn.1006-0863.2020.04.12
11. Dawei C (2018) Research on the construction of big data resources for government decision-making. Res Lib Sci 13:28–32. https://doi.org/10.15941/j.cnki.issn1001-0424.2018.13.005

Juan Liu born in April 1978, Doctor of Management, associate researcher, master supervisor. She graduated from China Agricultural University with Doctor's degree in 2006 and now works in the Agricultural Big Data Mining Laboratory of Agricultural Information Research Institute of Chinese Academy of Agricultural Sciences. She has been engaged in scientific research in data organization, data governance and agriculture-related big data applications for a long time, presided over and participated in more than 20 national and provincial projects, published more than 30 academic papers as the first author or corresponding author, published 6 academic monographs, obtained 2 invention patents and more than 10 software Copyrights.

Progress and Effectiveness of Informatization in the Scientific Observing Network of the Chinese Academy of Sciences in the Western Pacific Ocean

Fan Wang and Jianing Wang

Abstract The Institute of Oceanology, Chinese Academy of Science has established a scientific observing network in the western Pacific Ocean, with objectives to meet requirements of the national marine power strategy and to resolve the key scientific questions, such as formation and evolution of ocean energy and material confluence in the Indo-Pacific convergent area. The deep ocean data are real-time transmitted to the surface buoy through the wireless acoustic and inductive coupling communications, and then to the office through the Beidou satellite in an autonomous and controlled manner. We further build a data application system to help users acquire and analyze observation data efficiently. This network compensates for the deficiency of long-term continuous subsurface, intermediate, and deep layer data in the western Pacific, and contributes to many innovative research findings, including systematically revealing the temporal-spatial variations of three-dimensional ocean circulations in the western Pacific and their climate effect, proposing new mechanisms for the propagating of climate signal into the abyssal ocean, and providing supports for the improvements of the ocean environment and climate simulations and predictions.

Keywords Western Pacific Ocean · Scientific Observing Network · Subsurface mooring · Real-time transmission · BeiDou satellite

1 Introduction to the Scientific Observing Network in the Western Pacific

After more than 7 years of construction, the Scientific Observing Network of the Chinese Academy of Sciences (CASSON) in the western Pacific Ocean now has more than 30 sets of deep-sea subsurface moorings and has successfully acquired

F. Wang (✉) · J. Wang
Institute of Oceanology, Chinese Academy of Sciences (IOCAS), Qingdao, China
e-mail: fwang@qdio.ac.cn

China's e-Science Blue Book 2023, https://doi.org/10.1007/978-981-99-8270-7_14

temperature, salinity, and currents data for 6–7 consecutive years. These measurements constitute the first time long-term simultaneous observations of the equatorial current system, western boundary current system, and intermediate-deep ocean circulation system in the western Pacific (Fig. 1). Since 2014, we have organized a comprehensive expedition to the western Pacific each year using the national major scientific and technological infrastructure research vessel known as "*KeXue* (*Science*)". During the cruise, we performed repeated hydrographic section observations on board and realized synchronous observations over the whole water depth and covering multidisciplinary ocean elements; additionally, we recovered and redeployed the subsurface moorings and replaced the batteries and wearable connections of the subsurface mooring equipment.

Most of the subsurface moorings are equipped with measurement sensors in the full-water depth. This provides a strong database for understanding the three-dimensional circulation structure and the vertical exchange of substances and energy throughout the water column. Some subsurface moorings are equipped with measurement sensors only at depths deeper than 3000 m to monitor the deep ocean circulations. In addition to temperature, salinity, and velocity sensors, some subsurface

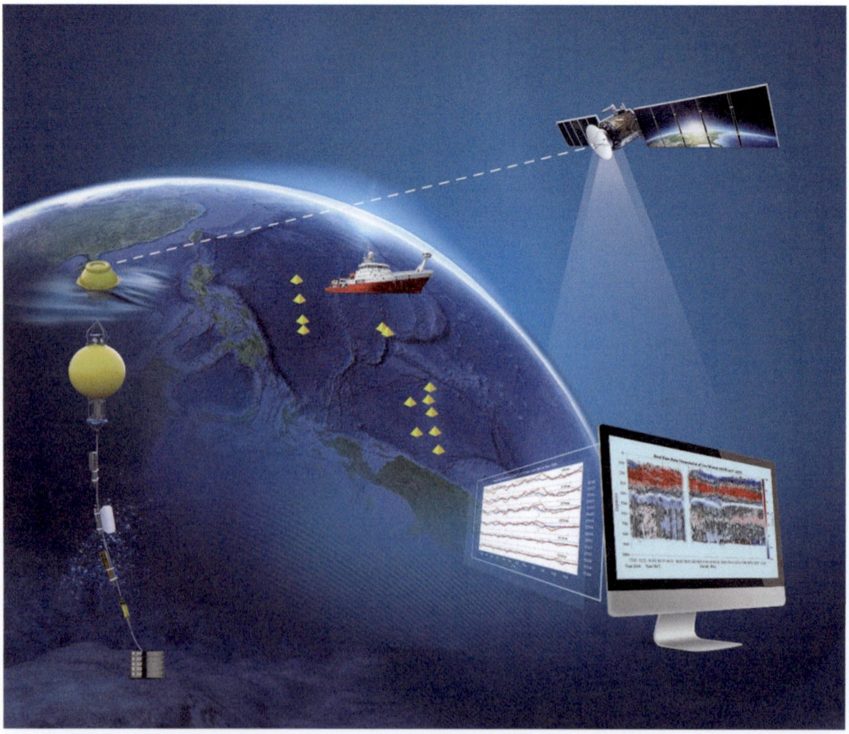

Fig. 1 Schematic diagram of the Scientific Observing Network of the Chinese Academy of Sciences (CASSON) in the western Pacific and real-time communication subsurface mooring

moorings are equipped with dissolved oxygen sensors, which provide data support for understanding the interactions between hydrodynamic and biochemical processes. All sensors return the records with an interval of less than 1 h. This can provide data for understanding multi-scale dynamical processes and their interactions, including internal tides and near-inertial internal waves [1, 2]. The subsurface moorings in the CASSON form several arrays to target different circulation systems. The subsurface mooring array along 130°E focuses on the characteristics and multi-time scale variabilities of the North Equatorial Current and North Equatorial Undercurrent along with their carrying water masses, the structure of mesoscale eddies and their influence on substance-energy exchange and transport, and the influence of the warm pool on typhoons. The subsurface mooring array along 140°E/142°E focuses on the characteristics and multi-time scale variabilities of equatorial current systems such as the North Equatorial Countercurrent, Equatorial Undercurrent, South Equatorial Current, New Guinea Coastal Current, and Equatorial Intermediate Current and their carrying water masses, the thermohaline structure of the warm pool, the interaction of circulation systems with climate events such as the El Niño-Southern Oscillation, and the water mass exchange between the Northern and Southern Hemispheres. The deep western boundary current array is mainly at the Yap Mariana Junction, and has recently been expanded westward to the Kyushu-Palau Ridge and the Philippine Basin, and eastward to the East Mariana Basin. Such an array targets to observe the pathways, structure, volume transport, and multi-time scale variabilities of the upper and lower limb of the deep western boundary current in the western Pacific, and explain the connections of the deep dynamical processes with the upper ocean and climate change.

2 Informatization of the Observing Network

2.1 Underwater-to-Sea Surface Data Communication Subsystem

Real-time transmission of deep ocean data is important for the ocean and climate forecasting, and deep-sea scientific research. Since 2016, we have been working on overcoming the technical problems of the real-time transmission of underwater data over a large water depth range and a long time series, leading to the development of a real-time subsurface mooring [3]. In 2016, through utilizing the wireless hydroacoustic communication, the velocity profiles from the surface to 1000 m recorded by two acoustic Doppler Current Profilers (ADCPs) were successfully transmitted to the surface communication float, and then to the office terminal through satellite (Fig. 2). In 2017, the maximum depth of data that could be transmitted in real-time was extended to 3000 m by upgrading the hydroacoustic communication system. In 2018, such maximum depth was further extended to 5000 m, and wired inductive

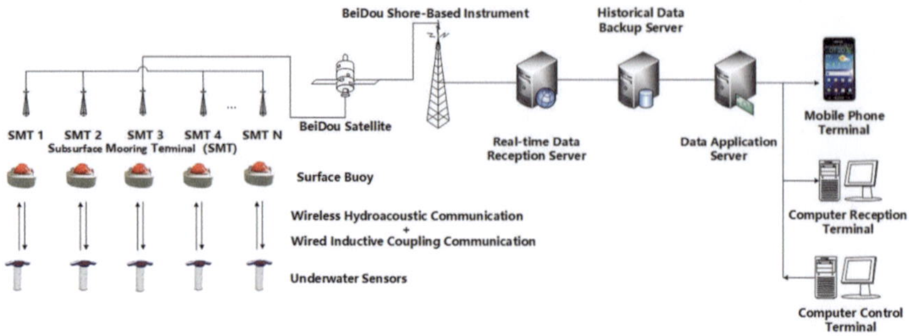

Fig. 2 Schematic diagram of the information system of the scientific observing network in the western Pacific

coupling and wireless hydroacoustic communication were integrated for simultaneous real-time transmission of data in the upper and deep layers. In 2019, two-way real-time transmission of data between underwater sensors and the office terminal was achieved. Through such an operation, scientists can adjust the working frequency of sensors in real-time. Since 2017, subsurface moorings with the real-time communication function have begun to work in a network, with more than 10 sets of moorings in operation each year. The average running time of more than 10 real-time subsurface moorings in one annual cycle is 6–8 months, and the stability and long-term operation of the system have gradually been improved.

2.2 Sea Surface-to-Office Data Satellite Communication Subsystem

Once the deep-sea data reach the surface floats, communication satellites must be used to continue sending the data back to the shore-based office (Fig. 2). To evaluate the overall performance and system reliability of the real-time communication subsurface moorings, we used the foreign satellite to transmit data in the early stages of constructing the observing network. Later, the application of China's self-developed BeiDou satellites became mandatory to ensure independent and controllable real-time deep-sea data transmission.

During the independent development of the BeiDou satellite communication subsystem, it was necessary to overcome three limitations. First, the payload space of the mooring is very small, and the volume for loading satellite modules and antennas is limited. Second, there is no independent system for generating power, and thus the power consumption of satellite modules must be low. Third, the communication data volume is much greater than the transmission capacity of BeiDou's single card. For example, the data volume of two ADCPs is more than 2700 bytes per hour, whereas the transmission rate of BeiDou's single card is only 78 bytes per minute. If only a single card is used for transmission, the hourly data transmission would require more than 30 min to be completed, and the variable environmental factors at sea make this almost impossible. A variety of innovative designs have been adopted to overcome these limitations. First, to address the small load volume, we designed a double-layer stacking scheme for the communication cardboard for the integration and simultaneous transmission of 16 communication cards. Second, to address the absence of an autonomous power generation system, we adopted a low-power microprocessor and duty circuit to greatly reduce energy consumption by activating the BeiDou communication module at regular intervals through the data acquisition board. Third, to handle a large amount of communication data, a multi-card switching operation, dynamic packetization, and sliding window packetization transmission were used to prevent data loss. Finally, a two-way interactive SMS function was implemented to realize the two-way communication function further. Using these innovative designs and through continuous optimization, data from two moored ADCPs can be transmitted in 2–3 min, achieving autonomous and controllable transmission of deep-sea data.

2.3 Office-to-User Data Application Subsystem

To more efficiently apply real-time deep-sea data and distribute user-oriented data, we constructed an observing network data application management system. The system fulfills three major functions: the structured storage and download of historical observation data; automatic reception, quality control, visualization, and transfer of the real-time deep-sea data to mobile phones, web pages, and other applications; and management and two-way configuration functions between subsurface moorings and office terminals. Overall, the system has a three-layer architecture based on C language, and the functions of receiving, backup, and application analysis of subsurface mooring data are separated to ensure the real-time transfer and integrity of the data. The server operating system uses the Linux kernel and Mysql custom database to avoid security concerns related to the Microsoft Windows system and improve data security. The self-developed data encryption algorithm eliminates the leakage loopholes during network transmission. Additionally, according to personalized customization, rich data interfaces are reserved for making it convenient to add functions for secondary development and interfacing with other data analysis systems. This application system enables scientific and operational personnel to examine the historical and real-time hydrographic and dynamical conditions of the deep ocean using their computers or mobile phone terminals.

2.4 Construction Effectiveness

The CASSON has strongly supported many innovative research results. First, the data revealed the three-dimensional circulation structure of the western Pacific Ocean, including the surface, subsurface, intermediate, and deep ocean circulations. Previous studies of circulations below the surface in the western Pacific Ocean relied on limited shipboard observations, with great discrepancies among each observation. Thus, there remains a large gap in understanding circulations below the surface. Long-term continuous data from the CASSON were used to directly confirm the existence and structure characteristics of the Mindanao Undercurrent, North Equatorial Undercurrent, and Equatorial Intermediate Currents [4, 5], as well as determine the pathway, volume transport, and structure of deep meridional overturning circulation into the western Pacific Ocean at depths deeper than 3000 m [6, 7]. Second, the multi-time scale variabilities of the circulations were revealed, and new mechanisms have been proposed for the propagation of climate change signals into the deep ocean at different timescales through the atmospheric intraseasonal variability, planetary and topographic Rossby waves, and eddies. Based on long-term continuous data, the intraseasonal-seasonal-interannual variabilities of subsurface, intermediate, and deep ocean currents, which were difficult to detect previously, have been revealed; particularly, a fast connection of the deep ocean currents with the upper ocean and climate change was discovered [7–14]. These innovative results have greatly accelerated the Northwestern Pacific Ocean Circulation and Climate Experiment (NPOCE) [15] and major frontier research on multi-layer interactions in the "Indo-Pacific Convergence Zone" initiated by Chinese scientists.

We established a platform for the regular sharing of real-time subsurface mooring data with scientific research and operational departments. The National Marine Environmental Forecasting Center considers that the real-time data helps improve data forecasting in the western Pacific Ocean, filling the long-standing gap of real-time data below the surface, improving the accuracy of underwater environmental forecasting, and providing strong support for improving the forecasting capability of the marine environment in China. The Institute of Atmospheric Physics, Chinese Academy of Sciences reported that the real-time data improved the accuracy of El Niño and climate forecast models in the western tropical Pacific. Particularly, the modeled three-dimensional structure of the equatorial current system in the western Pacific was significantly improved. The real-time subsurface mooring data effectively compensate for the lack of subsurface current observation in the existing international ocean observation system and provides strong support for the development and improvement of China's ocean dynamics and climate models. The 3-km multi-layer coupled prediction system covering the "two oceans and one sea" (Pacific Ocean, Indian Ocean, and the South China Sea), built by the Physical Oceanography Laboratory, Ocean University of China, has been providing the "Maritime Silk Road" regional sea and air prediction system with data as the main operating system for the Qingdao Pilot National Laboratory for Marine Science and Technology since 2018. Researchers at the laboratory stated that the real-time data provided by the CASSON improved the accuracy of subsurface temperature, salinity, and current velocity forecast products in the western Pacific as well as improved the simulation of the three-dimensional structure of the western boundary current system and equatorial current system in the western Pacific, overcoming the lack of real-time subsurface, intermediate and deep ocean circulation data in the existing international ocean observation system. The related research by the South China Sea Institute of Oceanology, Chinese Academy of Sciences showed that assimilating real-time data from the CASSON in the numerical model largely reduced simulation errors in low-frequency variabilities and mean axis positions of the North Equatorial Countercurrent and Equatorial Undercurrent, improved the simulation of the baroclinic structure at low-frequency time scales, and corrected the simulation estimates of the potential energy [16, 17].

3 Summary and Outlook

Construction of the Scientific Observing Network of the Chinese Academy of Sciences (CASSON) in the western Pacific Ocean has achieved integrated operation of the entire process, including scientific planning of the observing network, design of subsurface moorings, in situ operations, underwater and satellite real-time data transmission, data analysis and mining, and graphics reception on the computer and mobile phone terminals. The CASSON has accumulated long-term time series of ocean hydrodynamic observation data and enabled the two-way real-time transmission of data over 0–5000 m using the independent and controllable BeiDou satellite, demonstrating the global influence of China's construction and operation of the observing network. In addition, the construction and operation of the CASSON have resolved the long-standing lack of international long-term continuous observation of the subsurface, intermediate, and deep ocean environment in the western Pacific, which has widely impacted China and the world. Furthermore, the NPOCE international program supported by CASSON has been highlighted as a successful example of the Climate Variability and Predictability Program (CLIVAR) in the western Pacific over the past 20 years, which has historically established China's internationally leading research position in the tropical western Pacific.

To meet China's needs for deep-sea scientific research, equipment research and development, and promote cross-disciplinary and major scientific and technological achievements, we are steadily promoting the orderly opening and sharing of observation data from observation networks and subsurface mooring test platforms. For data sharing, we have developed a marine science dataset with regional advantages in the western Pacific by integrating subsurface mooring data, which led to the appearance of the first Chinese curve in the latest global climate change report. Additionally, we have provided direct data download links through relevant website platforms in an orderly manner. To enable sharing of subsurface mooring test platforms, we have added equipment to existing platforms, such as multidisciplinary observation equipment for specific scientific targets, the equipment that can be used for joint observations with existing observation network equipment, and the equipment that needs to be further evaluated in deep-sea trials.

We will continue to promote the construction of the CASSON; serve China's major national needs, such as the Maritime Silk Road, participating in comprehensive global ocean governance, and marine disaster prevention and mitigation; and undertake mission tasks proposed by national development for deep-sea scientific observation, research, and explorations.

Acknowledgements We are sincerely grateful to all the researchers, staff, and the crew of R/V *Kexue* (*Science*) for their efforts in the construction of CASSON. We thank the support from the Scientific Instrument Developing Project of the Chinese Academy of Sciences (Nos. YJKYYQ20170038, YJKYYQ20180057).

References

1. Wang F, Wang J (2016) Initial establishment of China's scientific observing network in western tropical Pacific Ocean. Bull Chin Acad Sci 31(2):258–262 (in Chinese with English abstract)
2. Wang J, Wang F, Zhang L (2017) Construction and operation of a deep-sea scientific observation network in the western Pacific. Oceanologia et Limnologia Sinica 48(6):1471–1479 (in Chinese with English abstract)
3. Wang F, Wang J, Xu L et al (2020) The development of a new real-time subsurface mooring. J Ocean Limnol 38(4):1080–1091
4. Zhang L, Hu D, Hu S et al (2014) Mindanao current/undercurrent measured by a subsurface mooring. J Geophys Res: Ocean 119(6):3617–3628
5. Zhang L, Wang FJ, Wang Q et al (2017) Structure and variability of the north equatorial current/undercurrent from mooring measurements at 130°E in the western Pacific. Sci Rep 7(1):1–9
6. Wang J, Ma Q, Wang F et al (2020) Seasonal variation of the deep limb of the Pacific meridional overturning circulation at Yap-Mariana Junction. J Geophys Res: Ocean 125(7):e2019JC016017
7. Wang J, Wang F, Lu Y et al (2021) Pathways, volume transport, and seasonal variability of the lower deep limb of the Pacific meridional overturning circulation at the Yap-Mariana Junction. Front Mar Sci 8:672199
8. Wang J, Ma Q, Wang F et al (2021) Linking seasonal-to-interannual variability of intermediate currents in the southwest tropical Pacific to wind forcing and ENSO. Geophys Res Lett 48(5):e2020GL092440
9. Song L, Li Y, Liu C et al (2019) Observed deep-reaching signatures of the Madden-Julian Oscillation in the ocean circulation of the western tropical Pacific. Geophys Res Lett 46(24):14634–14643
10. Ma Q, Wang F, Wang J et al (2019) Intensified deep ocean variability induced by topographic Rossby waves at the Pacific Yap-Mariana Junction. J Geophys Res: Ocean 124(11):8360–8374
11. Wang F, Song L, Li Y et al (2016) Semiannually alternating exchange of intermediate waters east of the Philippines. Geophys Res Lett 43(13):7059–7065
12. Hu S, Sprintall J, Guan C et al (2018) Spatiotemporal features of intraseasonal oceanic variability in the Philippine Sea from mooring observations and numerical simulations. J Geophys Res: Ocean 123(7):4874–4887
13. Ma Q, Wang J, Wang F et al (2020) Interannual variability of lower equatorial intermediate current response to ENSO in the Western Pacific. Geophys Res Lett 47(16):e2020GL089311
14. Zhang Z, Pratt LJ, Wang F et al (2020) Intermediate intraseasonal variability in the western tropical Pacific ocean: meridional distribution of equatorial Rossby waves influenced by a tilted boundary. J Phys Oceanogr 50(4):921–933
15. Hu D, Wang F, Sprintall J et al (2020) Review on observational studies of western tropical Pacific Ocean circulation and climate. J Ocean Limnol 38(4):906–929
16. Liu D, Zhu J, Shu Y et al (2018) Model-based assessment of a Northwestern Tropical Pacific moored array to monitor the intraseasonal variability. Ocean Model 126:1–12

17. Liu D, Wang F, Zhu J et al (2020) Impact of assimilation of moored velocity data on low-frequency current estimation in northwestern tropical Pacific. J Geophys Res: Ocean 125:e2019JC015829

Fan Wang Ph.D., Professor, Director of the Institute of Oceanology, Chinese Academy of Sciences (IOCAS). Chairman, Northwestern Pacific Ocean Circulation and Climate Experiment (NPOCE); Member, Climate Variability and Predictability Program (CLIVAR). He was awarded "National Excellent Scientific and Technological Worker" and "National Leader in Scientific and Technological Innovation".

Prof. Wang has led National Program on Key Basic Research Project (973 Program), the Key Research and Development Program, Major and Key Programs of the National Natural Science Foundation of China, the Strategic Priority Research Program of the Chinese Academy of Sciences, etc. He performs research on multi-scale dynamical processes in the Indo-Pacific convergence area and surrounding regions, and has achieved the important scientific discoveries and the theory innovation in the three-dimensional circulations in the western Pacific and the subsurface structure and variabilities in the warm pool. He realized the breakthrough in the real-time transmission of the deep ocean data, conducted pioneered frontier research on multi-layer interactions in the Indo-Pacific convergence area, and advocated the "artificial intelligence oceanography". Prof. Wang has published more than 160 scientific papers including Nature, National Science Review, and Science Advances, and 3 Academic Monographs. He was awarded "Outstanding Achievement Prize in Science and Technology of the Chinese Academy of Sciences", the first prize of "Natural Science of Shandong Province", the first prize of "Oceanographic Engineering Science and Technology Award", etc.

High-Performance Numerical Simulation Software for Life Extension of High-Speed Trains: Developments and Applications

Yujie Wei, Guowei Yang, Xianjia Chen, Pan Xiao, Bo Yin, Shan Liang, Xia- zhen Liu, Han Wu, Zhanling Ji, Zheng Yuan, Chunbao Zhou, Zhonghua Lu, Hongju Cui, and Long Ma

Abstract For the whole-vehicle-level life evaluation and optimization of high-speed trains (HSTs), we develop the theory and model for life extension optimization (LEO) of HSTs for exascale computing, thus improving the scale and efficiency of calculations. We realize a million-core parallel computing for aerodynamics/structure coupling analysis by breaking through the massively parallel algorithms for aerodynamic solving and overset grids that are applicable to domestic heterogeneous supercomputing platforms. From the perspective of system mechanics, we propose a hierarchical and collaborative LEO method for HSTs, which organically links the multi-physics analysis and optimization with the high-performance computing (HPC). For the needs of full-process simulations of LEO, we independently develop the core functional modules used for aerodynamic solving, structural parametric modeling, structural finite element analysis, fatigue life calculation and structural multi-objective optimization, thus integrating the high-performance numerical simulation software for LEO of HSTs ("Numerical HST Version 1.0"). We have completed

Y. Wei (✉) · X. Chen · P. Xiao · Z. Yuan
The State Key Laboratory of Nonlinear Mechanics (LNM), Institute of Mechanics, Chinese Academy of Sciences, Beijing 100190, China
e-mail: yujie_wei@lnm.imech.ac.cn

G. Yang · B. Yin · H. Wu · Z. Ji
Key Laboratory for Mechanics in Fluid Solid Coupling Systems, Institute of Mechanics, Chinese Academy of Sciences, Beijing 100190, China

S. Liang · X.-. Liu · C. Zhou · Z. Lu
Computer Network Information Center, Chinese Academy of Sciences, Beijing 100190, China

Z. Yuan
School of Mechanical, Electronic and Control Engineering, Beijing Jiaotong University, Beijing 100044, China

H. Cui · L. Ma
National Engineering Research Center for High-Speed EMU, CRRC Qingdao Sifang Co., Ltd, Qingdao 266111, China

China's e-Science Blue Book 2023, https://doi.org/10.1007/978-981-99-8270-7_15

the software deployment and testing on typical HPC platforms like "Sunway Taihulight" and "SunRising-1", as well as in the computing environment of CRRC Sifang Co., Ltd. Moreover, the developed software has been applied in the tasks for the safety and reliability design of HSTs, such as aerodynamic surface modification design for 400 km/h HSTs, prediction of the respective contributions of load components on the bogie frame to its structural fatigue life, and LEO for the key components of HSTs.

Keywords High-performance computing · High-speed train · Life extension optimization · Parallel algorithm · "Numerical high-speed train V1.0"

1 Introduction

High-performance computing (HPC) has evolved into a powerful tool for human beings to tackle major challenges in fields such as engineering, materials, energy, and environment, and its hard and soft capabilities are a key indicator of the country's level of scientific and technological development [1]. The United States launched the National Strategic Computing Initiative (NSCI) in 2015 to maintain and enhance its scientific, technical and economic leadership in high performance computing research, development and deployment [1]. As part of the NSCI, the U.S. Department of Energy is implementing the Exascale Computing (1 EFlops) Project (ECP) to deliver three exascale computers in succession by 2023, as well as to facilitate the synchronous research and development of supporting software and applications [1]. The EU and Japan have also launched several development plans to promote the development of supercomputing infrastructure, such as Horizon Europe and Digital Europe Programme (DEP) in the EU [1], and Flagship 2020 in Japan [2]. These projects all have as their primary goal the coordinated development of hardware, software, algorithms and applications for supercomputers.

After years of continuous efforts, China has developed a number of supercomputers, such as the Tianhe series, "Sunway Taihulight" and so on. Although China's hardware infrastructure for supercomputing is at the international leading level, there is still a lack of domestic industrial software suitable for HPC. According to the statistics of the "Comprehensive evaluation report on the development level of the national HPC environment (2015–2020)" released by China National Grid, among the 2,130 kinds of professional software for scientific research and education deployed on the supercomputing platforms of the nineteen national key sites for HPC (including six national supercomputing centers) in 2020, there are only 255 domestic software, and nearly 90% of the software is foreign [3]. The lack of application software adapted to the domestic HPC architecture in various disciplines has become a thorny issue restricting the development of applications of HPC in China. While proposing the construction of exascale and ten exascale supercomputing centers, the "14th five-year plan" of China also clearly pointed out the need to accelerate the filling of bottlenecks and deficiencies in basic software, improve the industrial level of key software, and promote the demonstration application of the first version of the software [4]. From the latest global supercomputer TOP 500 list released in May 2022, it can be seen that the Frontier supercomputer at Oak Ridge National Laboratory in the United States reached a peak performance of 1.1 ExaFlops (one billion-billion calculations per second) [5], indicating that supercomputing has entered the era of exascale computing. While the enormous computing power brings unlimited opportunities for scientific research, it also poses challenges for the development of corresponding engineering software.

Over the past two decades, China's high-speed railway technology has undergone the development journey of import, digestion and absorption, and independent innovation, and the high-speed railway network of China has been continuously extended and densified. Data from China State Railway Group show that by the end of 2021, the operating mileage of China's high-speed railway has exceeded 40,000 km [6], the "four vertical and four horizontal" high-speed railway network has completed ahead of schedule, the "eight vertical and eight horizontal" high-speed railway network is being formed, and the number of the high-speed electric multiple units (EMUs) in China has accounted for more than two-thirds of the world.

In the early stage of the development of high-speed trains (HSTs), people focused on how to ensure high-speed, stable and safe operation of the trains. However, with a large number of high-speed EMUs operating on vast lines for long periods of time, the safety of their service and the reliability of their key components are increasingly prominent. As the core equipment of the high-speed railway system, HST is a typical large-scale system with complex dynamics [7], and the safety and reliability design of HSTs needs to comprehensively consider the influence of wheel-rail, pantograph-catenary, and fluid–solid couplings as shown in Fig. 1 [8].

In recent years, the scholars and engineers in China have carried out extensive and in-depth researches on the key mechanical problems associated with the design and service safety of HSTs [9, 10]. On the aspect of aerodynamics, the aerodynamic performances of HSTs were calculated under complex operating conditions such as in crosswinds and passing through tunnels [11, 12]; a number of design methods were developed in order to optimize the aerodynamic shapes of HSTs [13]; the aerodynamic drag reduction measures for some unsmooth regions of an HST such as wind-shields, skirt plates, pantographs and deflectors were explored [14]; a moving model test platform of HSTs applicable to multiple operating conditions was established and the corresponding test techniques were developed [15]. On the aspect of safety and reliability design of the structures, the dynamics analysis model of the coupled vehicle-track-catenary system [7], the service simulation modeling and calculation method for HSTs [16], as well as the dynamics model of an HST considering crosswinds, wheel treads and track irregularities [17, 18] were developed. Moreover, in order to address the fatigue reliability problems of bogies, axles, bearings, gear boxes and other key load-bearing structures, a series of reliability assessment approaches and technologies have been developed, from load test [19, 20], load spectrum compilation [21] to load modeling [22–24], and so on. However, due to the inadequacy of systematic analysis methods as well as the limitation of computational scale and efficiency, the design of structural fatigue and reliability of HSTs now remains at the level of materials and key components, and it is difficult to

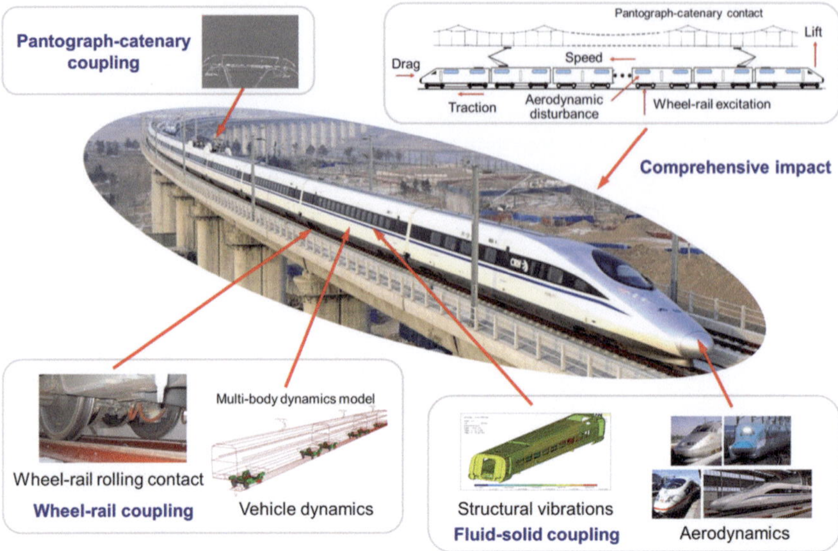

Fig. 1 Schematic diagram of the wheel-rail, pantograph-catenary, and fluid–solid couplings of a high-speed train (HST)

explain and tackle some holistic issues arising from the structural fatigue and reliability of HSTs, such as the interaction law between components. Therefore, there is an urgent need for novel interdisciplinary methods and HPC-oriented software to facilitate the whole-vehicle-level life assessment and design optimization for HSTs. In practice, supercomputers generally adopt a highly heterogeneous architecture with a high-speed interconnection internal network [25], so that the prevailing engineering computational software applied for the multidisciplinary analysis and optimization of HSTs cannot be compiled and run directly on the HPC platform, and a lot of work, such as algorithm transplantation and adaption, needs to be carried out in order to accelerate heterogeneous computing.

With the joint support of the "HPC" special project of the National Key Research and Development Program sponsored by the Ministry of Science and Technology of China, the Strategic Priority Research Program of the Chinese Academy of Sciences, and the Special Project for Informatization of the Chinese Academy of Sciences, the Institute of Mechanics, Chinese Academy of Sciences (IMECH, CAS), together with the Computer Network Information Center, Chinese Academy of Sciences (CNIC, CAS), Dalian University of Technology (DUT) and CRRC Qingdao Sifang Co., Ltd, has undertaken the research on "High-performance Numerical Simulation Software for Life Extension Optimization of High-speed Trains and its Applications". In this project, we start from the principles of system mechanics to realize the fatigue life analysis and optimization of HSTs at the whole-vehicle level by developing high-performance numerical calculation methods and engineering-specific software suitable for heterogeneous platforms. The followings are the specific objectives of our research: (1) Develop massively parallel algorithms and break through the million-core parallel techniques for the demand of multidisciplinary collaborative simulations such as structural finite element analysis, rigid multi-body calculation and aerodynamic solving at the whole-vehicle level; (2) Propose a systematic optimization scheme for HSTs with life extension as the goal that comprehensively takes factors, such as drag, mass and stability, into account, and develop exascale computing-oriented numerical simulation software for life extension optimization (LEO) of HSTs to solve the application problems arising from the lack of methods as well as the limitation in computational scale and efficiency for whole-vehicle-level life assessment and design optimization of HSTs; (3) Develop methods to predict the respective contributions to fatigue of complex loads, such as aerodynamics, gravity and wheel-track excitation, that the HSTs are subjected to; (4) Deploy the developed software in the national supercomputing environment to carry out application demonstrations for the CRH 380A HST, and propose schemes for life estimation and LEO of the train, so as to provide technical support for structural life management, maintenance program formulation as well as LEO design of HSTs. In this paper, we will elaborate on the major advances achieved from the following three aspects: (1) Models and methods of HPC for LEO; (2) Development of "Numerical HST Version 1.0" software platform; and (3) Typical application demonstrations for HSTs.

2 Models and Methods of HPC for LEO

The models used for simulation are a numerical representation of the real physical object under research, while the algorithms are the key to enable the rational, effective and efficient simulations. Here, we develop multi-physics models of a three-car marshaling HST derived from the CRH-380A electric multiple units (EMU) for aerodynamics analyses, structural finite element simulations and rigid multi-body calculations. For aerodynamics analysis, the most time-consuming part during LEO of a HST, we develop a million-core parallel algorithm for aerodynamic computing that is suitable for heterogeneous platforms, and combine it with the parallel overset grid technique to realize massively parallel simulations for the coupled aerodynamics/structure (rigid multi-body dynamics). Basing on the principle of system mechanics, we propose a hierarchical and collaborative method for LEO of HSTs that organically combines the multi-physics analysis and optimization with the HPC.

2.1 Multi-physics Computational Models for a Three-Car Marshaling HST

We consider a three-car marshaling HST, consisting of head, middle and tail cars, derived from the CRH-380A EMU as our research objective, and develop its aerodynamic model, structural finite element model and rigid multi-body dynamics model for aerodynamics analyses, structural response simulations and multi-body dynamics calculation during the process of LEO, respectively. Specifically, the computational domain of the aerodynamic model as shown in Fig. 2 is discretized using full hexahedral meshes, with a grid count of 13,438,772, or about 113 million. For the structural finite element model as demonstrated in Fig. 3, we use shell and tetrahedral elements to model the major components of the train such as vehicle bodies, bogie frames and traction pins, as well as various combinations of spring and damping units to achieve the simplified modeling of connectors like steel springs, air springs and couplers, resulting in a finite element model with approximately 4.56 million elements and 6.77 million nodes. As for multi-body dynamics modeling, we treat the major components, such as vehicle bodies, bogie frames and wheelsets, as rigid bodies, and connect these components via spring and damping members, while considering the track irregularities as well as the elastic support and nonlinear contact between the wheels and rails. All parameters are carefully chosen to match the real dynamics of the HST.

In terms of the computational efficiency of different disciplines involved in the LEO, it is found that the computational cost for aerodynamic analysis and optimization is the most expensive, accounting for more than 90% of the overall computational effort, the structural dynamic stress simulation is the second most time-consuming, and the multi-body dynamics analysis is the most computationally efficient. Therefore, we employ the incremental superposition method to parametrically model the head shape of the HST, and develop a cross-validation-based Kriging model for high-precision predictions of aerodynamic forces with few samples to significantly reduce the computational effort for aerodynamic design of HSTs (as shown in Fig. 4)

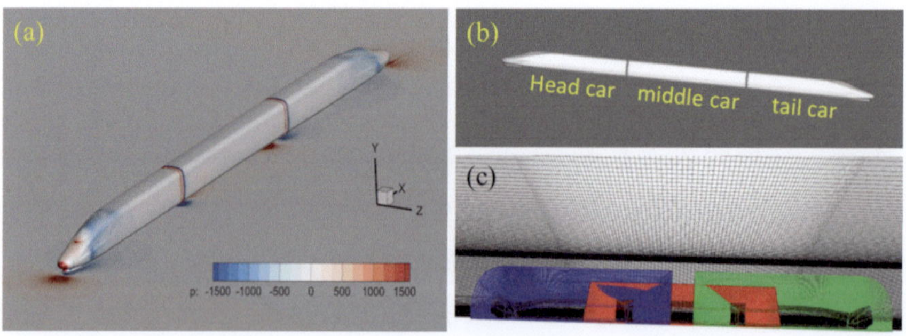

Fig. 2 Aerodynamic model of a three-car marshaling HST: **a** illustration of the static pressure distribution on the train surface; **b** geometric shape of the train; and **c** schematic diagram of the overset grid for the head, middle, and tail cars

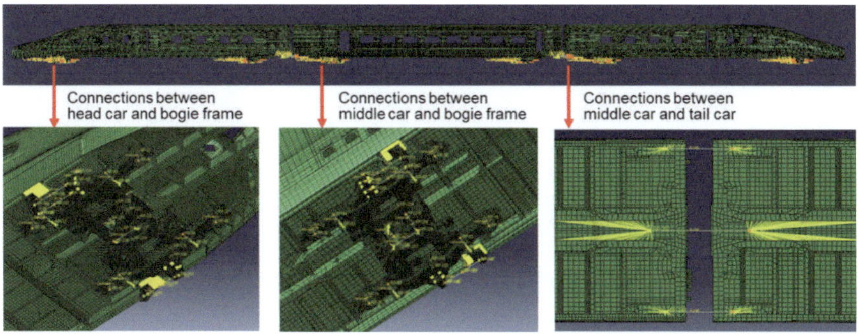

Fig. 3 Structural finite element model of a three-car marshaling HST including key connections

[13, 26]. Meanwhile, in order to improve the computational efficiency for dynamic stress simulations of critical structures, we propose an intelligent model in predicting the structural load spectra as demonstrated in Fig. 5. We take the force–time histories that the structure of our interest is subjected to and the stress–time histories at sites prone to fatigue from tests as the inputs and outputs of our intelligent model, respectively, and use the gated recurrent unit (GRU) neural networks to model the time-dependent intricate relationship between the inputs and the outputs. Our investigation indicates that the combination of the traditional finite element method with the novel intelligent approaches can considerably reduce the time required for solving the dynamic stresses of structures, thus dramatically saving the computational cost for the structural fatigue life of HSTs [27].

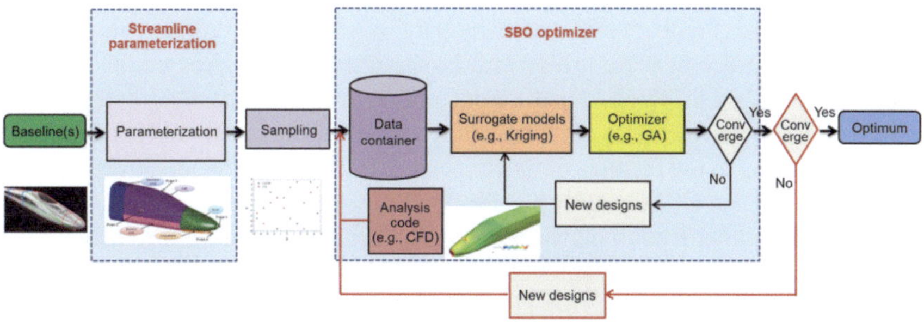

Fig. 4 Procedure for aerodynamic shape optimization of HSTs using cross-validation-based Kriging model

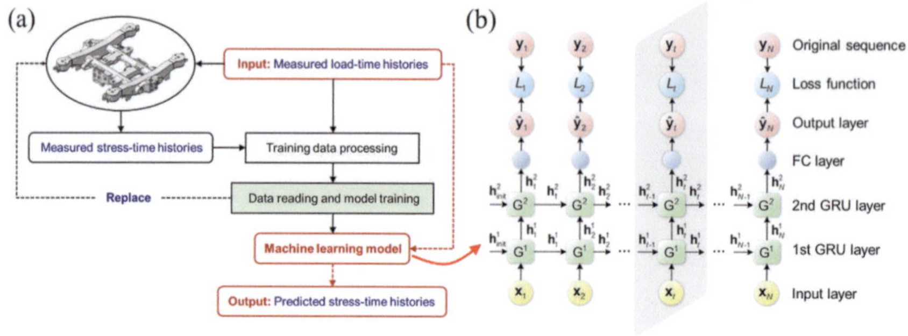

Fig. 5 Intelligent model in predicting the load spectra of critical components involved in HSTs: **a** inputs and outputs of the model; and **b** schematic diagram of GRU neural networks used in our model

2.2 *Efficient Parallel Algorithms for Aerodynamics/Structure Coupling*

In order to significantly improve the efficiency of aerodynamics analysis, the most time-consuming part during LEO of HSTs, we develop the parallel algorithms of CCFDv3.0 software, which is adapted to the domestic heterogeneous platforms in China such as the "Sunway Taihulight" supercomputer. In CCFDv3.0 software, the N-S equation is solved using the finite volume method, and the computational domain is discretized with multi-block structured grids. In practice, we need to utilize the domain decomposition techniques to divide the large-scale structured mesh blocks into multiple smaller sub-structured mesh blocks, and then allocate them to different processors for solving to enable massively parallel computing [28]. In this mapping and allocation process, there are several issues need to be addressed, such as the computation of mesh blocks on different compute nodes, the load balancing in communication, and the data communication between mesh blocks and inside a mesh block. According to the features of the domestic SW26010 many-core processor, the parallel algorithm of CCFDv3.0 software adopts the message passing interface (MPI) to keep data communication between core groups, and uses Athread to achieve the parallel acceleration of master and slave cores within a core croup. We take a series of optimization measures, such as direct memory access (DMA) for data transmission, thread-thread register communication, data storage and calculation process optimization, vectorization processing and instruction rearrangement, to improve the memory access and communication efficiency of the algorithm on different nodes, so that the CCFDv3.0 software can run efficiently on domestic HPC platforms. For instance, the structure of array (SoA) is employed to store the mesh block data to adapt to the vectorization and memory access characteristics of domestic processors; the double-buffer optimization is adopted to enable concurrent execution of computation and memory access in the cache, so as to realize the time overlap between computation and memory access and promote the memory access efficiency of the slave core;

efficient data exchange between groups of threads is implemented through same-row and same-column register communication paths supplied by the slave core array of the processor, thus enabling the high-efficiency Stencil calculation; by means of the shuffling instruction supplied by the SW26010 processor, the original data are reassembled to a data storage form that is stored continuously by variables or meets the actual computational requirements of the algorithm, thus fulfilling the optimization of replacing memory access with calculation and completing vectorization operations. Here, we take the aerodynamic analysis of a HST as an example to explore the acceleration performance of CCFDv3.0 on the "Sunway Taihulight" HPC platform as compared to the benchmark, a pure MPE version with a single process. We demonstrate in Fig. 6a the performance improvement in time stepping forward calculation due to different optimization measures for many-core heterogeneity. As shown in Fig. 6b, we further illustrate the speedup of each stage when using CCFDv3.0 to solve the N-S equation in parallel. It can be clearly seen that the aforementioned optimization measures could effectively enhance the computational performance of CCFDv3.0 when operating on domestic heterogeneous platforms.

To meet the demand of stability analysis of HSTs under aerodynamic loads, we develop a massively parallel algorithm for aerodynamics/structure (rigid multi-body dynamics) coupling based on the breakthrough in parallel overset grid techniques [29]. It is essential for an aerodynamics/structure coupling algorithm to enable the data transfer between the two disciplines as well as the moving of aerodynamic meshes. The basic process of the coupled aerodynamics/structure analysis in each time step for a HST is as follows: (1) Calculate the aerodynamic forces and moments on each car via aerodynamic analysis using CCFDv3.0 and then apply them to the center-of-mass position of each car in the dynamics model; (2) Carry out the multi-body dynamics analysis based on the calculated aerodynamics and track excitations to achieve the translational and rotational displacements of each car, which are in

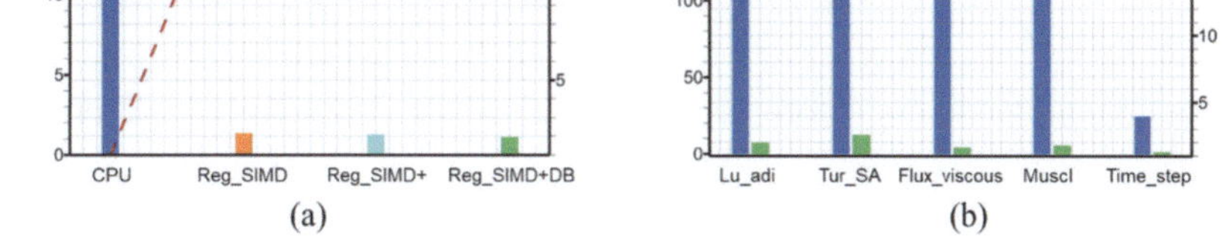

Fig. 6 Acceleration capability of CCFDv3.0 on "Sunway Taihulight" supercomputing platform: **a** speedups of time stepping forward calculations with respect to different many-core heterogeneous optimization strategies; **b** time consumption and speedup of each stage during parallel solution of N-S equation using CCFDv3.0

Table 1 Parallel efficiency of aerodynamic and coupled aerodynamic/structural calculations for the head car model with about 360 million aerodynamic grids in a strongly scalable pattern

Number of core groups	Number of cores	Number of grids per core group	Aerodynamic analysis		Aerodynamics/structure coupling analysis	
			Average computing time for each step (s)	Parallel efficiency	Average computing time for each step (s)	Parallel efficiency
800	52,000	449,700	3.82	–	6.40	–
1,600	104,000	224,850	1.94	98.5%	3.21	99.8%
3,200	208,000	112,400	1.14	83.9%	1.84	87.0%
6,400	416,000	56,200	0.62	76.8%	0.99	80.1%
16,000	1,040,000	22,486	0.46	41.5%	0.72	44.3%

turn transferred to the aerodynamic model; (3) Obtain the aerodynamic model with a modified rigid position of vehicle by rotating the mesh blocks and updating the overset relationships between mesh blocks, and then return to the first step for further aerodynamic calculations. This cycle is iterated until both the aerodynamics and multi-body dynamics analyses converge or reach a specified number of iterations. Here, we chose the head car from our three-car marshaling HST model for aerodynamics analysis and aerodynamics/structure coupling calculation to investigate the large-scale scalability of our algorithm on the "Sunway Taihulight" platform. We present in Table 1 the average single-step computing time and parallel efficiency of the two computational ways with different number of parallel cores in a strongly scalable pattern. As the result shows, our algorithm achieves an efficiency of 41.5 and 44.3% for 1.04 million cores in parallel for aerodynamic and coupled aerodynamic/structural calculations, respectively, using 52,000 cores in parallel as the benchmark.

2.3 System-Mechanics-Based Method for LEO

The HST is a typical complex mechanical system, and the LEO of its key components requires a system-level multi-physics coupling analysis as well as a hierarchical and collaborative optimization. More specifically, the fatigue life of a HST is dependent on the dynamic characteristics of the vehicle structures under external loads, such as aerodynamic loads and wheel-rail excitations. Hence, it is necessary to simultaneously use the aerodynamics, multi-body dynamics and structural finite element methods to calculate the stress distribution and stress–time histories of the vehicle structures of our interest, and then determine their fatigue life distribution in combination with the fatigue model and material fatigue properties. In terms of the coupling relationship, there is a bidirectional strong coupling between the aerodynamics and the multi-body dynamics. However, the variations of aerodynamics arising from the elastic deformation of the vehicle can be ignored compared to the those induced

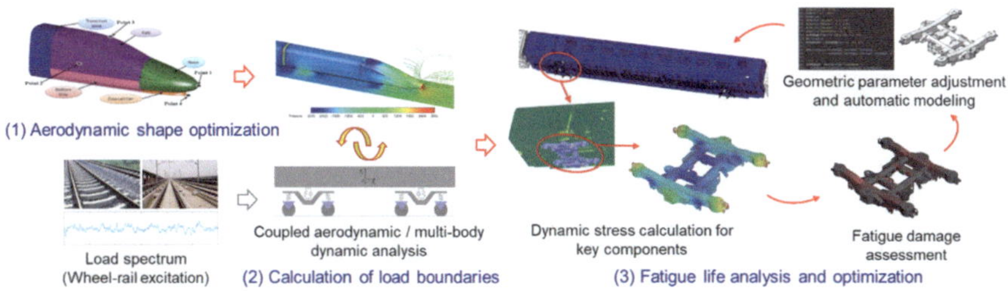

Fig. 7 Framework for life extension optimization (LEO) of high-speed trains (HSTs) based on system mechanics

by the rigid body displacement of the vehicle, and the relationship between the aerodynamics and structural dynamics could be treated as a unidirectional weak coupling. Thus, the aerodynamics/multi-body dynamics coupling analysis is first required to supply load boundary conditions, such as forces and accelerations, to the structural finite element simulation of a HST for fatigue life evaluations. As for the computational expenses of the individual disciplines, they are, in a descending order, aerodynamics analysis, structural dynamic stress calculation and rigid multi-body dynamics solving, and there are orders of magnitude differences between them. Consequently, it is much needed to improve the efficiency of aerodynamic analysis through massively parallel computation, and on the contrary, the simulations of other disciplines, such as multi-body dynamics analysis, structural dynamic stress solving and fatigue life calculation, may be carried out at a much smaller parallel scale, as appropriate.

Based on the mentioned considerations, we draw on the principle of system mechanics to propose a hierarchical and collaborative framework for LEO of HSTs as shown in Fig. 7, which organically integrates the multi-physics analysis and optimization with the HPC. The following is the detailed procedures using our framework, in which the aerodynamic optimization and structural LEO are performed in a hierarchical manner.

(1) Aerodynamic shape optimization. Firstly, we create a parametric aerodynamic model of a HST and choose m key parameters determining the aerodynamic shape, namely $\boldsymbol{X}_a = (x_{a1}, x_{a2}, \ldots, x_{am})^T$, as our aerodynamic design variables. Then, we calculate the aerodynamic forces, such as lifts and drags, under varying values of design variables to construct a surrogate model $F_a(\boldsymbol{X}_a)$, for aerodynamics with respect to the design variables. Finally, we adopt the constructed surrogate model to carry out optimization designs for getting the optimal aerodynamic shape $\boldsymbol{X}_a^*$. The entire process for aerodynamic optimization can be described as follows:

$$\begin{cases} find \boldsymbol{X}_a \\ \min F_a(\boldsymbol{X}_a) \\ s.t. \boldsymbol{X}_a \in \boldsymbol{\Omega}_a \end{cases} \tag{1}$$

where $\boldsymbol{\Omega}_a$ is the design domain of $\boldsymbol{X}_a$.

(2) Calculation of load boundaries. We develop a computational fluid dynamics (CFD) model of a three-car marshaling HST with the optimal shape $\boldsymbol{X}_a^*$, and combine it with the multi-body dynamics model for coupled aerodynamic / structural calculations to achieve the dynamic responses of the HST under the joint effect of aerodynamics and track excitations, thereby determining the dynamic load boundary conditions $\boldsymbol{F}_b(t) = \left(F_{b1}(t), F_{b2}(t), \ldots, F_{bp}(t)\right)^T$ for key components like vehicle bodies, bogie frames and traction pins, where p is the number of load boundary conditions.

(3) Fatigue life analysis and optimization. We first construct the geometric parametric models of key components of a HST and choose n geometric parameters $\boldsymbol{X}_s = (x_{s1}, x_{s2}, \ldots, x_{sn})^T$ as our structural design variables. We then use the automatically generated finite element models for different design variable values of key components to calculate their stress–time histories, which are converted into load spectra, namely the magnitude—frequency information of stresses, by using the rainflow counting method, under the dynamic load $\boldsymbol{F}_b(t)$, so as to obtain the corresponding fatigue life distributions of key components with the help of linear cumulative damage theory and S–N curves of materials. We eventually utilize the optimization algorithms to find the optimal structural solutions for key components of the HST with the largest fatigue life. The aforementioned life extension problem of structures could be expressed as follows:

$$\begin{cases} find\boldsymbol{X}_s \\ \max Life(\boldsymbol{X}_s, \boldsymbol{F}_b(t)) \\ s.t.\boldsymbol{X}_s \in \boldsymbol{\Omega}_s \\ G(\boldsymbol{X}_s) \le 0 \end{cases} \tag{2}$$

where $\boldsymbol{\Omega}_s$ is the design domain of $\boldsymbol{X}_s$; $\boldsymbol{G}(\boldsymbol{X}_s) = \left(G_1(\boldsymbol{X}_s), G_2(\boldsymbol{X}_s), \ldots, G_q(\boldsymbol{X}_s)\right)^T \le 0$ denotes the constraints, such as mass and stiffness constraints, and q is the number of constraints.

Bear in mind that when the aerodynamic shape and structural geometry remain the constant, the above framework for LEO of key components of HSTs degenerates into a system-mechanics-based procedure for fatigue life analysis.

3 Development of "Numerical HST Version 1.0" Software Platform

Based on the breakthrough of million-core parallel algorithms for aerodynamic and coupled aerodynamic/structural analyses, we independently develop the core functional modules used for aerodynamic solving, structural parametric modeling, structural finite element analysis, fatigue life calculation and structural multi-objective optimization according to the workflow of system mechanics-based LEO, and then integrate the high-performance numerical simulation software for LEO of HSTs ("Numerical HST Version 1.0") [30] and construct the corresponding industrial database.

3.1 General Architecture Design of Software Platform

As illustrated in Fig. 8, the general architecture of the "Numerical HST Version 1.0" software platform is composed of four layers, which are, from bottom to top, the fundamental resource layer, application support layer, application function layer and user interface layer. In particular, the fundamental resource layer provides the most basic hardware facilities and software resources for the whole platform, including the basic algorithm library, standard solvers and other modules. The application support layer offers requisite supports for the application function layer, enabling it to more conveniently and securely use the underlying infrastructures, such as the job operation, data handling, resource scheduling, and in-process monitoring. The application function layer is the pivotal link to perform massively parallel numerical simulations, consisting of modules for pre-processing, multi-physics field solving, and post-processing. The user interface layer assumes the human–computer interaction function of the whole software platform and provides an integrated solution for each module, which can be seamlessly integrated with the pre- and post-processing modules and conveniently call all major solvers with the characteristics of simplicity, ease of use, customizability and specialization.

The development of the "Numerical HST Version 1.0" follows the design philosophy of modularity, service-oriented component model, dynamization, and scalability. The application functions of the software are strictly divided into modules, which are implemented in the form of plug-ins. The plug-ins are uniformly governed by the plug-in manager, which enables non-invasive extension of new functions by adding plug-ins to the system. We demonstrate in Fig. 9 the user interface of the "Numerical HST Version 1.0" software, which is developed using the JAVA language and composed of functional domains such as the toolbar, shortcut button, project navigation, text editing, wizard view for control files, file list, and job monitoring window.

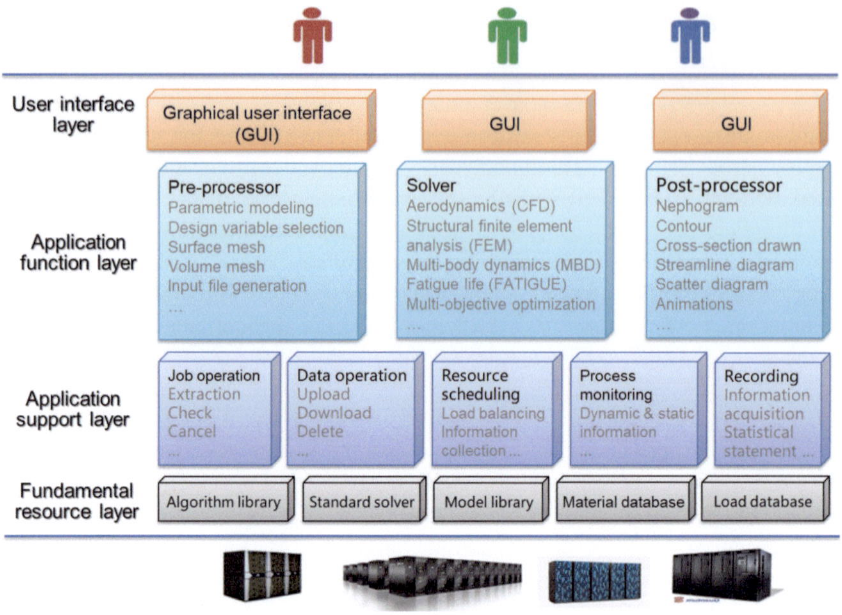

Fig. 8 Schematic diagram of the general architecture of the "Numerical HST Version 1.0" software

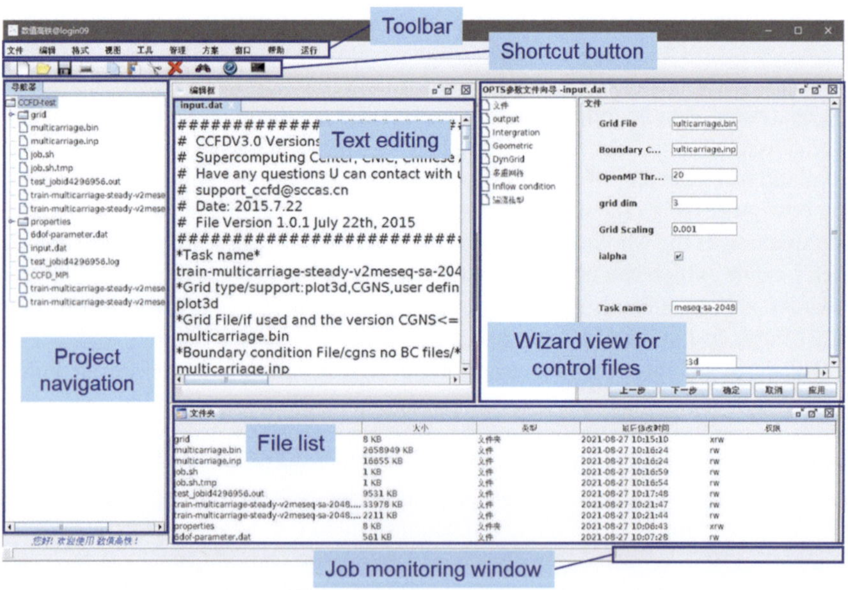

Fig. 9 User interface of "Numerical HST Version 1.0" software and its functional layout

3.2 Development of Core Functional Modules

To meet the full-process needs for LEO design of HSTs, we have independently developed modules for pre-processing, multi-physics analysis and optimization, as well as post-processing (as shown in Fig. 10), which constitute the core of the application function layer of the "Numerical HST Version 1.0" software.

(1) **Pre-processing module**

In order to accommodate the changes in structural geometry during LEO, we develop the xFreeCAD extended functional function library for geometric parametric modelling and finite element mesh generation of major components using Python language, and integrate it into the "Numerical HST Version 1.0" software platform as a pre-processing module. The main functions of this extended library include simplifying the modelling process and setup, adding complex modelling functions, defining design variables, adding meshing control functions, creating node and element groups, defining force elements and constraints required for finite element calculations, and automatically generating the input files for finite element calculations. In practice, we could utilize the pre-processing module to rapidly tune the values of design variables and automatically generate the finite element models of the major load-bearing structures, such as the vehicle body, bogie frame and axles, for fatigue life analysis and optimization tasks.

(2) **Multi-physics analysis and optimization module**

In terms of multi-physics analysis and optimization, we develop four functional modules, namely CCFDv3.0 aerodynamic parallel solver, FEAP structural finite

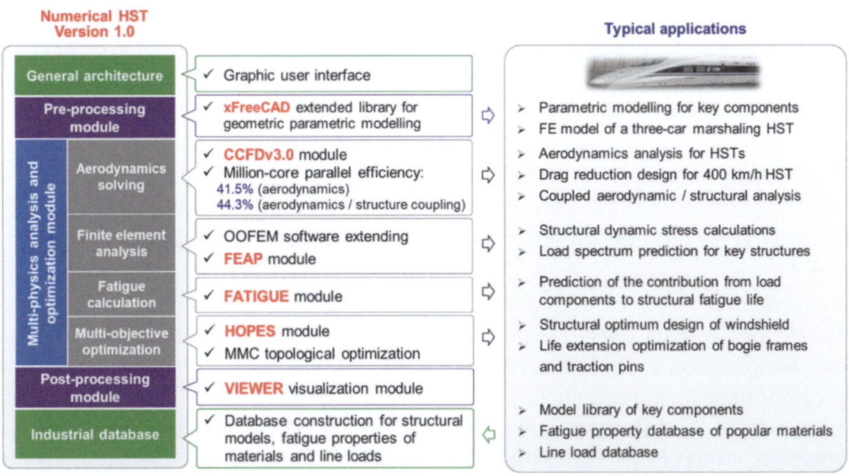

Fig. 10 Core functional modules of "Numerical HST Version 1.0" software and their applications in the safety and reliability design of HSTs

element solver, FATIGUE structural fatigue life calculator and HOPES structural multi-objective parallel optimizer, which are used to enable the multi-physics collaborative analysis and optimization required by the life extension design of HSTs. Among them, the **CCFDv3.0 module** [31] is used for aerodynamics solving, which is the most computationally expensive aspect in the LEO task. By designing the underlying data structure, computing mode, parallel algorithms, load balancing model and so on, we refactor the source code of CCFDv3.0 to make it available for domestic heterogeneous platforms with multi-layer storage structure and multi-layer nested computing cores, thereby enabling the efficient computational capability with millions of cores in parallel. Furthermore, to tackle the kinematic boundary problem involved in coupled aerodynamic/structural analysis, we create efficient and robust algorithms for hole digging and hole surface optimization under a distributed environment, thus breaking through the parallel overset grid technique. In combination with a multi-body dynamics calculation procedure, the CCFDv3.0 module possesses the capability of massively parallel solving for aerodynamics/structure coupling [32].

The **FEAP module** is intended to fulfill the functionality of structural finite element simulations and supply stress/strain data for the structural fatigue life analysis module. At present, the FEAP module is able to handle more than 10 types of elements, such as solids, beams, trusses, springs and mass points, and can support common constraints and loading conditions, as well as the solution of linear, nonlinear, explicit/implicit dynamics and frequency problems. This module is developed with the object-oriented programming language C++ and support more than 10 types of extensions including elements, loading, solvers and materials. The solver integrates linear algebra libraries such as PETSc and SPOOLES to cover the needs of finite element calculations for key load-bearing components of HSTs.

The **FATIGUE module** [33] uses the stress/strain data obtained from the structural finite element analysis to generate the load spectra, which are combined with the material fatigue model to calculate the fatigue life of key components for their LEO. This module is developed using the C++, and the input control takes place in the form of command line and script files to adapt to HPC. Currently, the FATIGUE module can read in files of finite element models, load histories and stresses in a variety of formats, and support commonly used 2D and 3D elements, the combination, association, editing and interpolation of load history and stress files, as well as different rainflow counting methods, fatigue stress data forms and fatigue calculation models.

The **HOPES module** is a generalized multi-objective parallel optimization framework for engineering structures. This module is developed in Python, and for various optimization problems, it defines a unified input interface and form to facilitate data exchanging between different modules. The HOPES module calls the script files from pre-processing, FEAP and FATIGUE modules in conjunction with optimization algorithms to automate the iterative process of the LEO for critical components. Moreover, the module also integrates a variety of optimization strategies such as COBYLA, BOBYQA, MMA, PRAXIS and Nelder-Mead to meet the requirements of applications for optimization problems in different disciplines.

(3) **Post-processing module**

In order to display the numerical simulation results in a more intuitive, visual and vivid way, we specifically develop the VIEWER visualization module [34] for post-processing of results from multi-physics solutions, such as aerodynamics, structural responses and fatigue lives, generated by each module during the LEO process. This module is developed in C++ with excellent scalability, and the interface, which can be used across platforms, is developed using wxWidgets. With regard to display and operation, the post-processing module supports the grouping of data in various presentation styles, the flexible setting of point, line and surface showings for models, the concurrent display of multiple models and model parts, the modification and analysis of model data through expressions, the dynamic visualization over multiple time steps, the 3D display in various modes, and the export of the display results as images or animations.

With the help of the above modules, users can conveniently accomplish many simulation and optimization tasks relating to HSTs, such as aerodynamic analysis and optimization, aerodynamics / multi-body dynamics coupling calculation, structural dynamic stress simulation, structural fatigue life estimation, LEO and so on.

3.3 Industrial Database Construction

With our years of accumulations in the research of HSTs, we have constructed a model library for major components of HSTs, a fatigue property database of popular materials as well as a line load database to facilitate our "Numerical HST Version 1.0" software platform to better serve the industrial applications.

(1) **Model library of key components**

We demonstrate in Fig. 11 the structural model library created for the key components of HSTs, including the major structures such as the coupler box, vehicle body, cross-beam, luggage rack, sleeper beam, windshield, air conditioner mounting bracket, pantograph, gearbox, bogie frame, motor hanger, anti-snake shock absorber seat, wheelset, cowcatcher, sanding device, primary spring, brake hanger, brake disc, traction pin, axle-box body, together with the structural finite element model of a three-car marshaling HST composed of these key components.

(2) **Fatigue property database of popular materials**

To meet the needs of fatigue life prediction for structures with different configurations and materials, we establish a fatigue property database of popular materials for users to choose from when using the FATIGUE module. This database gathers various theoretical models of mean stress correction for fatigue life prediction such as linear correction, FKM guideline correction, Smith–Watson–Topper correction,

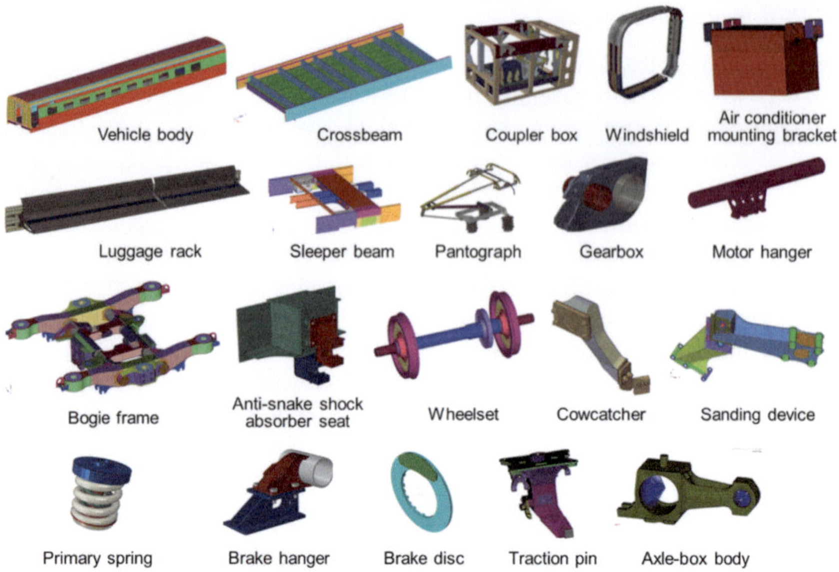

Fig. 11 Structural model library for major components of HSTs

Gerber Parabola correction, Soderberg correction, Modified Goodman correction and bilinear correction, as well as fatigue property data files for more than 50 commonly used materials defined in JSON format.

(3) **Line load database**

We sort out the force–time histories acting on the bogie springs, bogie rotary arms, traction rods and gearbox suspender, the stress–time histories at the fatigue-prone spots of bogie frames, and the track irregularity spectra from field tests to form a line load database. All these load data are obtained from the daily operations of both CRH 380A and CR 400AF EMUs on various routes such as Beijing~Guangzhou, Beijing~Shanghai, Beijing~Chengdu, Beijing~Tianjin and Beijing~Taiyuan, which may offer abundant load boundary conditions for the fatigue life estimation and optimization of key structures of an HST.

4 Typical Application Demonstrations for HSTs

We have completed the deployment and testing of the "Numerical HST Version 1.0" software on supercomputers, such as "Sunway Taihulight" and "SunRising-1", and in the computing environment of CRRC Sifang Co., Ltd. As shown in Fig. 10, the software has been applied to the safety and reliability design of HSTs. Here, we will focus on the applications of the software in several typical tasks such as aerodynamic surface modification design for 400 km/h HSTs, prediction of the respective contributions of load components on the bogie frame to its structural fatigue life, and LEO of the key components of HSTs.

4.1 Design of Aerodynamic Surface Modification for 400 km/h HSTs

As contrasted with an overall aerodynamic shape optimization, it is much more economical to improve the aerodynamic performance of an existing HST by reshaping the surface of the localized area. Hence, we attempt to utilize the CCFDv3.0 module provided by the "Numerical HST Version 1.0" software to carry out an aerodynamic analysis of the localized surface modification of a 400 km/h HST. Drawing on the mechanism of drag reduction on non-smooth surfaces of living organisms, the aerodynamic drag of the train may be effectively reduced by arranging pits or grooves on the surface of the bulkheads near the bogie frame. The aerodynamic drag reduction effect of a non-smooth surface structure is heavily dependent on its topology, such as the shape, size and distribution of the microstructure. We examine the drag reduction capacity of various patterns of surface pit and groove structures, and find that the greatest drag reduction could be achieved when hemispherical pit structures with a diameter $D = 20$ mm and a distribution spacing $R = 40$ mm, as illustrated in Fig. 11, are arranged on the surfaces of all bulkheads near the bogie frames. With this scheme, the overall drag of a three-car marshaling HST is reduced by 2.79% in comparison to the prototype, and is even expected to reach 5.34% if this modification is applied to an eight-car marshaling HST. In particular, our research also reveals that surface modifications in the bogie region may result in much greater drag reduction than those in the windscreen and pantograph regions. Our research may serve as a guidance for advancing the engineering application of the drag reduction technology of surface biomimetic modification to 400 km/h HSTs (Fig. 12).

4.2 Prediction of the Contribution of Load Components to Structural Fatigue Life

In order to identify the contribution of the three dominant external loads, namely aerodynamics, gravity and wheel-rail excitation, to the fatigue life of the bogie frame during the operation of an HST, we follow the procedure for fatigue life analysis supplied by the "Numerical HST Version 1.0" software to calculate the dynamic responses and the resulting fatigue damages of the bogie frame under multiple loads, thereby determining their respective contributions to the structural fatigue life. Here, we consider the following six working conditions for a train running at a speed of 300 km/h for investigation: (1) G—gravity, (2) GT—gravity + wheel-rail excitation,

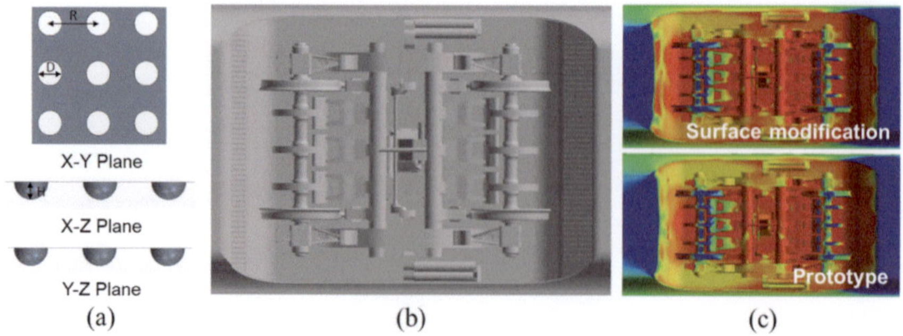

Fig. 12 Biomimetic modification design on the surface near the bogie frame: **a** size and distribution spacing of the hemispherical pit structure; **b** set-up of pit structures on the bulkhead surface near the bogie frame; **c** comparison of aerodynamic static pressure distributions in the vicinity of the bogie frame before and after surface modification

(3) AmG—aerodynamics in the open air + gravity, (4) AmGT—aerodynamics in the open air + gravity + wheel-rail excitation, (5) AsG—aerodynamics through the tunnel + gravity, and (6) AsGT—aerodynamics through the tunnel + gravity + wheel-rail excitation. We employ the developed multi-physics computational model of the three-car marshaling HST to obtain the stress–time histories along with the resulting fatigue damages at the positioning node of the rotary arm, a vulnerable point for fatigue of the bogie frame under the aforementioned loading conditions, through a series of numerical analyses involving aerodynamics solving, multi-body dynamics calculations, structural finite element simulations and fatigue life estimations.

We show in Fig. 13 the stress–time histories at the positioning node of the rotary arm for different operating conditions. As can be observed from Fig. 13, the gravity load is the main determinant of the magnitude of the mean stress, while the amplitudes of stress fluctuations due to aerodynamics in the open air and wheel-rail excitations are relatively small. As the degree of structural fatigue damage is mainly reliant on the stress amplitude and its cycling number, in terms of load-induced stress fluctuations, the sharpest fluctuations in stresses are caused by the wheel-rail excitations, followed by the aerodynamics in the open air, while the gravity load can hardly lead to stress fluctuations. As a result, the wheel-rail excitation is the dominant contributor to the number of strain cycles at the position of our concern. In Fig. 13b we further compare the impacts of aerodynamics in the open air and through the tunnel on the dynamic stress levels at the investigated position. It can be seen that the dynamic stress level caused by the aerodynamics through the tunnel is greater than that arising from the aerodynamics in the open air, but the fluctuating frequencies are similar for both. In Table 1 we present the relative values of fatigue damages and fatigue life estimations regarding the positioning node of the rotary arm for six working conditions, indicating that the load components, ranked from high to low in terms of their respective contributions to the fatigue life at the position of our interest, are wheel-rail excitation, aerodynamics through the tunnel, aerodynamics in the open air and gravity (Table 2).

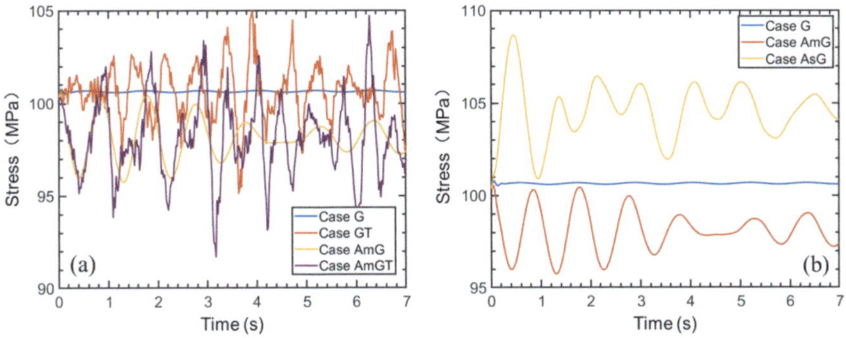

Fig. 13 Stress–time histories at the positioning node of the rotary arm of the bogie frame under various working conditions: **a** gravity, gravity + wheel-rail excitation, aerodynamics in the open air + gravity, and aerodynamics in the open air + gravity + wheel-rail excitation; **b** gravity, aerodynamics in the open air + gravity, and aerodynamics through the tunnel + gravity

Table 2 Fatigue damages and lives at the positioning node of the rotary arm of the bogie frame under different operating conditions

Case	Load	Fatigue damage	Fatigue life (cycles)
G	Gravity	6.6×10^{-13}	2.41×10^{18}
GT	Gravity + wheel-rail excitation	2.7×10^{-5}	6.72×10^{11}
AmG	Aerodynamics in the open air + gravity	5.4×10^{-7}	2.4×10^{12}
AmGT	Aerodynamics in the open air + gravity + wheel-rail excitation	1.05×10^{-4}	1.48×10^{11}
AsG	Aerodynamics through the tunnel + gravity	3.8×10^{-6}	3.71×10^{11}
AsGT	Aerodynamics through the Tunnel + gravity + wheel-rail excitation	1.13×10^{-4}	1.28×10^{11}

4.3 LEO of Key Components

The fatigue reliability of the bogie frame, a critical load-bearing component connecting the wheelsets and the vehicle body, is crucial to the operating safety of HSTs. Here, we take the bogie frame of the CRH 380A EMU as the prototype, and adopt the "Numerical HST Version 1.0" software to perform LEO for it. Based on the analysis of the dynamic stresses and fatigue damage distributions of the bogie frame subjected to an actual load spectrum from the line load database, we identify two sites prone to fatigue, the upper plate of the steel spring seat and the positioning node of the rotary arm, and choose the structural thicknesses of these two sites as our design variables. Following the LEO, we suggest a bogie frame structure with increased thicknesses of 23 mm for the upper plate of the steel spring seat and 36 mm for the positioning node of the rotary arm as the optimized scheme, which results in a calculated fatigue life of 1.73×10^{12}, 168% longer than the initial one. We show the clouds of the calculated fatigue damage distributions for the bogie frame before and after optimization in Fig. 14.

The traction pin is another key load-bearing component linking the vehicle body and the bogie frame. Here, we still use the traction pin of the CRH 380A EMU as the prototype, and calculate the dynamic stresses and the resulting fatigue damages of the initial structure under a given load spectrum from the indoor experiment with the help of the "Numerical HST Version 1.0" software. We take the thicknesses of the front and rear side plates, and the right and left side plates of the traction pin, which have a significant influence on the damage of the structure according to our preliminary analyses, as the two design variables to improve the fatigue life of the traction pin by optimizing their values. As a result of the LEO, we recommend a traction pin structure with an increased thickness of 12 mm for the right and left side plates and an unchanged thickness for the front and rear side plates as the optimized scheme, which results in a calculated fatigue life of 8.30×10^6, 54% longer than the initial one. We demonstrate in Fig. 15 the calculated distributions of fatigue damage for both initial and optimized traction pins. To further validate the feasibility of our scheme, we carry out fatigue strength tests on the optimized traction pin. The results of our tests show that the optimized traction pin is found no cracks by a non-destructive detection after 1.3×10^7 load cycles, and thus meets the requirements for fatigue strength.

The above work clearly indicates that the "Numerical HST Version 1.0" software is capable of performing LEOs of structures. By refining the modelling of structural welds and joints, enlarging the design variables and tailoring the optimization model, we could realize the life-extension design of bogie frames and traction pins for different demands. The proposed optimization framework may then be applied to the fatigue life analysis and optimization for many other engineering structures.

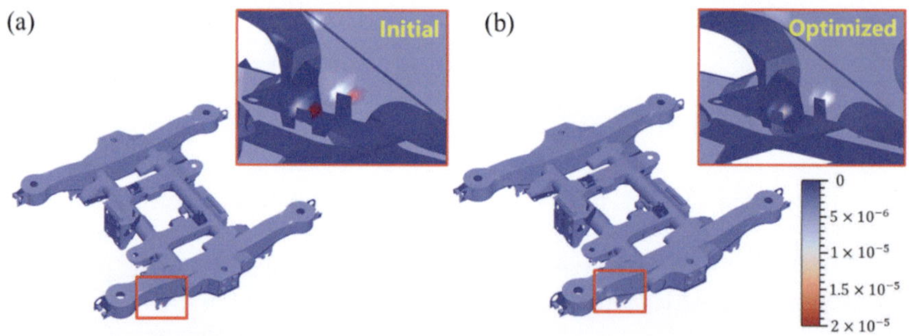

Fig. 14 Distributions of fatigue damage for the bogie frame: **a** before optimization; **b** after optimization

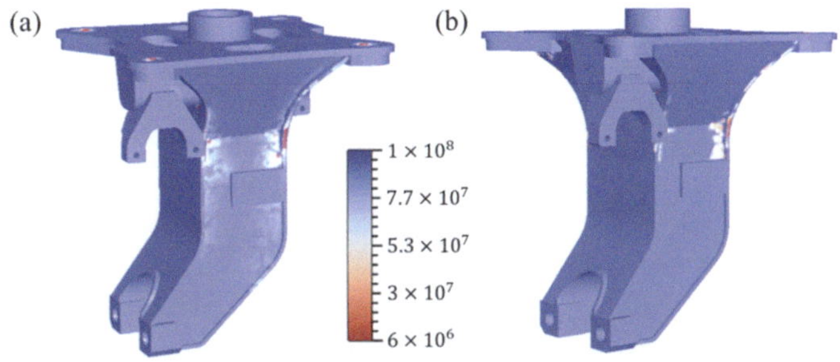

Fig. 15 Distributions of fatigue damage for both **a** initial and **b** optimized traction pins

5 Conclusions and Prospects

With the joint support of the "HPC" special project of the National Key Research and Development Program sponsored by the Ministry of Science and Technology of China, the Strategic Priority Research Program of the Chinese Academy of Sciences, and the Special Project for Informatization of the Chinese Academy of Sciences, we develop the high-performance computational theories and models for LEO of HSTs oriented towards exascale computing, and achieve a million-core parallel computing for coupled aerodynamic/structural analyses by breaking through the massively parallel algorithms for aerodynamics solving and parallel overset grid techniques applicable to domestic heterogeneous platforms. In a strongly scalable pattern, our algorithm achieves an efficiency of 41.5 and 44.3% for 1.04 million cores in parallel for aerodynamic and coupled aerodynamic/structural calculations, respectively, using 52,000 cores in parallel as the benchmark. From the perspective of system mechanics, we propose a hierarchical and collaborative LEO method for HSTs, which organically links the multi-physics analysis and optimization with the HPC, to solve the application problems arising from the insufficient methods as well as limited computational scale and efficiency for whole-vehicle-level life assessment and design optimization of HSTs. We independently develop the core functional modules used for aerodynamic solving, structural parametric modeling, structural finite element analysis, fatigue life calculation and structural multi-objective optimization according to the workflow for a hierarchical and collaborative LEO of HSTs, and then integrate the numerical simulation software for LEO of HSTs ("Numerical HST Version 1.0"), and construct a model library for major components of HSTs, a fatigue property database of popular materials as well as a line load database. We have completed the deployment and testing of the software on typical supercomputers like "Sunway Taihulight" and "SunRising-1" as well as in the computing environment of CRRC Sifang Co., Ltd. Moreover, the developed software has been applied in the tasks for the safety and reliability design of HSTs, such as the design of aerodynamic surface modification for 400 km/h HSTs, the prediction of the respective contributions of load components on the bogie frame to its structural fatigue life, and the LEO for the key components of HSTs.

The development of the “Numerical HST Version 1.0” software follows the design philosophy of modularity and scalability from the outset, allowing for non-invasive extension of new functions by adding plug-ins to the system. As is known to all, the development of software is a process of repeated iterations and upgrades. Starting from the current version, we will continue to improve the processing capability of existing modules, strive to improve user experience and extend new functional modules, such as (1) developing non-traditional and efficient computational modules for engineering structures, that combines machine learning with traditional computational methods to enable efficient and high-precision mechanical analysis and optimization for complex structures; (2) improving the parallel calculation module for coupled aerodynamic/structural analysis by extending the functions of interface data information transfer and mesh deformation, thus achieving efficient solving for aerodynamics/structure coupling problems; (3) expanding the libraries of constitutive models and elements for novel materials and structures, thereby enhancing the corresponding simulation and analysis capabilities. We aim to gradually develop the “Numerical HST Version 1.0” software into a general-purpose engineering computing software capable of HPC through continuous investment, so as to provide an efficient simulation tool for multi-disciplinary analysis and optimization of large-scale engineering issues.

Acknowledgements This work was supported by the National Key Research and Development Program of China (2017YFB0202800), the Strategic Priority Research Program of the Chinese Academy of Sciences (XDB22020200), and the Special Project for Informatization of the Chinese Academy of Sciences (XXH13506-204).

References

1. Qian DP, Wang R (2020) Key issues in exascale computing (in Chinese). SCIENTIA SINICA Inform 50(9):1303–1326
2. Computer Network Information Center, Chinese Academy of Sciences. The supercomputing development report (2016–2020) of Chinese Academy of Sciences was released (in Chinese). http://www.cnic.cas.cn/zhxw/202112/t20211229_6329885.html. Last accessed 13 May 2022
3. China National Grid (2015) Comprehensive evaluation report on the development level of the national high-performance computing environment (2015–2020) (in Chinese). http://www.cngrid.org/yjcg/fzbg/201808/W020211014388411373175.pdf. Last accessed 13 May 2022
4. The Central People’s Government of the People’s Republic of China. The 14th five-year plan for economic and social development of the People’s Republic of China and the long-range goals for 2035 (in Chinese). http://www.gov.cn/xinwen/2021-03/13/content_5592681.htm. Last accessed 30 March 2022
5. China National Grid (2022) Latest supercomputer TOP500 ranking: AMD processor-powered US Frontier takes the first place with a performance of over one ExaFlops (in Chinese). http://www.cngrid.org/xwtz/xw/202205/t20220531_703157.html. Last accessed 31 May 2022
6. China State Railway Group Co., Ltd. (2022) In the first year of the 14th five-year plan, the railway handed over a dazzling transcript (in Chinese). http://www.china-railway.com.cn/xwzx/ywsl/202201/t20220109_119362.html. Last accessed 30 March 2022
7. Zhang WH (2015) Study on dynamics of coupled systems in high-speed trains (in Chinese). Strategic Study CAE 17(4):42–52

8. Yang GW, Wei YJ, Zhao GL et al (2015) Research progress on the mechanics of high speed rails (in Chinese). Adv Mech 45(1):201507
9. Zhai WM, Jin XS, Zhao YX (2010) Some typical mechanics problems in high-speed railway engineering (in Chinese). Adv Mech 40(4):358–374
10. Ding SS, Chen DW, Liu JL (2021) Research, development and prospect of China high-speed train (in Chinese). Chin J Theor Appl Mech 53(1):35–50
11. Xu HM, Zhang P, Guo Z (2019) Study review on operation safety of high speed train in gales (in Chinese). China Railway 5:17–26
12. Wang MZ, Mei YG, Jia YX (2017) Simulation of aerodynamic effects generated by a high-speed train passing through a tunnel with overset grid method (in Chinese). Chin J Appl Mech 34(3):589–595
13. Sun ZX, Yao YF, Guo DL et al (2021) Research progress in aerodynamic optimization of high-speed trains (in Chinese). Chin J Theor Appl Mech 53(1):51–74
14. Li T, Dai ZY, Liu JL et al (2021) Review on aerodynamic drag reduction optimization of high-speed trains in China (in Chinese). J Traffic Transp Eng 21(1):59–80
15. Guo Y, Guo DL, Yang GW et al (2021) Moving model analysis of the slipstream of a long grouping high-speed train (in Chinese). Chin J Theor Appl Mech 53(1):105–114
16. Zhang WH (2021) Research on service simulation modeling and calculation method of high-speed train (in Chinese). Chin J Theor Appl Mech 53(1):96–104
17. Xu L, Zhai WM (2018) The vehicle-track stochastic model considering joint effects of cross-winds and track random irregularities (in Chinese). J Vibration Eng 31(1):39–48
18. Xu L, Zhai WM (2020) Stochastic analysis model for vehicle-track coupled systems under joint random irregularities of wheel treads and track irregularities (in Chinese). J China Railway Soc 42(2):79–85
19. Zhang YY, Sun SG, Wang BJ et al (2021) Study on the classification and verification of bogie frame loads of the high-speed EMU (in Chinese). J Mech Eng 57(6):156–163
20. Zhang YY, Sun SG, Yang GX et al (2020) Load characteristics and fatigue damage assessment of high speed train bogie frame (in Chinese). J Mech Eng 56(10):163–171
21. Sun JJ, Sun SG, Li Q et al (2021) Compilation method of test load spectrum for bogie frame (in Chinese). J China Railway Soc 43(6):29–36
22. Wang X, Hou Y, Sun SG et al (2021) Advances in key mechanical parameters for reliability assessment of high-speed train bearings (in Chinese). Chin J Theor Appl Mech 53(1):19–34
23. Xu SS, Huang GH, Jiang HB et al (2017) Discussion on the method to evaluate the vibration severity of the gear box in high-speed trains (in Chinese). Railway Qual Control 45(11):30–32
24. Yuan Z, Chen XJ, Wei YJ et al (2021) A segmented load spectrum model for high-speed trains and its inflection stress as an indicator for line quality. Int J Fatigue 148:106221
25. Jin Z, Lu ZH, Li HY et al (2019) Origin of high performance computing—current status and developments of scientific computing applications (in Chinese). Bull Chin Acad Sci 34(6):625–639
26. Yao SB, Guo DL, Sun ZX et al (2012) Multi-objective optimization of the streamlined head of high-speed trains based on the Kriging model. Sci China Technol Sci 55(12):3494–3508
27. Chen XJ, Yuan Z, Wei YJ et al (2023) A computational method on the load spectra of large-scale structures using data-driven learning algorithm. Sci China Technol Sci 66(1):141–154
28. Liang S, Zhang J, Liu XZ et al (2019) Domain decomposition based exponential time differencing method for fluid dynamics problems with smooth solutions. Comput Fluids 194(15):104307
29. Hu XD, Liang S, Yuan W et al (2019) Communication optimization algorithm of hybrid overlap grids (in Chinese). Comput Syst Appl 28(12):146–151
30. Computer Network Information Center, Chinese Academy of Sciences, Institute of Mechanics, Chinese Academy of Sciences. High-performance numerical simulation software for life extension of high-speed trains (NHST) V1.0: 2021SR2042068. 10 Dec 2021
31. Computer Network Information Center, Chinese Academy of Sciences. Massively Parallel CFD Simulation Software (CCFD) V3.0: 2019SR0764891. 23 July 2019

32. Computer Network Information Center, Chinese Academy of Sciences. Massively parallel CFD solver software based on multigrid/overlapped grid V3.0 (CCFDv3.0-MGOG): 2020SR1885027. 23 Dec 2020
33. Institute of Mechanics, Chinese Academy of Sciences. SIMECH fatigue life calculation software (SIMECH Fatigue) 1.0: 2018SR443875. 12 June 2018
34. Institute of Mechanics, Chinese Academy of Sciences. SIMECH visualization software for multiscale mechanics (SIMECH Viewer) 1.0: 2020SR0592824. 09 June 2020

Yujie Wei is a professor and doctoral supervisor at the Institute of Mechanics, Chinese Academy of Sciences (CAS). He is currently the director of the State Key Laboratory of Nonlinear Mechanics, Institute of Mechanics, CAS and the head of the Department of Modern Mechanics, School of Engineering, University of Chinese Academy of Sciences (UCAS). He was the recipient of the National Science Fund for Distinguished Young Scholars in 2014 and the Excellent Supervisor Award of CAS. He received his Ph.D. in Mechanical Engineering in 2006 from the Massachusetts Institute of Technology, his M.S. in 2000 from the Institute of Mechanics, CAS, and his B.S. in 1997 from Peking University. From 2006 to 2008, he held a postdoctoral position at the Engineering Division of Brown University, and was Assistant Professor (Tenure Track) from 2008 to 2009 at the Department of Mechanical Engineering, University of Alabama. Since 2010, he has been professor at the Institute of Mechanics, CAS. His research focuses on the deformation mechanisms of metallic polycrystalline and amorphous on the bulk-, micro and nano-multi-scale level. He has also done research on the interface mechanics of the nano crystalline, two-dimension materials, and service reliability of engineering structures. He has led research on several major projects such as the National Key Research and Development Program of China and the Strategic Priority Research Program of the CAS. As the corresponding author, he has published 1 monograph and more than 120 papers in international academic journals, such as Nature, Rev Mod Phys, Nature Mater, Nat Sci Rev, Nature Comm, PRL, PNAS, Nano Lett, Acs Nano, Adv Mater, J Mech Phys Solids, Acta Mater, PRB, and so forth.

Open Source Software Supply Chain for RISC-V Ecosystem

Yanjun Wu, Guanyu Liang, Siyang Tian, and Chen Zhao

Abstract Open source software has gradually become the mainstream mode for software co-development and distribution. The rapid growth of the number of open source software and its increasingly widespread applications have made it an important need in the current information industry to build high-quality open source software supply chains, and also a core competitiveness of the instruction set ecosystem. RISC-V, as a new generation of reduced instruction set architectures, compared with the mature and prosperous X86 and ARM ecosystems, needs to quickly build a high-quality RISC-V open source software supply chain for its ecosystem to be sustainable and prosperous. We first introduce the definition and characteristics of open source software supply chains, as well as the current status of the rapid development of the RISC-V instruction set. Then, the critical links in the open source software supply chain for RISC-V instruction sets, such as the general open source IP core, general basic software, open source code migration tools, open source software upstream communities, and ecological fragmentation management, are deeply analyzed. At the same time, we point out the potential intellectual property risks, software vulnerability risks, vulnerability propagation risks, and continuous maintenance risks that need to be addressed in building an open source software supply chain, as well as possible solutions under RISC-V. We believe that the combination of the RISC-V instruction set and the open source software supply chain will bring a new and more prosperous hardware and software ecosystem.

Keywords Open source software · Open source software supply chain · RISC-V · Instruction set ecosystem

Y. Wu (✉) · G. Liang · S. Tian · C. Zhao
Institute of Software, Chinese Academy of Sciences, Beijing, China

G. Liang
University of Chinese Academy of Sciences, Beijing, China

China's e-Science Blue Book 2023, https://doi.org/10.1007/978-981-99-8270-7_16

1 Open Source Software Supply Chain

Software has two significant characteristics: firstly, software is the product of human intellectual activities, and its creators are human beings. Automation cannot be achieved by importing, imitating, or modifying production equipment to increase production efficiency in the same way as a production line. Secondly, once the software product is developed, distribution only requires simple copying, with almost zero marginal cost. However, it usually requires continuous updating and maintenance to adapt to changing needs and maintain competitiveness. The former determines that human is the most critical factor in software production process, and that human resources consumption is often the biggest cost. The latter determines that the software industry's investment in human resources is long-term rather than one-time. Therefore, we can derive the first principle of software development: software should reuse code as much as possible with continuous iteration.

Open source software is the inevitable outcome of the first principles of software development mentioned above. In 1983, Richard Stallman launched the Free Software Movement, and open source software officially stepped onto the stage of software history. In 1998, Eric Raymond et al. founded the Open Source Initiative (OSI), gradually providing an official definition of open source software [1]. Open source software inherits the characteristics of software, but at the same time extended features such as free copying, distribution, modification of source code and large-scale distributed collaboration in the development process. This largely solved problems such as code reuse, coordination of human resources, and continuous product evolution in the software industry.

Open source software has become an unstoppable trend sweeping the world and is becoming the mainstream mode for software development, release, deployment and operation. According to the statistics of Forrester Research [2], more than 80% of commercial software uses open source components, and in fields such as the internet and finance, this number reaches over 95%. With this trend, the software development model has changed from independent development by individuals/organizations to collaboration by multiple individuals/organizations. With the support of software modular design, the development model of building customized software based on open source software modules published by others has gradually become mainstream. Open source operating systems are typical products of this model.

Due to the increasing scale of open source software, there are more and more obvious dependency and combination relationships between open source projects. For example, a typical open source operating system (such as Ubuntu and Android) is composed of tens of thousands of open source software from which is called upstream. These upstream open source software are widely reused and become the "raw materials" and "meta-components" of the software world. We refer to this supply chain relationship that exists in the field of open source software as an open source software supply chain. The process of building an operating system is essentially an integrated optimization process of the open source software supply chain.

The concept of a supply chain was first proposed in the field of enterprise management, and it can be traced back to Material Requirement Planning (MRP) in the 1960s. At that time, due to the low production capacity of enterprises, the supply–demand contradiction mainly focused on resources. MRP was mainly proposed to solve the planning problems between raw material inventory and product parts production with the minimum input and critical path as its basic starting point. Today, there are many different definitions of supply chain according to different target needs, application environment, etc. Among them, Martin Christopher, a British supply chain management expert, gave a highly accepted definition in 1998. He defined supply chain as "a network composed of multiple organizations, in which the organizations are interconnected with their upstream and downstream relationships, and they contribute value to the final customer product in the form of products or services in different production activities or processes".

Corresponding to the concept of traditional supply chain, an informal definition of open source software supply chain (OSSSC) can be expressed as follows: The open source software supply chain is an actual business system that involves all upstream open source software communities, source packages, binary packages, package managers, repositories, developers, maintainers, communities, foundations, etc. that are involved in the development and operation process, forming a supply chain network based on dependency, combination, hosting, guidance, and other relationships.

This is a relatively broad definition, which can also be called the definition of a generalized open source software supply chain. In contrast, the narrow definition of an open source software supply chain is equivalent to a Software Bill of Materials (SBOM), which only involves the dependencies, combinations and other relationships of the open source software itself (source packages or binary packages).

According to the first principles of software development mentioned earlier, the pursuit of maximum software code reuse and rapid iteration has resulted in tremendous changes in both software composition and development process. Nowadays, software development is more focused on agility and efficiency. Basic functions usually prioritize reuse of related open source software and developers only need to make necessary extensions and improvements. It has been proved that this mode can effectively reduce the cost of software development, accelerate software upgrade, lower the development barrier and speed up the demand response. This has also led to the emergence of a vast supply chain for open source software. According to Octoverse annual report [3], as of the end of 2019, more than 3.6 million open source repositories with dependencies on the top 50 open source projects, with an average of 180 package dependencies per open source project. Taking the operating system as a typical example of a large complex system software, we can show the organizational structure and complexity of the open source software supply chain. In terms of organizational structure, an operating system is a collection of system programs that manage computer hardware resources and software resources, including kernel and other system level tools. Operating system distributions derived from the Linux kernel, such as Ubuntu, CentOS, Android, are collectively known as Linux distributions. They organically integrate a large number of open source software with different functions into the Linux kernel in the form of software packages to meet the different requirements from end users. DistroWatch [4] and Repository statistics [5] show that commonly used Linux distributions require tens of thousands of software packages to support their functionality and ecosystem. For example, Ubuntu 18.04 involves 29,207 packages and Debian Unstable 32,453 packages. Even leaner systems built through customization also contain nearly several hundred packages. From the above data, it is obvious that under the open source collaboration mode, the supply chain relationship between software is both widespread and complex globally. The construction and maintenance of open source software supply chain has become a common challenge in the global open source field.

2 RISC-V Development Status

Instruction Set Architecture (ISA), is the boundary and interactive standard between hardware and software in a computer system. It is the basis and ecosystem origin of system software and hardware processors, and determines the way of software and hardware collaboration. The current mainstream instruction sets are X86 and ARM. One of the most significant characteristics of X86 processors, led by Intel, is a rich and comprehensive set of instructions covering a wide variety of possible operations. When programmers build software based on one instruction set, each instruction in which has a basic hardware implementation that enables high computational efficiency. However, this also leads to the complexity of CPU design. In addition, power consumption increases the complexity of CPU design, making it difficult to optimize for specific embedded scenarios. As a representative of the Reduced Instruction Set Computer (RISC), ARM instruction set effectively makes up for the gap of X86. It starts from embedded scenarios and gradually extends to the server field through the explosion of smartphones.

Both X86 and ARM benefit from open source software. X86 processors, for example, are well suited to Linux, making them the dominant player in the server processor market. Similarly, before the emergence of the open source Android operating system, ARM processors were concentrated in the embedded fields such as microcontrollers. However, with the birth of Android, ARM processors become the indisputable mainstream of the mobile terminal market, and the market scale of the entire ecosystem expanded from approximately tens of billions of dollars to trillions of dollars.

However, both X86 and ARM instruction sets face a common problem in that they are proprietary instruction sets owned by commercial companies. Many technologies in the computing field, such as TCP/IP protocol, are developed based on open standards, forming a pattern of open competition and cooperation. But the ISA is proprietary, meaning that others cannot access or modify them without high licensing fees. Given that ISA is the starting point of an ecosystem, it is challenging to imagine a future where the open source software ecosystem will be unrestricted on proprietary ISA. Likewise, an open source software supply chain based on private instruction sets cannot fundamentally address risks such as intellectual property violations, suboptimal code quality, security vulnerabilities, and long-term maintenance. Open source developers will face a variety of technical obscurity, or the commercial terms of the company's intellectual property and confidentiality.

Under this background, the RISC-V instruction set stepped onto the historical stage and achieved rapid development. In 2011, the University of California, Berkeley released RISC-V, the fifth generation of RISC instruction set for the design and innovation demands of the computing field in the next 50 years. The most significant feature of RISC-V is open source. The RISC-V specification adopts the "Creative Commons Attribution 4.0 International License". The RISC-V International (RVI, formerly the RISC-V Foundation) is responsible for the definition and maintenance of the specification, which has now moved to the permanently neutral country of Switzerland. Although there might be patents in microarchitecture design and peripheral circuit design under RISC-V instruction set specification, RVI is gradually solving this problem by constructing a shared patent pool.

In addition, RISC-V also features simplicity and modularization. RISC-V instruction sets are divided into base instruction sets, standard extended instruction sets, and other custom instruction sets. Among them, base and standard extended instruction sets are reviewed, ratified, and maintained by RVI, keeping the number of instructions as few as possible. This leaves plenty of room for optimization in chip design and deployment. For example, RV32I, the 32-bit integer instruction set in the base instruction set, contains only 47 instructions and will never be changed. Other modules, such as multiplication instruction set module M and single precision floating point instruction set F, only need to be implemented according to the actual need of the hardware. Custom instructions for some complex instruction sets can be combined with multiple simple instructions, eliminating the need for separate hardware design. Based on these features, RISC-V is highly suitable for flexible scenarios, which can be customized according to specific needs of functions, performance, and power consumption.

In recent years, the global RISC-V ecosystem has witnessed rapid growth. As of December 2021, RVI enterprise members come from more than 70 countries and regions, and the number of enterprise members has increased to over 2,000, more than 50 times the number compared to the initial startup in 2015, covering education, research, chip, software, security, finance, and other industries. By December 2021, more than 2 billion RISC-V cores have been shipped globally, and nearly 100 different types and models of RISC-V chips have been applied in multiple industries, such as the Internet of Things, industrial control, and machine learning. According to the survey report "RISC-V Market Analysis: Emerging Markets" released by Semico in 2019, it is estimated that the total consumption of RISC-V chips will reach 62.4 billion by 2025, with a growth rate of 146% [1]. RISC-V has become one of the three major instruction set ecosystems alongside X86 and ARM.

In China, several research institutes and companies have started the development of RISC-V. As an active force of RISC-V ecosystem, the Institute of Software of Chinese Academy of Sciences (ISCAS) has made continuous and core contributions to RISC-V compilers, simulators, JavaScript runtime and fundamental libraries, and more [6]. The "Xiangshan" RISC-V processor [8] was proposed in June 2021 by the Institute of Computing Science of the Chinese Academy of Sciences (ICT-CAS) [7]. The first-generation architecture "Yanqi Lake" has a clock frequency of up to 1.3 GHz, and the chip design adopts an out-of-order processing core with 11-stage pipelines, 6 dispatch units and 4 access components, mainly targeting ARM's A72 product. At the 2021 Yunqi Conference in October, Ali T-Head open sourced 4 types of RISC-V XuanTie processors, and simultaneously with a full set of software tool chain [1]. The largest open source community in China, openEuler, began to support RISC-V architecture at the beginning of its establishment, with the RISC-V SIG group to carry out RISC-V software packages porting, compilation and construction. The open source project OpenHarmony for the Internet of Things also initiated a migration adaptation for RISC-V. According to the website of RVI, as of August 2021, 11 of the 13 premier members in RVI were from China, and 2 were from the United States. Among the 208 strategic members, 72 were from China, 78 from the United States, and 47 from Europe. Nine of the 17 board members are from China and 6 from the United States, while three of the four development partners were from China and one was from the United States.

In the broad research and business prospects of RISC-V, many opportunities waiting to be explored. Building a high-quality open source software supply chain on top of the RVI ecosystem processes is important and timely.

3 Critical Factors in Building RISC-V Open Source Software Supply Chain

Software is the catalyst and glue of the instruction set ecosystem. The RISC-V instruction set, which originates from the open-source philosophy, naturally complements open source software. As early as 2012, in the second year of RISC-V's birth, UC Berkeley researchers have ported common system software to RISC-V, such as GCC, Clang, Linux. In order to improve the programming efficiency of the CPU IP core, Chisel, a hardware programming language with high-level logic abstraction, was also developed. After being compiled, Chisel can be compatible with Verilog. More importantly, it achieved the goal of code reuse, by introducing the traditional programming concepts such as function and object-oriented programming. While improving the programming efficiency, it also lowers the technical barrier, make it easier for developers to get started.

However, an open source software supply chain of an instruction set ecosystem is much more than that. The RISC-V open source software supply chain is essentially a large-scale software knowledge graph starting from instruction set. This graph shows connections between various components in the open source world. Each package entity is a vertex, and each edge represents an association between two packages. Just as highways connect different cities, many highways together form the open source world's transportation network. Whether developers, users, maintainers or managers, as long as they can drive according to the traffic rules on the highway, they can quickly and efficiently shuttle through each traffic node. As a result, the originally scattered and fragmented world of open source is glued together. The construction and evolution of software for large complex systems also become well-organized.

There are some critical factors in the RISC-V ecosystem that are of utmost importance in building an open source software supply chain.

(1) A Landmark Open Source IP Core for General-purposed CPU

Although the modular design philosophy of RISC-V allows unlimited combination of instruction modules, the design of the general-purposed processor also has a high technical barrier, and most of the software and hardware development companies still need one or more "classical" chips. When such a landmark and excellent chip appears, it will bring obvious domino effect, and the subsequent embedded hardware, supporting software and the entire commercial process will be generated. UC Berkeley's Rocket and BOOM provide the basics for getting started, but there is still a long way to go from a truly functional, general-purposed, classical open-source IP core. It should be emphasized here that due to the widespread use of programming languages such as Chisel and Verilog, the development process of open-source IP cores can fully adopt mature models and platforms of open source software, such as DevOps, continuous integration (CI), continuous testing (CT). In some sense, IP core source code also constitutes an important part of the open source software supply chain.

(2) General Fundamental Software

After RISC-V IP cores, there should be supporting fundamental software, such as assembler and compiler. At present, multiple compilers and fundamental libraries have been supporting for RISC-V, such as GCC, LLVM, QEMU and GLIBC. Similarly, more and more operating systems support RISC-V, such as Fedora, Debian and openEuler. Even so, there are still many chanllenges in the actual use. Take the openEuler operating system being migrated and adapted by ISCAS as an example. Among more than 8,000 community-maintained software packages (which can be compiled, built and run on X86 and ARM), only more than 2,400 software packages have been successfully compiled, built and run on RISC-V after nearly a year of efforts. If considering the over 30,000 software packages of Fedora and Debian, the workload of subsequent migrations can be estimated, which is by no means an easy task. In this situation, it is necessary to prioritize the transplantation of the most universal and important software packages. Such as OpenCV, a library used extensively in the field of machine vision and artificial intelligence, or OpenJDK, a programming toolkits and runtime environments used extensively in enterprise-level commercial software.

(3) Open Source Code Migration Tools

For software running on the RISC-V, it is not feasible or cost-effective to implement everything from scratch. A more rational and efficient way is to automate or semi-automate the migration of existing broadly available software to the RISC-V platform. The open source community currently has a large number of open source projects that are potential candidates for migration to the RISC-V platform. Therefore, it is necessary to have some quick and practical source code level migration tools to improve the efficiency of software migration and the reuse rate of open source code.

(4) Upstream Community of Open Source Software

A significant sign that an instruction set has become a mainstream instruction set, is that it is supported by a large number of upstream open source software. Upstream open source software is usually an independent, self-contained open source project adopted by major operating system distributions. And often each upstream open source software represents an integral part of the supply chain. Currently, the level of support for RISC-V in upstream software has not yet reached scale. In terms of customization and optimization, it is far from reflecting the unique advantages that RISC-V should have.

(5) Ecological Fragmentation Management

The modular design of RISC-V makes it easy to diversify but also easy to fragment. A typical example of fragmentation is Google's Android operating system. Android's open source strategy allows users to customize its open source version (AOSP version, under the Apache 2 license) according to their own needs, and then fork to their own product line. In order to avoid fragmentation, Google has successively used agreements to restrict phone manufacturers, used JetPack/AndroidX to comb third-party libraries in the supply chain, and even switched the APP packaging format that has been used for 12 years from APK to AAB. The RISC-V ecosystem needs to be proactive, from instruction set specification, compilation of toolchains, software development kits (SDKS), distribution package management tools, and even through RVI membership agreements to avoid similar problems.

The above constitutes the five critical factors of RISC-V open source software supply chain construction. These were inspired by the open source ecosystem builds on X86/Linux and ARM/Android.

4 Major Risks of RISC-V Open Source Software Supply Chain

Building an open-source software supply chain not only requires focusing on critical links but also facing complex and diverse risks. These risks also exist in the existing X86/ARM instruction set ecosystems. However, due to the historical limitations of open source development, there is almost no systematic preparation, resulting in the accumulation of risks, which has become a heavy burden on the open source industry. For this reason, when RISC-V builds its own supply chain as a new ecosystem, these risks must be considered in advance to avoid the same mistakes as possible.

(1) Intellectual Property Risk

Any open source software must have at least one open source license. An open source license states the extent to which users can use, modify, and distribute source code. Some licenses are very lenient, such as the MIT license, which allow modification of the source code while also allowing for it to be made proprietary. These licenses also allow to use the project name of the source code for promotional endorsements when the software needs to be advertised. Some licenses are very strict, such as the GPL, where when the source code is modified, the new code must also be open sourced and must be licensed under the GPL.

When a large complex open source software contains a large number of open source packages, it means that it may contain a variety of open source licenses. If the management of licenses is neglected, it is easy to leave intellectual property risks. For example, LINKSYS, a company specializing in developing routers, released the WRT54G router in 2003. Later, it was discovered that the firmware of the router was based on the Linux kernel, which used the GPL license, meaning that the router firmware code had to be open sourced as well. In the same year, Cisco acquired LINKSYS, and was forced to open source WRT54G software under the pressure of Linux lawsuit, which in result spawned the open source router operating system OpenWRT. This meant another enterprise-level open source router operating system for the community, but possibly the breaking of the company's hard-built technological barriers. Therefore, commercial companies must pay attention to the license, slight carelessness may result in huge losses.

The open source license rules of RISC-V instruction set itself cannot represent the intellectual property agreements of other open source software in its ecosystem. However, the instruction set is the starting point of the ecosystem. Starting from RISC-V, it is possible to build a more commercial-friendly open source software supply chain, thereby greatly reducing the intellectual property risks for commercial applications.

(2) Software Vulnerability Risk

Vulnerabilities (also known as exploits or bugs) are ubiquitous in every piece of software. Some bugs cause widespread malign effects or even malicious attacks, while others may go undiscovered until the end of the software's life cycle. Just like license issues, massive software packages can bring a lot of defective code. It is impossible for a user to screen every software package for vulnerabilities. However, these open source software with vulnerabilities are likely to be a squib covering dust for years, and will explode without warning at any time, thus bringing down the entire business system and suffering losses.

A typical and serious example is the HeartBleed bug in 2014 [8]. This vulnerability is found in a version of the OpenSSL, an open source software package widely used around the world. In particular, a large number of internet sites are using the OpenSSL component, resulting in the theft of numerous of usernames, passwords and related sensitive data. When the vulnerability was exposed, millions of hosts worldwide were affected. In fact, this vulnerability was introduced into the OpenSSL code in 2012 but was not discovered until two years later by a Google engineer. It is hard to know how many servers were hacked during this period. Currently, there are some vulnerability scanning tools, such as Vtopia, FOSSID, SONATYPE. They provide early warning by integrating vulnerability information published by upstream communities and proactively scanning to discover vulnerability information.

The RISC-V instruction set itself cannot reduce the probability of software vulnerabilities, but an open source instruction set can provide a more transparent binary-level method of vulnerability scanning, avoiding the "black hole" of debugging and detection that previously comes from proprietary instruction sets, especially so-called *undocumented instructions*.

(3) Vulnerability Propagation Risk

When a large and complex software discovers a new vulnerability, there must be precise repair instructions down to the individual level to improve the speed of vulnerability repair. For commercial software maintained by the company, there are usually specialized testers who are responsible for finding vulnerabilities. When they find a vulnerability, they will directly request the corresponding programmer to fix it. However, in the supply chain system of open source software, testers cannot always quickly and accurately feedback vulnerabilities to the designated developers, and the feedback channels are not smooth, which ultimately leads to the failure of timely repair of vulnerabilities. When a widely used open source software module is repaired, we also hope that it can be quickly synchronized to all other open source projects using the module, rather than the same vulnerability repeatedly found in different software, repeatedly repaired, wasting manpower and resources, and even being repeatedly expoited by attackers. Whether open source maintainers can respond stably and timely, and whether bug fixes can be pushed on a large scale and with full coverage, are the special risks of vulnerability propagation in the open source software supply chain.

Let's take the recent Log4j2 outbreak as an example of why maintaining open source software modules is important but difficult. Apache Log4j2 zero-day vulnerability (Log4Shell, CVE-2021–44,228) was exposed on December 10, 2021. Its "Lookup" mechanism had parsing problems, resulting in JNDI injection vulnerability. The triggering condition of this vulnerability is simple, but the harm is enormous. An attacker can enter a specific attack string into the program, and when the program logs, the vulnerability is triggered, which can be further used to execute malicious code. Log4j2 is an open source logging component widely used in Java projects, so the vulnerability quickly became a security crisis in the open source software supply chain. According to incomplete statistics, more than 8,600 open source software on GitHub directly rely on log4j2 components. And tracing back through these open source software, eventually more than 200,000 open source software affected. Furthermore, even a week after the first official release of the patch, more than 80% of indirectly linked open source software has not been fixed yet. It can be seen from the data that once serious vulnerabilities are found in such open source components, they will directly or indirectly affect the upper open-source software that depends on them. Directly dependent on these components can easily troubleshoot and repair the component's vulnerability. However, through the hierarchy of complex dependencies between open source software, the spread of the flaw is not easy to be found in the application with deep dependence, and will bring immeasurable and uncontrollable influence on the global supply chain, and will last for a long time.

Earlier, we mentioned that OpenSSL, which caused the HeartBleed vulnerability, also belongs to this type of component. In addition, similar open source components include Apache Struts2, Linux Kernel, and other widely used open source software. Apache Struts2 is a web application framework based on the MVC design pattern. A series of security vulnerabilities in this component directly affected multiple large websites in China, including Baidu, Tencent, and Taobao, and even national-level government websites. The Dirty Cow is also a notable vulnerability that existed in the Linux Kernel for 9 years before being fixed. As Android is an operating system based on the Linux Kernel, the Dirty Cow was indirectly spread, affecting most Android phones on the market at that time. For a system, the closer the vulnerability is to the lower level of the system, the greater the range of impact, the more difficult it is to be fixed, and the higher the risk of propagation.

The open property of RISC-V instruction set is expected to form a vulnerability information sharing mechanism in which different processor manufacturers can participate under the same instruction set. This will enable different supply chains to be fixed in a shorter period of time and even possible to introduce a standard extended instruction set specifically for vulnerability repair. This will further bring the OTA (over-the-air) upgrade mechanism commonly used in current smart terminal devices down to the instruction set level. Once this instruction set is enabled, remote repair will be automatically performed when the devices are connected to the Internet.

(4) Continuous Maintenance Risk

The long-term obligation to maintain an open-source project may lead to a series of unfair phenomena. For example, commercial companies earn huge profits through open source software, but do not give any feedback to the maintainer, and even deliberately avoid talking about the use of open-source software, which causes the aversion of open source maintainer and even some extreme behaviors. The recent vandalism of the open source libraries of faker.js and colors.js by their authors is a typical example. Faker.js and colors.js are widely used, with weekly downloads approaching 2.5 million and 22.4 million respectively, which are critical upstream nodes in the open source software supply chain. Faker.js is under the MIT, which is a very loose open source license, making so many commercial companies are not paying anything to use this project. As one of the best open source projects in the field of fake data, the enormous workload of faker.js and colors.js was largely done by Marak, the author of faker.js, and received no support or reward from commercial companies. In a vicious cycle that had been building for a long time, the author poisoned the supply chain by submitting malicious code to these two packages, published it to GitHub and NPM package managers, and then emptied the project repository of all code and stopped maintenance altogether, leaving thousands of projects that depended on those two libraries unviable.

In fact, when OpenSSL had a vulnerability, it was discovered that only two part-time employees maintained the OpenSSL library, and were completely voluntary. Log4j is also maintained by several part-time community open source enthusiasts. It is conceivable that when a vulnerability was found, people cannot expect the timely and comprehensive repairment.

Due to the diversity of scenarios, maintenance of the open source software supply chain in RISC-V ecosystem will be more challenging. The traditional "volunteer" model needs to be broken down and replaced perhaps with a "limited liability" model, where open source developers assume some maintenance responsibility, matched by award or income. The organization mode of supply chain will also change from the traditional "horizontal stratification" mode to "vertical penetration" mode, which makes the upstream and downstream coordination in a single supply chain even closer.

The above revelations about the risks in the RISC-V open source software supply chain do not overturn the general trend of open source as the future software mode. On the contrary, systematically identifying and resolving these risks will bring new opportunities for the development of the RISC-V industry and even the entire information technology ecosystem. Furthermore, this may become the core competitiveness of RISC-V compared with other proprietary instruction sets in the future.

5 Conclusion

The combination of open source software and RISC-V instruction set will have great potential. Open source software with RISC-V can solve intellectual property, vulnerabilities and other problems from the origin, without being bound by historical burden. On the other hand, RISC-V can also develop a full-stack open source ecosystem differentiated from X86/ARM with the help of open source software. With the further popularization of the open source mode and RISC-V, the RISC-V ecosystem's open source software supply chain will become increasingly clear to the industry, penetrating not only the entire information technology industry but also various other industries, becoming the fundamental element of production and life. However, the risks existing in the open source software supply chain will also be unavoidable and need to be solved jointly by global developers, open source communities and even governments. In summary, although there is a long way to go to build RISC-V open source software supply chain, a prosperous future can be expected.

References

1. Segall RS (2021) What is open source software (OSS) and what is big data? [M]. In: Research anthology on usage and development of open source software. IGI Global, pp 817–857
2. Qi Y, Liu J, Li N (2021) Security risk analysis of open source software supply chain [J]. Inf Secur Res 7(9):5 (in Chinese)
3. Forsgren N, Alberts B, Backhouse K et al (2021) 2020 state of the octoverse: securing the world's software [J]. arXiv preprint arXiv: 2110.10246
4. Searls D (2011) Eof: what does Linux want? [J]. Linux J 2011(201):12
5. Legay D, Decan A, Mens T (2021) A quantitative assessment of package freshness in linux distributions[C]. In: 2021 IEEE/ACM 4th international workshop on software health in projects, ecosystems and communities (SoHeal). IEEE, pp 9–16
6. Liu C, Yanjun W, Jingzheng W et al (2021) Survey of RISC-V ISA Research [J]. J Softw 32(12):3992–4024 (in Chinese)
7. The Chinese Academy of Sciences announcd the open source RISC-V IP core "Xiangshan" [J]. Inf Syst Eng 2021(07):2 (in Chinese)
8. Hu Z, Chen P, Zhu M et al (2021) A co-design adaptive defense scheme with bounded security damages against Heartbleed-like attacks [J]. IEEE Trans Inf Forensics Secur 16:4691–4704

Yanjun Wu Chief engineer of Institute of Software, Chinese Academy of Sciences. He is also the vice President of openEuler Community, the largest operating system community in China, and TSC member of RISC-V International (RVI). He also received the Outstanding Member Award of Youth Promotion Association of Chinese Academy of Sciences. His research interests include operating system and RISC-V ecosystem software.

Research Progress and Prospects of Strategic Consulting Intelligence Support System

Yuan Zhou, Yufei Liu, and Wenjiang Zheng

Abstract Relying on the experience and wisdom of the academician of the Chinese Academy of Engineering, a series of strategic consulting research achievements have been made in the field of engineering science in China. However, with the explosive growth of knowledge and the advent of the era of big data, technology represented by a new generation of artificial intelligence has become an important tool to support consulting research. Data analysis support has become an urgent demand for engineering science and technology consulting research. At the same time, the development of information technology and data environment makes it easier to obtain and analyze data, and it is possible to support the strategic decisions of academicians and experts with objective data analysis results. On this basis, the Intelligent Support System (iSS for short) for engineering science and technology strategic consulting emerged. At the same time, it has built characteristic databases for engineering technology such as future technology library, technology list library, consulting report library, etc., and provides support services for strategic consulting researchers through standardized processes and methods. Focusing on the construction goal of taking experts as the core, process as the norm, data as the support, and interaction as the means, the iSS platform has provided intelligent decision support for several strategic consulting projects of the Chinese Academy of Engineering and more than 100 project groups. At the same time, it provides support for the teaching of relevant courses for undergraduates, masters and international students in Tsinghua University, Nanyang Technological University, Huazhong University of Science and Technology, Beijing University of Posts and Telecommunications and other universities. On this basis, this research discusses the construction background, construction plan, functions and services of the intelligent support system for engineering science and technology strategic consulting in detail from four parts: the overview of the research content, the platform structure, the support process, and the

---

Y. Zhou (✉)
School of Public Policy and Management, Tsinghua University, Beijing, China
e-mail: zhou_yuan@mail.tsinghua.edu.cn

Y. Liu · W. Zheng
Center for Strategic Studies, Chinese Academy of Engineering, Beijing, China

China's e-Science Blue Book 2023, https://doi.org/10.1007/978-981-99-8270-7_17

summary and outlook, so as to provide process guidance and methodological tool support for readers to use the iSS platform.

Keywords The iSS platform · Intelligent decision support · Technology foresight · Strategy consulting

1 Introduction

With the advent of the era of big data and the development of artificial intelligence technology, technologies such as massive data storage, multi-source data mining, and cloud computing have developed to maturation, which brings new development opportunities for a new round of technological revolution and industrial transformation. Big data technology promotes the transformation of data resources into valuable information through the rapid collection and in-depth mining, timely judgment and sharing of big data. It has carried out many application services in optimizing production, reducing energy consumption, accurate delivery and other aspects. In this way, big data has become a powerful means to support scientific decision-making and accurate prediction of social governance. Cloud computing technology can realize the convenient access of network data and the on-demand sharing of computing resources, improve the collaborative ability between different users, lower the technological threshold of digital resource management, and play an important role in creating a new social governance model of ecological green and intelligent management. Artificial intelligence technology enforces the ability of machine interactive learning. It generates rich industrial application scenarios in intelligent manufacturing, drug research and development, unmanned driving and other fields, providing specialized technologies for the innovation and optimization of social governance mode to keep pace with the times. With the explosive growth of knowledge and rapid growth rate of data generated by scientific and technological management and related scientific research activities, any individual, whether a senior scholar or an experienced expert, cannot fully grasp all the information and data in a specific field. However, strategic consulting research represented by technology foresight needs full interaction between experts and data. In the interaction process, it needs not only the support of data and analysis tools but also the guidance of scientific and standardized processes. Therefore, applying the most cutting-edge data analysis technology, objective data analysis results, scientific and effective technology forecasting process, building a data analysis interactive platform to assist experts in decision-making consultation for improving the quality of strategic consulting services is of great significance.

Strategic consulting research needs to solve industrial or technological development planning problems from different perspectives, which involves many stakeholders and is difficult and complex. Moreover, relying only on experts' experience and knowledge, it is often difficult to obtain a comprehensive understanding of the field, and experts' opinions are difficult to converge and prone to subjective bias.

However, the development of information technology makes the acquisition and analysis of big data more convenient and efficient.

Therefore, engineering science and technology strategy consulting research need to provide objective scientific data measure for experts with experts as the core, supported by big data and artificial intelligence methods. At the same time, it makes the expert knowledge and objective data analysis results are better able to interact and integrate, to improve the effect of consulting research and quality. At the same time, a process framework that can better interact and integrate expert knowledge with objective data analysis results is needed to improve the effect and quality of consulting research. In this context, relying on the China Knowledge Center for Engineering Sciences and Technology, Tsinghua University, Inspur Group Co., Ltd, Huazhong University of Science and Technology and Hunan University jointly built a strategic consulting intelligent support system (Intelligent Support System), provides intelligent and process online support services for high-end think tanks in the field of engineering technology. The iSS platform was officially launched in 2017 to provide services and all users can register for free. The main interface is shown in Fig. 1.

The construction of the iSS platform follows the concept of simultaneous research, construction, promotion and application, with authoritative data, forward-looking thinking and credibility of conclusions. Under the overall goal of "experts as the core, process as the norm, data as the support, and interaction as the means," iSS platform's position is to build a strategic consulting intelligent support system that integrates process, method, tool, case and operation manual for strategic consulting research. The overall functions of the platform include designing and improving the academician consultation process, providing technology foresight methods, tools,

Fig. 1 The iSS platform homepage

cases and manuals, assisting and supporting projects to complete policy analysis, technological analysis, market and product analysis, research reports and roadmaps and other data support reports.

The iSS platform focuses on the development trend of world science and technology and judges the new direction of world science and technology revolution as the application scenario. It's an intelligent support platform combined with modern information technologies such as cloud computing, big data, artificial intelligence, etc., with experts as the core, data as the support, interaction as the means, and embedded consulting research. This platform supports scientific consultation with scientific methods, supports scientific decision-making with scientific consultation, and leads scientific development with scientific decision-making. With the help of the iSS platform, it can effectively support the scientific development of engineering science and technology strategic consultation and effectively improve the quality and level of strategic consultation of national high-end science and technology think tanks.

2 Platform Structure

In this study, the consulting process and analysis framework are systematically designed based on high-quality data, objective cognition, expert analysis, and consulting results. On the one hand, this consulting process can better play the supporting role of scientific and technological literature, patent data, industry reports, and other heterogeneous data to strategic consulting. On the other hand, it can better give full play to the core role of academicians in the whole process of strategic consulting. Focusing on the actual needs of the research group, following the principles of practicality, ease of use, and effectiveness, and focusing on technological trend analysis, technology system construction, technology list, Delphi method, and technology roadmapping, build a multi-dimensional, process-oriented, intelligent and interactive technology foresight characteristic product. This research design uses technology foresight as the characteristic service. It uses the database, method base, and index base as the main functional modules to support the data analysis to complete the data collection, mining, and research better. The development of the data analysis module carries out the framework design from the functional modules based on the database, method base, and index base. Based on these data analysis modules, an strategy consulting intelligent support system for engineering science and technology is built, and the application of scientific and technological innovation and strategic planning is completed based on the system (Fig. 2).

(1) **Multi Source Database**

The iSS system combines the data resources of the Chinese Academy of Engineering and Tsinghua University and the data uploaded by users. It summarizes and sorts scientific and technological documents, patent documents, scientific research projects, business reports, industrial data, and other data. The KGO patent database,

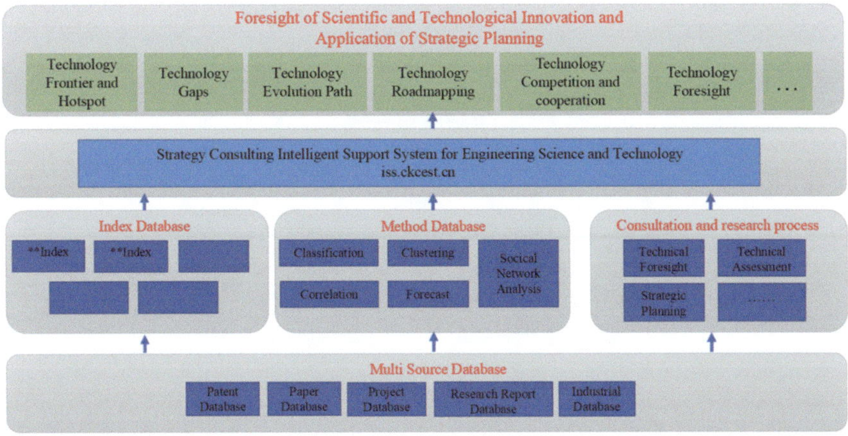

Fig. 2 iSS platform structure

SCI database, and scientific research projects have been completed. In addition, the construction of the database (Fig. 3), consulting report database, global technology inventory database, Chinese enterprise investment relationship, etc., is now completed. The collection of literature data mainly comes from Thomson innovation, China National Knowledge Infrastructure, and Web of Science, and displays the information of each dimension of the current literature task to users in the way of data visualization, mainly including time trend, country distribution, theme River, keyword co-occurrence, correlation analysis, etc.; The analysis database of scientific research projects mainly collects the data of fund projects of various countries reveals the future direction of technology development from multiple dimensions such as area, project, keyword, undertaking unit, undertaking person, field direction, support type and amount, reveals the funding of discipline development and the distribution of regional disciplines, constructs the relationship between critical fields and funding, and finally provides support for high-end think tanks to select experts, do Delphi survey, draw up technology mapping and the like; The Consulting Report Database mainly collects the research reports issued by institutions in various fields worldwide. After manual sorting and division of areas, interpretation is performed, and lots sort summary keywords to build a database. Users can search for corresponding reports through keywords to analyze technology development status; The Global Technology Inventory Database mainly collects reports on the prospect and foresight of future technological development around the world, and manually sorts out the crucial technological development directions, forms technology list entries, organizes technology entries by field, and builds a database to support strategic consulting research; The China Enterprise Investment Relationship Database is based on the investment relationship and supply chain relationship between Chinese listed companies to explore the impact of inter-enterprise cooperation and investment on

Fig. 3 Query interface of multi-source database on iSS platform

the flow of multi-dimensional knowledge such as markets and products better to support the strategic consulting research work of this project.

(2) **Index Database**

At present, the index database has accumulated some research results, mainly including patent analysis index [1], paper analysis index [1], and industrial economic index [2–4].

(3) **Method Database**

The iSS platform needs to use data mining and analysis methods to clean and process the data in the database and perform correlation analysis on the data from different sources during the construction process. The quality of the data source determines the accuracy of an analysis result. Therefore, natural language processing, machine learning, deep learning, and other methods are used to process the data in the data integration process. Then data mining methods such as association rules, clustering, classification, and prediction are used to comprehensively analyze different types of data and generate technological trend analysis reports for the reference of academicians and experts. Meanwhile, with the increase of users, the platform will face the situation of insufficient computing power when processing massive data. Therefore, it is necessary to develop a highly concurrent deep learning algorithm base and its architecture to support the simultaneous use of many users to support the application and promotion of the platform in a broader range.

(4) **Consulting Research Process Database**

The process base of consulting research divides the strategic consulting process into three categories according to the research characteristics of strategic consulting by investigating the current situation and development trend of the domestic and foreign strategic consulting: technology foresight, technology assessment, strategic planning. Technology Foresight is to analyze the past data in the field of a consulting project, summarize its development trends technological routes, release related results, and predict candidate hot technologies for expert reference; technological assessment is assessing the trend of changes in the results of past projects. Strategic consulting planning is to make a comprehensive analysis of multi-source data in the field, define the time required for the transformation of research results into the industry, use the data to find the critical points of industrial development, and provide support for expert judgment.

(5) **System Layer**

Firstly, based on the above data analysis module, the cutting-edge technologies of extensive data analysis and new generation artificial intelligence are applied to data mining. Then, combined with the distributed computing platform of artificial intelligence 2.0, the platform is used as the underlying computing framework to support the application of various analysis functions of the platform. Finally, establish an online platform with experts as the core and data support, integrating consultation process, analysis methods, analysis tools, consultation cases, and operation manuals.

(6) **Application Layer**

Based on the iSS system, it can provide more scientific decision support for strategic consulting services and multi-dimensional analysis tools for scientific researchers. For example, use the data analysis tools provided by the system to support the research of identifying technology frontiers and hotspots [5], technology gaps [6], technology evolution path [7, 8], technology roadmapping [9], technology foresight [10, 11], emerging technologies [12], technology convergence [13].

In the process of construction, the platform conducts in-depth research and analyzes the current situation and development trend of domestic and foreign strategic consulting, and explores the internal relationship between methods such as the objective, extensive data analysis, new generation artificial intelligence technology, technology roadmapping [6] and expert knowledge. This research's distributed data computing platform encapsulates a series of new-generation artificial intelligence algorithms represented by deep learning. The computing platform realizes the in-depth mining and disclosure of data resources, improves the ability of extensive data analysis, expands the function of data mining, and enhances the ability of association analysis between multi-source heterogeneous data. In addition, this study further optimizes the decision support process with Delphi expert analysis as the core and guides experts to conduct efficient discussion by strengthening the function of human–computer interaction to improve the quality of consulting service.

The iSS platform takes the concept of research, construction, promotion, and application to study the above-embedded technology foresight analysis framework and process. On the one hand, it studies the convergence of multi-source heterogeneous technology, deep learning technology, knowledge graph technology, human–machine collaborative intelligent interaction technology, technology roadmapping method, and technology foresight. On the other hand, it discusses how to innovate, develop and apply specific techniques and tools of new technologies and new approaches in technology foresight and continuously iterates platform functions to optimize the quality of decision-making consultation.

In general, integrating a large number of multi-source data and integrating the big data intelligence technology and human–machine collaborative intelligence technology in the new generation of artificial intelligence technology, the platform integrates multiple characteristic products and available tools for auxiliary analysis. These tools have been applied to practical case studies in the fields of "stratègic emerging industries," "intelligent manufacturing," and "high-quality development of manufacturing." It also provides scientific data and decision support (decision-making) for capturing the future development trend and possible evolution path of crucial technologies as soon as possible, formulating reasonable technology strategies and innovation policies, grasping the general trend of scientific and technological development, and clarifying the main direction and breakthrough of development.

3 Support Process

The design and formulation of the technology roadmapping is the most important link in strategic planning and the core and key of the consulting work. Therefore, during the construction of the iSS platform, a long-term and in-depth exploration of strategic management theory and consulting practice was carried out with the technology roadmapping project as the lead. On this basis, the technology roadmapping research process supported by the iSS platform is designed, which has made beneficial exploration and solid accumulation for the strategic consulting research of engineering science and technology in China.

3.1 Technology Foresight and Technology Roadmapping Research

In order to grasp the technology development trend in the field, look forward to the future of technology development, and obtain technological and industrial competitive advantages, countries around the world have carried out technology foresight activities to do technology foresight and layout. As a flexible strategic management tool, technology roadmapping can support technological innovation and strategic planning and is one of the core technology foresight methods. At the same time, in the strategic research of engineering science and technology consulting, the technology roadmapping is an important output of technology foresight work.

The technology roadmapping consists of a hierarchical time-based chart that provides a visual analysis framework for exploring and communicating strategic planning [14]. As a technology management method that can be used to analyze the dynamic relationship between organizational goals, technological resources, and changing environment, technology roadmapping is helpful to identify potential development opportunities and technological gaps. Besides, technology roadmapping can deeply support national mid-and long-term scientific and technological development plans, major projects, and other scientific and technological development strategies. The roadmap is widely used in a macro, national strategic planning and other aspects. On the one hand, the roadmap can help identify key market and technology knowledge gaps, analyze market, competitive and technological intelligence, and develop technology strategies.

On the other hand, the roadmap supports strategic planning at the national macro level, supports resource allocation, risk management and investment decisions, provides critical information on technology direction, and identifies cross-institutional synergies and barriers. In addition, technology roadmappings facilitate communication across industry and inter agency networks, such as between leading companies, core suppliers, key customers, associations, and government agencies. In developing one roadmap, participating experts will build consensus and trust, which will be disseminated among experts, academics, key enterprises, associations, and government agencies, facilitating their understanding and collaborative development of consensus-based future development strategies.

In the face of complexity, uncertainty, knowledge gaps, and information asymmetry in innovation and strategy research, the technology roadmapping provides an intuitive and logical framework to help answer three key questions: (1) where are we now? (2) Where do we want to go? (3) How can we get there? Further, it helps to describe potential and preferred technology or innovation development paths and effectively support innovation management and strategic planning [15].

The development process of technology roadmapping affects the quality of technology roadmapping to a great extent, so it's very important for the development of technology foresight and strategic consulting. The early technology roadmapping development mostly adopted the process-driven workshop method. The relatively famous t-Plan [16] and S-Plan [17] technology roadmapping fast-start method developed by Professor Robert Phaal of Cambridge University in the UK. T-plan provides a streamlined approach to bringing together those involved in roadmap development in the form of workshops to leverage group strengths, explore options and opportunities, and reach consensus on the initial building of the roadmap.

Based on the T-Plan method, the development process of a technology roadmapping suitable for strategic planning called S-Plan is proposed. Compared with T-Plan, S-Plan mainly focuses on industry, policy, and market level. Through 1–2 days of workshops, the brainstorming method quickly determines the current situation, development goals, opportunities, barriers, and future development strategies. Based on this, the development strategies of industry and innovation are formulated. The T-plan and S-Plan methods provide standardized and process-based methods for other organizations and institutions to develop technology roadmapping. On this basis, scholars integrate scenario analysis, Quality Function Deployment, QFD), Morphology Analysis (MA), Portfolio Assessment (PA), and a series of innovative management tools that assisted the construction of the technology roadmapping. The standardization process based on expert discussion and the application of qualitative research methods provide a reference for constructing the technology roadmapping. However, due to experts' limited time and energy, the information presented by the technology roadmapping is not comprehensive enough because experts only rely on expert knowledge as the input of the technology roadmapping and lack the support of objective data and quantitative research methods. At the same time, expert decision-making is prone to subjective errors.

With the rapid development of machine learning and big data technology, it is possible to analyze the exponentially growing literature data. Some scholars have begun to combine expert knowledge and machine learning methods to identify emerging technologies [18], analyze technology evolution paths [7] and technology convergence trends [13] while strengthening the interaction of data and expert knowledge to improve the quality of technology roadmappings. Under this research background, the strategic consulting services provided by the iSS platform are mainly based on experts and supported by data. Various functions and processes are constantly being improved and perfected.

3.2 Technology Roadmapping Research Process Supported by the iSS Platform

The iSS platform import the technology foresight, quantitative methods, and technology roadmapping drawing tools, which support rich and detailed data for looking forward to the development direction and critical tasks of China's engineering science and developing technology roadmappings of different fields. In this way, the systematization, completeness, and standardization of technology roadmapping research have been improved. Specifically, the iSS platform can carry out strategic consulting and technology foresight research according to the four steps shown in Fig. 4.

The First Step: Engineering Technology Demand Analysis

Combined with key events in recent years (such as sino-US trade war, AI technology, etc.), demand analysis is conducted from political, economic, social, technological, market and other factors to analyze changes in technological development trend and macro environment, and identify future development needs of the field. At the same time, the iSS platform's build in report database can also be used to provide important data support for the future prediction technology demand in the research field from multiple dimensions, such as engineering technology classification and

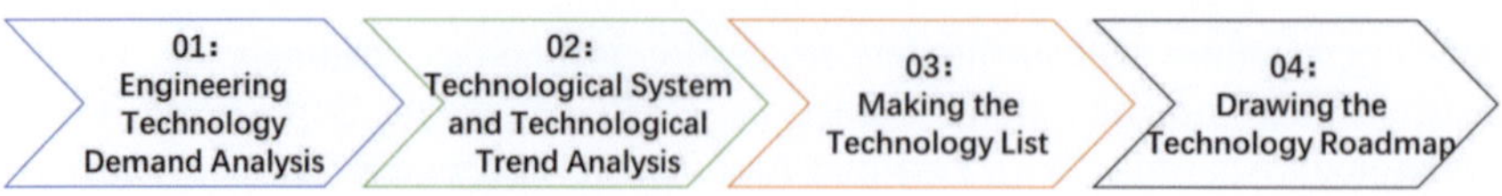

Fig. 4 The support process with iSS platform

ranking, report time, high-frequency word cloud and release trend. Finally, the expert group determines the field development needs, goals and tasks. On this basis, the content of the demand layer of the technology roadmapping is determined, that is, to what extent the target technology or industry should be developed, what goals need to be achieved at the technology and industry level in this field, what technological tasks need to be completed in order to achieve the technology and industry goals, and how to define the strategic priorities.

The Second Step: Technological System and Technological Trend Analysis

Technology is systematic and can be described by formulating the technology system. "Technology A," "Technology B," and the relationship between the two technologies are the basic elements of the technology relationship. These elements conform to the triplet pattern of knowledge graph in the new generation of artificial intelligence research. The technology system can be formed with technological support personnel and domain experts. While the technological support personnel can use the advanced technology of big data, artificial intelligence, to collect and sort the technologies rapidly, these intermediate technologies can support experts to formulate the final technology system. The technology system can guide experts to participate in technology foresight and help experts grasp the content and direction of analysis through a structured system. At the same time, the technology system provides comprehensive technological information, which is conducive to ensuring the comprehensiveness and uniformity of the technology list and can also help the subsequently participated experts quickly understand the work results of the previous experts and reach a consensus quickly. After the expert discussion, the final version of the technology system can help experts efficiently extract valuable information and then be used as a reference for the technology layer of the technology roadmapping.

The retrieval keywords can be made according to the technology system formulated in the previous step. Multi-source data such as patents, papers, and funds can be retrieved and accomplish the technology trend analysis. The iSS platform provide patents, papers, research projects, such as analysis function, can help users from the global, countries, institutions, the patentee and the key information to complete the thesis and the trend of patent analysis, comparative analysis, and situational analysis, which support team to complete the opening material selection, macro analysis, technology trend analysis report, Fig. 5 is a schematic diagram of technological trend analysis results in the field of numerical control machine tools.

The Third Step: Making the Technology List

The formulation of the technology list is the selection of key technologies for the future and the key to the success of technology foresight. There are three primary sources for getting the technology list: (1) Based on the results of the domain technology trend analysis, natural language processing method, and aggregation and classification algorithm are used to deep into domain knowledge, form domain knowledge clustering graph, analysis the main research topics. Then after manual sorting, forming the relevant technological items; (2) Using the technology list database of iSS platform to retrieval future-oriented key research and sort out the technology items

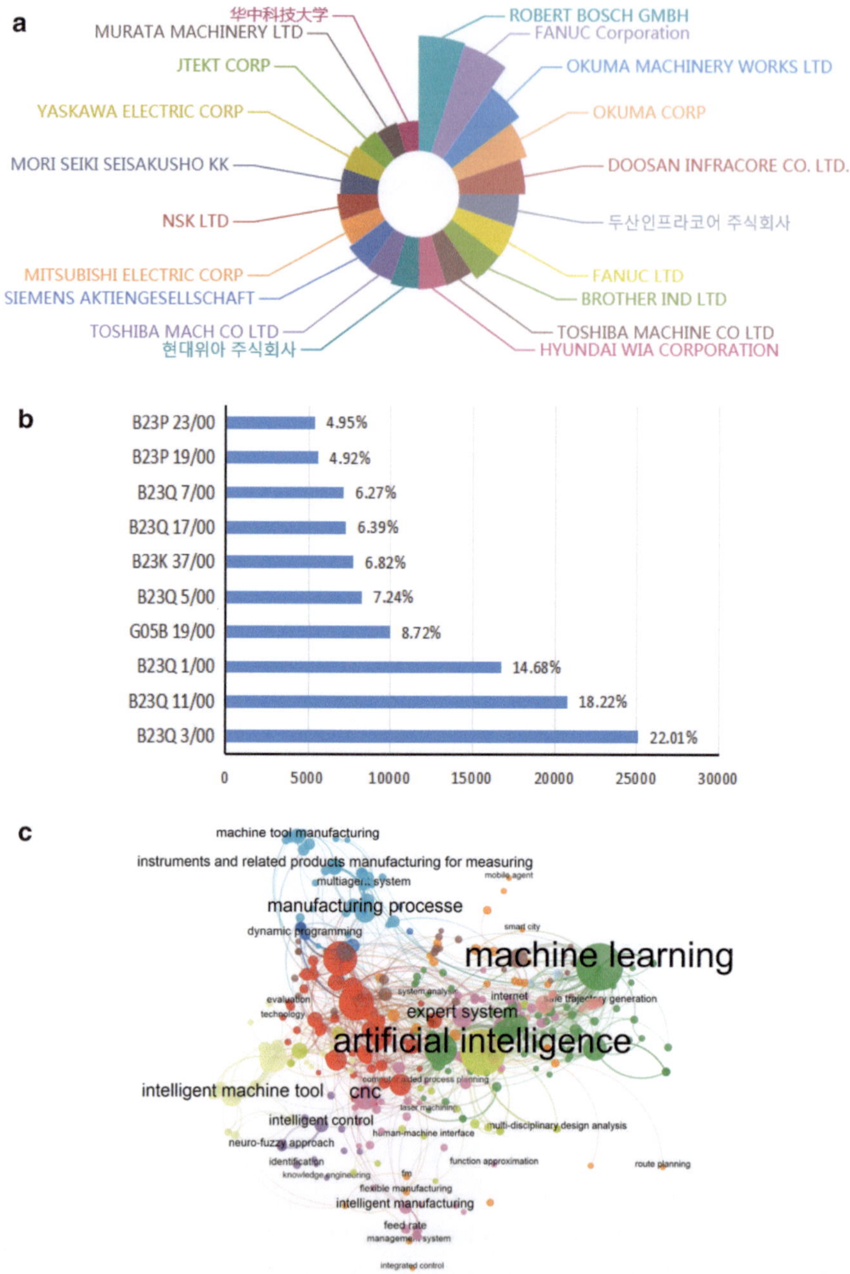

Fig. 5 An example of technological trend analysis of numerical control machine tools. **a** The patent applicant organization analysis. **b** IPC analysis. **c** Keyword co-occurrence analysis

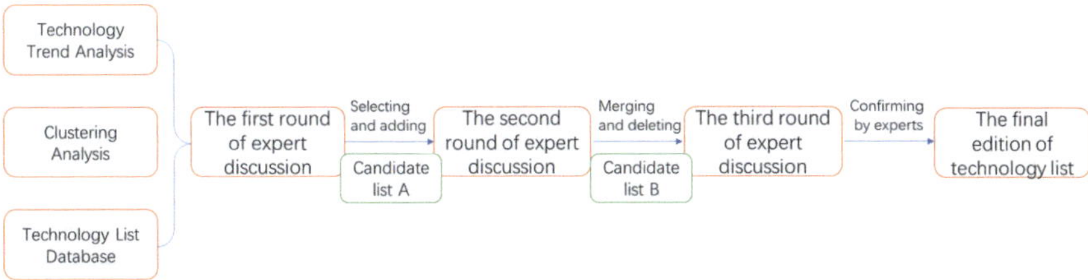

Fig. 6 Formulation process of technology list

suitable for the development of this field in China; (3) Experts will propose key technologies for the future through expert discussions and interviews. The technology list formulating process is shown in Fig. 6.

The specific development process of the technology list is as follows: First, based on the technology trend analysis report, users use clustering analysis and natural language processing methods to mine the core research topics in the field. After manual sorting by researchers, several key technology items are summarized, and the initial technology list is formed. Then, retrieving and screening future-oriented key technology items to supply the initial technology list and get the candidate technologies. Finally, three rounds of expert seminars will be held. In the first round, experts supplement the list of candidate technologies and add the missing technologies. In the second round of the workshop, technology items with inappropriate content or too small particles will be deleted, and similar technology items will be merged. In the third round of the workshop, the experts adjust the granularity of the technology items in the list. So that all the technology items in the list keep the same granularity, experts will write every technology items' category and connotation. At this point, the final technology list is generated.

The Fourth Step: Drawing the Technology Roadmapping

Technology roadmapping has the characteristics of highly summarized, integrated, and forward-looking, which can put the market, product, and technology information into one graph and display the connections between them. Based on this, the technology roadmapping can help users clarify the development direction of the target field and the key technologies needed to achieve the goal and clarify the relationship between products and technologies. The final technology roadmapping comes from the comprehensive results of demand analysis, technology system, technology trend analysis, technology list results, etc. Then the reasonable phased development goals are formulated by experts. In the process of developing a technology roadmapping, according to the Delphi survey, it can help experts clarify the current status and level of technology development recognize the key competitor, important technologies, important products, and projects to aid experts in drawing the technology roadmapping. Finally, the iSS platform provides online and offline versions of the technology roadmapping tool to help draw the technology roadmapping. Figure 7 is an example of a technology roadmapping drawn with the iSS platform.

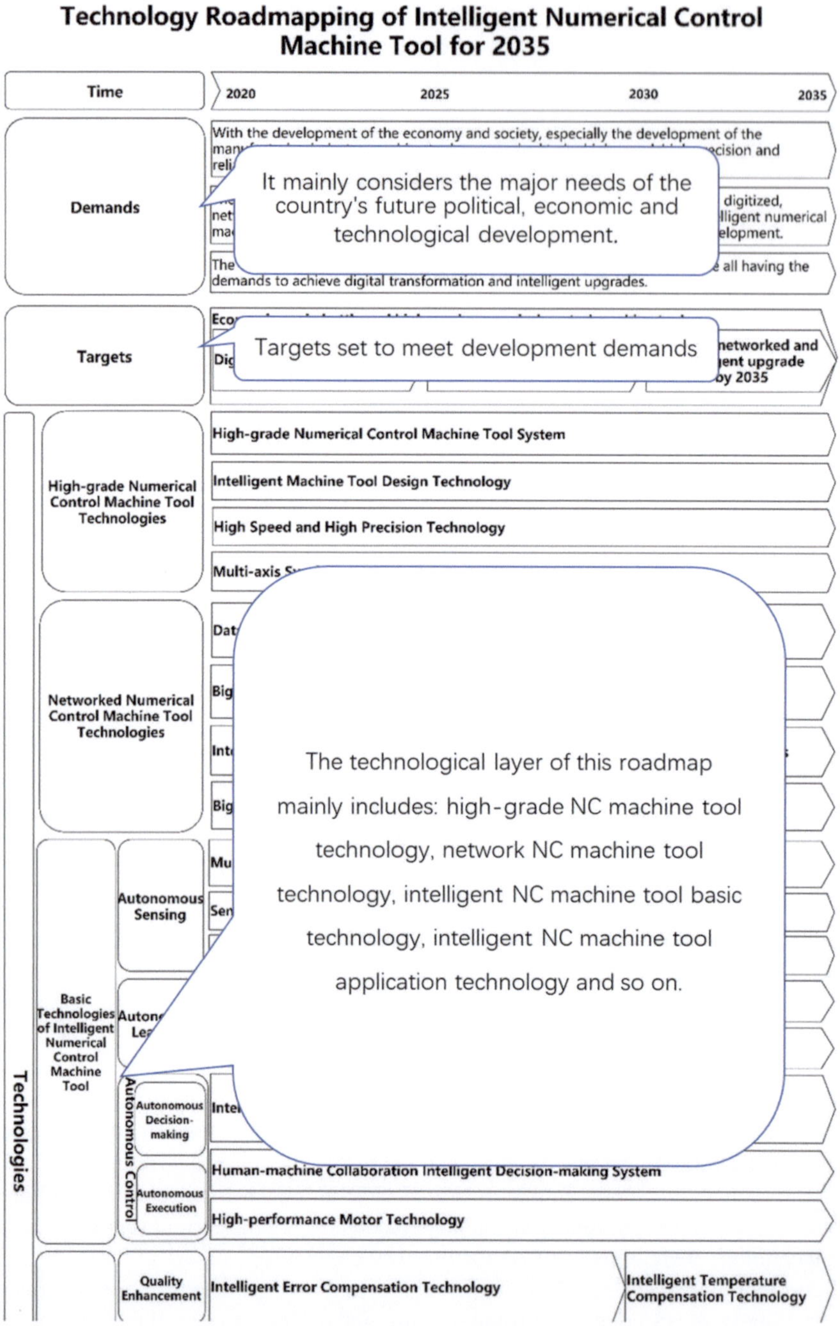

Fig. 7 Example of technology roadmapping

4 Conclusion and Discussion

The iSS platform focuses on the actual needs of the strategic consulting research group and takes "experts as the core, the process for the specification, the platform for the tool, the interaction of" as the guiding ideology. After years of construction, it has completed the data concentration, application integration, infrastructure integration, standard specification, safe and efficient advisory services, and information technology support platform of tool system. The platform integrates technology trend analysis, technology list, Delphi questionnaire, technology roadmapping, and other characteristic products with technology foresight as the core, and eight available tool sets such as project cloud disk, topic opening assistant, voice assistant, relevance analysis, social network analysis, citation network analysis, knowledge base, and literature map, to provide researchers with multi-dimensional analysis tools. By 2020, iSS has supported "Research on 2035 development strategy of China's engineering science and technology", that is, "Research on the medium and long-term development strategy of China's engineering science and technology," "intelligent manufacturing under the guidance of a new generation of artificial intelligence," "frontier research of Intelligent Manufacturing," "global engineering frontier" and "Research on Technology Roadmapping in key fields of manufacturing industry" "Research on the development strategy of strategic emerging industries under the new situation" and other projects, nearly 100 research groups completed the strategic consulting task.

At present, technology foresight is the main type of consulting research supported by iSS. In the process of consulting research, except for technology foresight topics, the scale, development status, development trend and other data of each industry and segment market are comprehensively studied and analyzed based on the industry map. With the help of industry distribution map, industry chain panorama map, industry heat map and list of key enterprises, multi-dimensional analysis is carried out on the upstream, middle and downstream of the industry chain, and the development focus of the industry is put forward from the perspective of "complementary chain", "strong chain" and "chain extension". Perfecting and optimizing the industry chain is also an important topic. Technology foresight is the primary type of consulting research supported by iSS. In consulting research, except for technology foresight topics, the scale, development status, development trend, and other industry and segment market data are comprehensively studied and analyzed based on the industry map. With the help of an industry distribution map, industry chain panorama map, industry heat map, and list of key enterprises, multi-dimensional analysis is carried out on the upstream, middle, and downstream of the industry chain. The development focus of the industry is put forward from the perspective of "complementary chain," "strong chain," and "chain extension." Perfecting and optimizing the industry chain is also an important topic. In addition, the iSS platform has accumulated some user data after years of construction and application. The knowledge contained in these data will be more fully mined in the service process to provide users with more effective support. At the same time, with the increase of users, the amount of data accumulated by the platform is becoming larger and larger, and the demand for high concurrent use of users is becoming stronger and stronger. The forum will face insufficient examples when dealing with massive data. Therefore, it is necessary to develop a high contemporary deep learning algorithm library and its architecture to support the concurrent use of many users and ensure the smooth operation of the platform. In the future, we will also strengthen the research on considerable data computing ability and comprehensively apply multi-source heterogeneous technology, deep learning technology, and human–computer collaborative intelligent interaction technology. Furthermore, explore the application of new processes, new methods, and new tools in the field of Engineering Science and technology strategic consulting to support the application and promotion of the platform in a broader range and provide a better user experience for the project team in the field.

References

1. Liu Y, Zhou Y, Liao L (2016) Application of big data analysis method in technology foresight for strategic emerging industries. Eng Sci 18(4):8
2. Yuan H, Zhang H, Zeng H (2014)Measuring localization of manufacturing industries and its dynamics—using Beijing firm-level data. China Ind Econ (9):13
3. Zhang L (2015) Research on the measurement methods of industrial agglomeration. Technoeconomics & Manag Res (6):6
4. Wang Y, Zhang H (2020) Comparative analysis on the G20 countries intelligent manufacturing development level. J Quant & Tech Econ 37(9):21
5. Lin H, Zhou Y, Liu Y (2017) Technology hot spots and frontier identification support to 2035: the technology list adjustment method, using robot technology as an example. Eng Sci 19(1):9
6. Liu Y, Liu H, Siyuan W (2021) Technology gap analysis of middle and high-grade computer numerical control machine tools at home and abroad based on patent document measurement. Sci Technol Manag Res 41(02):1–8
7. Liu H, Chen Z, Tang J et al (2020) Mapping the technology evolution path: a novel model for dynamic topic detection and tracking. Scientometrics 2:1–48
8. Yuan Zhou Y, Junfei D, Liu Y et al (2018) Identifying technology evolution pathways by integrating citation network and text mining. J Intelligence 37(10):76–81
9. Yufei L, Yuan Z, Heng C et al (2021) Expert interaction technology roadmapping approach: driven by engineering science and technology knowledge graph science of science and management of S.&.T. 42(03):29–47
10. China Engineering Science and technology 2035 development strategy research project team (2020) Project team of intelligent support system for engineering science and technology strategic consulting research strategic consulting center of Chinese academy of engineering. In: Research on 2035 development strategy of China's engineering science and technology-technology roadmapping volume (I). Electronic Industry Press
11. China Engineering Science and technology 2035 development strategy research project team (2020) Project team of intelligent support system for engineering science and technology strategic consulting research strategic consulting center of Chinese Academy of Engineering. In: Research on 2035 development strategy of China's engineering science and technology-technology roadmapping volume (II). Electronic Industry Press
12. Zhou Y, Dong F, Liu Y et al (2021) A deep learning framework to early identify emerging technologies in large-scale outlier patents: an empirical study of CNC machine tool. Scientometrics 126(2):969–994

13. Kong D, Yang J, Li L (2020) Early identification of technological convergence in numerical control machine tool: a deep learning approach. Scientometrics 125:1982–2009
14. Vinayavekhin S, Phaal R, Thanamaitreejit T et al (2021) Emerging trends in roadmapping research: a bibliometric literature review. Technol Anal & Strat Manag 1–15
15. Zhou Y, Phaal R, Farrukh C, Probert D et al (2021) Innovation and strategy roadmap-theory. Science Press, Methods and Applications
16. Phaal R, Farrukh C, Probert DR (2021) T-plan: the fast-start to technology roadmapping–planning your route to success. University of Cambridge, Institute for Manufacturing
17. Phaal R, Farrukh C, Probert DR (2007) Strategic roadmapping: a workshop-based approach for identifying and exploring strategic issues and opportunities. Eng Manag J 19(1):3–12
18. Zhou Y, Dong F, Liu Y et al (2020) Forecasting emerging technologies using data augmentation and deep learning. Scientometrics 123(1):1–29

Yuan Zhou is an associate professor with tenure, doctoral supervisor, and special researcher at School of Public Policy and Management, Tsinghua University (China). He also serves as the Assistant Director of China Institute for Engineering Development Strategies, the adjunct researcher at the Institute of National and Global Governance, Tsinghua University. He received his Ph.D. degree (technology management) from the University of Cambridge (UK) in 2010 and B.Eng and M.Sc degree in Nanyang Technological University (Singapore). His research interests include public policy, innovation policy, and innovation management.

Featured Technologies and Progress of BDS Spatiotemporal Parameter Services

Junping Chen and Jianhua Zhou

Abstract The BeiDou Global Navigation Satellite System (BDS-3) started providing fully operational global services on July 31, 2020. It provides global positioning, navigation and timing (PNT) services and serves as key infrastructure for the determination and maintenance of the spatiotemporal reference frame in China. Challenges in the production of the spatiotemporal parameters of BDS exist due to its unique hybrid satellite constellation and regional-tracking-only capability. Innovative approaches are applied in the BDS ground control segment to ensure its high accuracy, high availability and high reliability. This paper introduces parts of these innovative approaches and discusses some key concepts of the ongoing comprehensive PNT system, which is regarded as the next-generation of BDS.

Keywords Beidou satellite navigation system (BDS) · Spatiotemporal information processing · Spatiotemporal reference · Precise correction information

1 Introduction

Since the great navigation period, astronomical navigation technology has provided tools for human beings to explore the earth. People determine the user's position by measuring the right ascension and local time angle of natural celestial bodies. In the middle of the twentieth century, navigation technologies such as Loran-C system and satellite doppler navigation system were developed to determine the user's location information by measuring and processing radio signal propagation and frequency shift.

J. Chen (✉)
Shanghai Astronomical Observatory, Chinese Academy of Sciences, Shanghai, China

School of Astronomy and Space Science, University of Chinese Academy of Sciences, Beijing 100049, China

J. Zhou
Beijing Satellite Navigation Center, Beijing, China

China's e-Science Blue Book 2023, https://doi.org/10.1007/978-981-99-8270-7_18

Since the 1970s, radio satellite navigation technology based on high-precision atomic time and frequency has been developed. At present, it forms the so-called 'Big 4' global navigation satellite systems (GNSS) including Chinese BeiDou (BDS), European Galileo, US-American GPS and Russia's GLONASS.

The development of BDS experiences three steps, i.e. the so-called BDS-1, BDS-2 and BDS-3, where BDS-1 and BDS-2 provides service mostly in China and neighboring regionals while BDS-3 realizes global positioning, navigation and timing (PNT) services. Current BDS constellation is a mixture of BDS-2 and BDS-3 satellites, which forms unique hybrid orbit constellation consisting of GEO (Geosynchronous Orbit), IGSO (Inclined Geosynchronous Orbit) and MEO (Medium Earth Orbit) satellites [1]. Besides the satellites, ground control center (GCC) and user equipment are the other two components of the GNSS. The coordinate and clock information of satellites and ground tracking sites calculated by the GCC is the core spatiotemporal parameters of BDS. The BDS ground monitoring stations are only distributed in the mainland China, and the coverage of ground measurement and control network is extremely limited. In addition, MEO satellites experiences periods that cannot be observed by regional network with invisible period covers more than 3/4 of its revolution arc. As a result, the spatiotemporal parameters of the satellite have strong correlation, and observation error characteristics are complex, consequently the spatiotemporal parameters are difficult to be accurately determined. To solve this difficult problem, it is necessary to innovate new technologies and adopt new strategies to accurately determine the orbit and clock parameters of BDS satellites.

BDS system innovatively integrates the three services in GCC processing facility: Legacy PNT, SBAS (Satellite-Based Augmentation Services) and Precise Point Positioning (PPP). Its theoretical methods for PNT performance improvement and wide area augmentation solve the above mentioned problems and Fig. 1 shows the key performance indicators of the three main services. The accuracy of BDS satellite clocks is better than 1.5 ns, the accuracy of radial orbit component of broadcast ephemeris is better than 10 cm. The signal-in-space range error (SISRE) of Legacy PNT is about 50 cm [1], which corresponds to the navigation positioning error of about 1 m. Furthermore, the SISRE of BDS SBAS is around 30 cm by superposing the real-time orbit and clock corrections, and ionospheric grid corrections on the broadcast ephemeris [2–5]. The SISRE of BDS PPP service is around 0.2 m by further broadcasting the partition comprehensive corrections (PCC) [6].

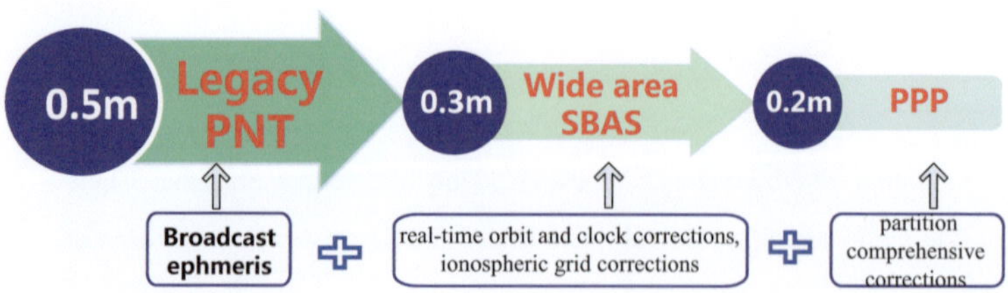

Fig. 1 Quadruple parameter superposition scheme of the three type of PNT services of BDS, showing also their SISRE accuracy

2 Spatiotemporal Information Processing Technique

The spatiotemporal information processing technology of satellite navigation system mainly includes the definition and maintenance of spatiotemporal reference frame, precise orbit determination and time synchronization (ODTS), signal propagation error correction etc.

2.1 *Spatiotemporal Reference Frame*

The principle of satellite navigation observation is to measure the time difference between local time and satellite time. Each GNSS system maintains its unique time datum through combination of ground and space facilities. The time system compatibility among GNSS is realized by tracing to common datum, i.e. UTC (Coordinated Universal Time).

The definition of spatial datum (i.e. reference frame) includes elements of frame origin, coordinate rotation and its scale. The international unified spatial benchmark is the international terrestrial reference frame (ITRF), which integrates the precise geodetic solutions of four space geodetic technologies: VLBI (very long baseline interferometry), SLR (satellite laser ranging), DORIS (Doppler Orbitography and Radio Positioning Intergrated by Satellite) and GNSS. The reference frame of each GNSS system has independent definitions, but they all realize the regular alignment with the ITRF by deriving the coordinate of monitoring stations under the same benchmark ITRF, so as to realize the coordinate compatibility of each system.

2.2 *Precise Orbit Determination and Time Synchronization*

The orbit and clock parameters of the GNSS satellites are the spatiotemporal reference of the navigation system to serve users. The ODTS processing in the GCC processing facility uses the observation data of the ground tracking network, solves the orbit-dynamic and observation equations.

The theory of precise ODTS mainly includes perturbation model, observation model and parameter estimation method. Due to the complex constellation configuration of BDS system and regional-tracking-only ground network, the ODTS processing in BDS GCC is different from other satellite navigation systems in terms of observation model and parameter estimation model. By combining ground-satellite and inter-satellite time synchronization technique, the parameter correlation problem between satellite orbit and clock parameters is largely solved in ODTS processing, and the parameter estimation accuracy of BDS system is greatly improved [7–9], and the orbit accuracy of centimeter accuracy is realized (Fig. 2).

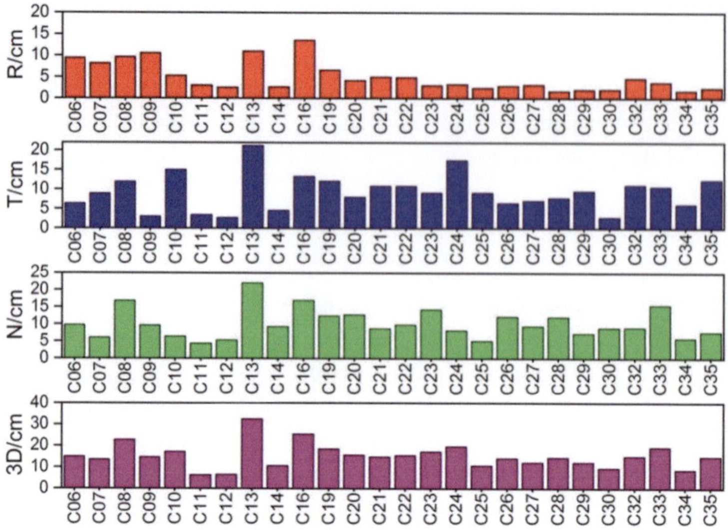

Fig. 2 BDS orbit precision (in RMS) in three components and 3D of BDS (R: radial, T: transverse, N: normal), where X-axis shows the PRN code of satellites

2.3 Signal Propagation Error Correction

The principle of user PNT function requires data of satellite orbit and clock, range between user and satellites. Errors of input data are among the factors affecting PNT accuracy. Errors in range observation occurs in the signal propagation path and user receivers. The signal propagation error correction technology aims to accurately calculate, encode and broadcast the following corrections, including satellite orbit, satellite clock, tropospheric delay, ionospheric delay etc. The main challenges for providing these corrections in BDS GCC include: high dynamic processing in real-time, limited and constrained broadcast bandwidth resources. BDS adopts the real-time quadruple wide area augmentation approach aforementioned, where satellite orbit and clock corrections, grid ionosphere and partition comprehensive corrections are superposed on the broadcast ephemeris [2–5]. Figure 3 is the schematic diagram of BDS spatiotemporal parameter broadcasting architecture. Through the quadruple wide area augmentation approach, and associated message structure and information encoding scheme design for the three type of PNT services, the challenges of high accuracy PNT service is largely solved. The real-time full-set of message parameters is broadcasted with bandwidth of less than 150bps.

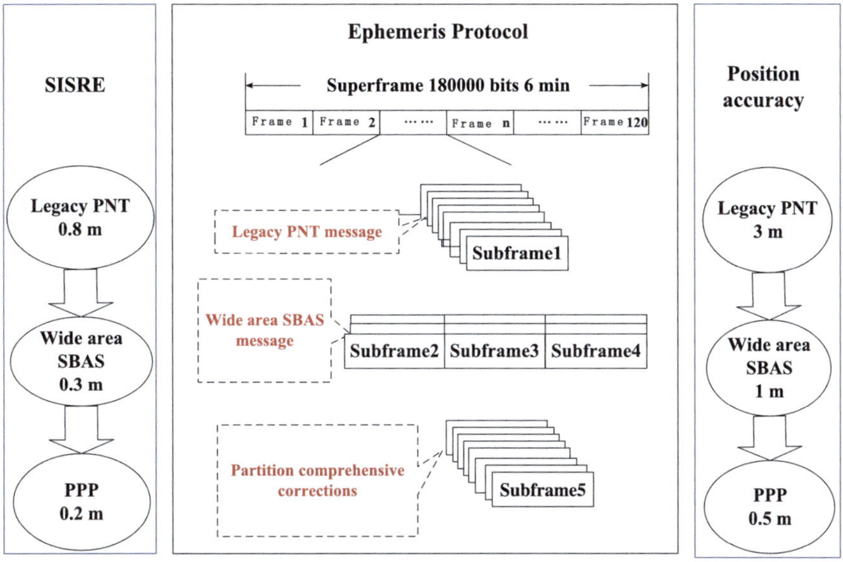

Fig. 3 Schematic diagram of BDS spatiotemporal parameter encoding architecture

3 Spatiotemporal Information Processing System

3.1 BDS GCC Processing Facility

The above innovative spatiotemporal information processing technology is implemented in BDS GCC processing facility. Figure 4 shows the basic structure and operation architecture of the processing system. The system adopts the architecture of 'duplex hot spare', which ensures high reliability of the real-time online running system and meanwhile provides additional platform for system performance optimization. The most important strategies are the running-fault prediction of the processing system based on background data, overall time synchronization mechanism among online and spare systems, and real-time digital twin environment of the whole system.

Under this structure, the output control strategy between the duplex units is based on overall system performance and service availability. The mechanisms of parallel processing of multi-channel data and cross verification through integrity monitoring are proposed. The overall system performance considers the comprehensive weighted performance such as positioning accuracy, SISRE etc. This operation architecture realizes not only the independence of operation service and performance optimization, but also the flexibility of seamless management and control in system switching operation.

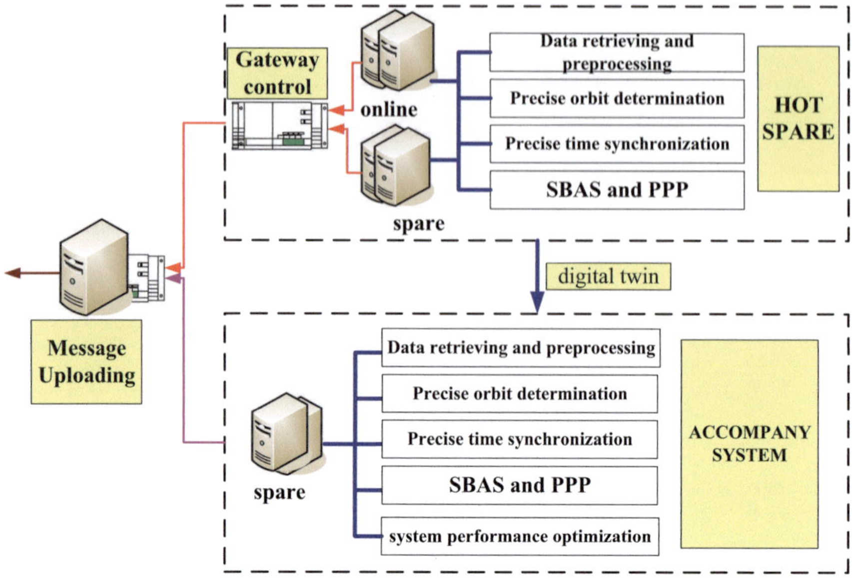

Fig. 4 BDS operation architecture of the GCC processing system

3.2 *SISRE of GNSS*

The purpose of the above spatiotemporal information processing technology and related processing system is to provide various type of spatiotemporal parameters and broadcast them to users via navigation messages. SISRE is the comprehensive index used to evaluate the GNSS performance, which mainly reflects the overall quality of satellite orbit and clock parameters. Montenbruck et al. (2020) systematically evaluated the SISRE of the 'big 4' as shown in Fig. 5. As it illustrates the European Galileo system has the minimum SISRE of smaller than 0.2 m, showing the most accurate PNT service; BDS has the second smallest SISRE of better than 0.4 m; SISRE of GPS is about 0.6 m, while GLONASS is about 1.7 m.

3.3 *PPP Performance of BDS*

The BDS satellite-based wide area augmentation corrections has improved the user positioning performance. Figure 6 shows the distribution of three-dimensional kinematic position accuracy of single frequency positioning in the two scenarios, where observation of BDS-2 satellites only and of BDS-2/BDS-3 full constellation is respectively used. In general, the positioning accuracy of BDS kinematic single frequency is mostly better than 0.6 m. The percentage of 3D position accuracy better than 0.1 m in the BDS-2/BDS-3 combination scenario has greatly increased against the

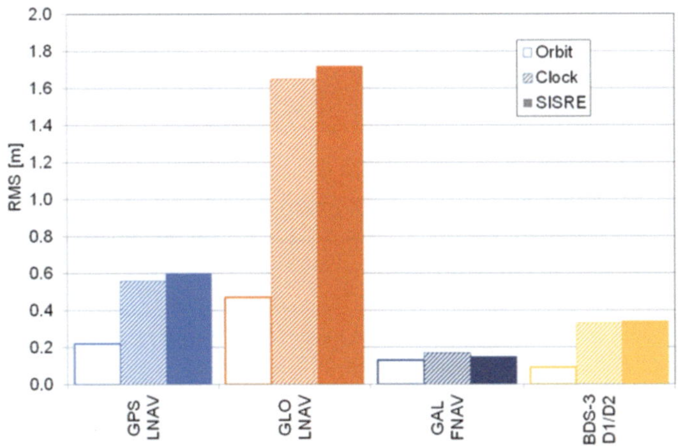

Fig. 5 SISRE of the 'big 4' [10]

BDS-2 only solutions. Figure 7 shows the 3D position accuracy distribution of dual frequency (i.e. B1/B3) kinematic positioning in the above two scenarios. Compared with the single frequency case, the positioning error under the two scenarios has been significantly reduced.

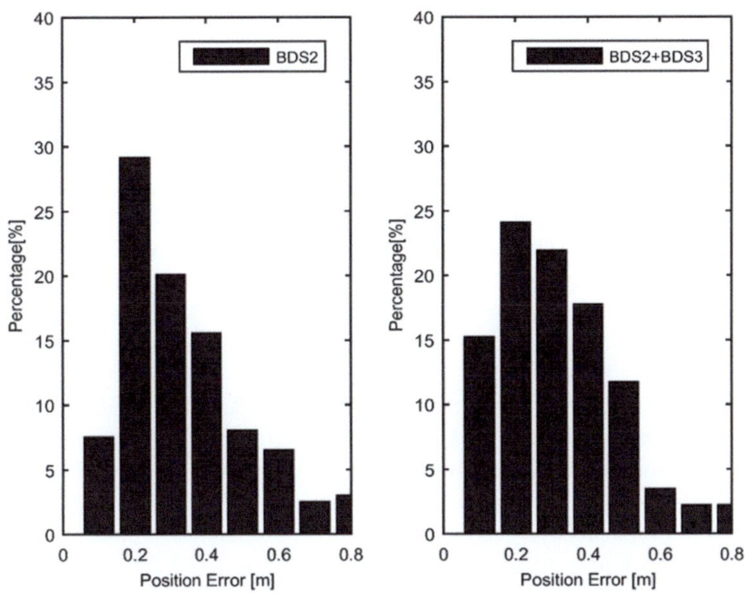

Fig. 6 Magnitude distribution of 3D coordinate errors of B3 single-frequency kinematic PPP using ionospheric grid corrections in the scenario of BDS-2 only (left) and BDS-2/BDS-3 combination (right)

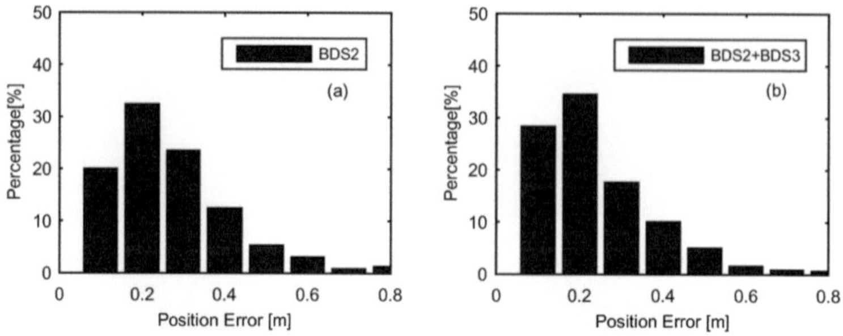

Fig. 7 Magnitude distribution of 3D coordinate errors of dual-frequency kinematic PPP. **a** BDS-2 only scenario; **b** BDS-2/BDS-3 combined scenario

3.4 Performance of BDS Location Reporting Function

RDSS (Radio Determination Satellite Service) short message function is a characteristic service of BDS, which realizes both the short-message communication and location reporting. Starting from its regional System (BDS-2), BDS has developed new technology of the location reporting in the Master Control Center (MCC) by integrating the RNSS(Radio Navigation Satellite System) and RDSS observations. In the new development, the RNSS range and carrier phase observations of user stations are transmitted to the MCC through RDSS link, and the RNSS observations are used to determine users' precise positions in the MCC. Supported by the various types of precise spatiotemporal data, e.g. precise satellite orbits and clocks, the precision of user positioning is thus greatly improved [11, 12]. The location reporting processing system consists of two parts: positioning processing and reporting unit in the MCC, RDSS and RNSS dual-mode user terminal. The main function of the positioning processing and reporting unit is to receive all kinds of real-time data from the MCC and carry out user position estimation and location reporting for all requesting RDSS users. The user terminal is the application service unit of the location reporting system, which carries out real-time tracking of each satellite within the tracking range, sends the observation data back to the MCC through the transmission link, and receives the positioning results calculated by MCC through the same transmission link.

BDS location reporting processing algorithm includes three modes: differential pseudo-range positioning (SPP), precision point positioning (PPP) and double difference relative positioning. The designed system positioning accuracy is better than 1 m with consecutive tracking period of shorter than 2 min. Figure 8 shows the histogram of horizontal and height positioning error under three positioning modes of SPP, PPP and double difference. As shown in the figure that the error distribution of the three modes is basically the same. The proportion of height error less than 1 m of the three modes reaches 67%, 67% and 65%, respectively. The mean 3D root mean square (RMS) of the precision positioning results of the three modes are 1.29 m, 1.07 m and 1.14 m, respectively.

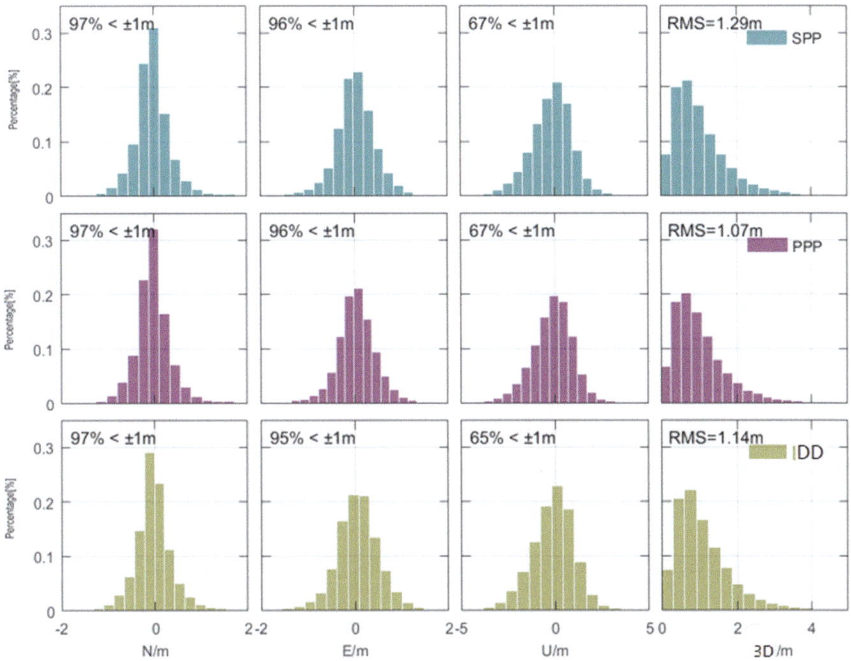

Fig. 8 Histogram of horizontal and height positioning error under three positioning modes of SPP (upper), PPP (middle) and double difference (bottom), each subplot shows the statistics of each coordinate component

4 Further Development of GNSS Spatiotemporal Information Processing

The 'big 4' GNSS systems are exploring upgrades and future designs. Taking GPS as an example, the United States continues to optimize the functions of GPS system, meanwhile it explores navigation modes that do not rely on or rely less on GPS system. In order to realize ubiquitous PNT under indoor and outdoor environment, the United States upgraded the iridium system to be provided with capability of signal enhancement to the GPS system. In recent years, many LEO (Low Earth Orbiter) constellation plans are under construction, constellations like 'oneweb' and 'Starlink' designs thousands or even more than 10,000 LEO satellites to explore the low-cost integrated service of network communication, PNT and remote sensing. Meanwhile, Europe focuses on solving the impact of navigation satellite clock error on service performance, and explores the use of LEO satellites to carry high-precision atomic clocks to realize the on orbit maintenance of time system.

China is also working on the design of LEO satellite constellation, in view of the future development of BDS, it is proposed that 'a more ubiquitous, integrated and intelligent comprehensive spatiotemporal service system will be built before 2035.'

LEO constellation has short satellite revisiting time and better ground-satellite geometry. New challenges emerge in the work of processing of spatiotemporal parameters of the new developments: (1) the constellation configuration is much complex, including high orbit and medium orbit satellites, as well as massive low earth orbit constellations; (2) the types of observation are complex, including the observation of ground-satellites link and between-satellite of high, medium and low orbit satellites; (3) The parameter types and parameter models are complex. There are dramatic increasing spatiotemporal parameters of satellites and ground receiving equipment, which needs to be accurately estimated.

Focusing on the sustainable development of satellite navigation system, it is necessary to refine the BDS spatiotemporal reference frame, including ground datum and timing datum. Key technologies of spatiotemporal information processing of 'Space-Earth Integration Network' need to be further invented and implemented to improve navigation service ability with BDS as the core. Topics of the real-time autonomous POD of LEO satellites, user positioning algorithms using the observation of BDS/LEO combined constellation need to be studied to improve the consistency between BDS spatial reference and the ITRF to be better than 1 cm, and the BDS SISRE may be improved from 0.4 m to be better than 0.2 m.

4.1 Accuracy and Reliability Enhancement

Accuracy and reliability improvement involves refined modeling and processing strategies of spatiotemporal parameters e.g. satellite orbit and clock parameters, monitoring station coordinates, velocity and other parameters etc. Particularly, the determination of earth orientation parameters (EOP) based on regional tracking network needs to be investigated and service systems will be developed.

(1) **Comprehensive Clock Based on Atomic Clock Assembly**

According to the different characteristics of BDS-3 onboard rubidium clock, hydrogen clock and possible cold atomic fountain clock in the future, comprehensive functional model integrating the best physical characteristics of each atomic clock type can be established to improve the overall performance timing datum. The comprehensive atomic clock function includes timing signals not only from onboard satellite but also ground-based. By developing sophisticated filtering, the long-term, medium-term and short-term characteristics of different clock groups are well modeled, thus to achieve better timing performance. By utilizing the comprehensive clock approach, the fragile situation of BDS timing keeping system, depending on single node clock assembly, can be greatly improved.

(2) **Spatial Reference Datum Based on Regional-Tracking Network**

The regional ground VLBI/SLR tracking network can be integrated with BDS monitoring network to realize the combined reference frame; EOP parameters can be determined based on the joint BDS inter-satellite link and ground VLBI/SLR/BDS tracking network.

(3) **Space Clock Quality Improvement and Onboard Space Timing Datum**

Through the improvement of physical and circuit system, the day frequency stability of hydrogen clock may enter the order of 10^{-16}, and the independent time keeping datum based on the space clock assembly can be realized. Alternative strategies and design for ground-satellite and inter-satellite links should be studied to independently maintain the timing datum in the situation that satellite constellation loses ground support.

4.2 Improvement of Inter-satellite Data Processing

BDS-3 achieves the break-through global PNT service, in which the innovative inter-satellite link technique plays a key role. The inter-satellite link is actually the first realization of its type, refinement and improvement of its hardware and software can be further performed. Besides, the inter-satellite link technique can be extended from between navigation satellites to between satellites orbiting at high, medium and low orbit planes. Methods for reference datum unification, system offsets calibration etc. will be developed.

(1) **ODTS Based on Inter-Satellite Link and Ground-Satellite Link**

Works regarding POD based on multi-source data, e.g. observation from LEO satellite, inter-satellite link etc. have to be performed. The algorithms in the combining of the time-division and continuous two-way measurement technique needs to be developed. Functional and stochastic models should be investigated, which covers satellite real-time attitude determination during satellite shadow periods, phase center correction of inter-satellite antennas, orbit and clock estimation and long-term orbit prediction etc.

(2) **Clock Prediction Based on Inter-Satellite Link and Ground-Satellite Link**

The related researches includes: calibration algorithm of time delay of multi-type equipment, function model of time variant errors of inter-satellite link, stochastic models of space borne and ground-based atomic clock assembly.

(3) **Model Refinement in Spatiotemporal Information Processing**

Aiming at the improvement of spatiotemporal information processing system, further studies includes: antenna phase center offsets and variation (PCO/PCV) calibration at each frequency of satellites and different types of receivers; satellite attitude, solar radiation pressure model and perturbation models etc.

4.3 Future ODTS and PNT Positioning Modes

With LEO constellation to be deployed in the future navigation, new PNT modes and related algorithm topics are emerging during the progress of the future navigation system.

(1) **Future User PNT Scenarios**

How to realize the rapid convergence of carrier ambiguity parameters owning to the fast motion speed of LEO satellites? How to improve the positioning accuracy and efficiency at the mean time? Considering the special topological structure and constellation evolution law of the future satellite constellation, new user positioning method with the integration of satellites spacing in the high, medium and low earth orbit planes need to be studied and developed.

(2) **Future ODTS Technique**

How to realize the real-time and high-precision solution of the spatiotemporal parameters of the whole constellation, considering the huge number of LEO satellites? The huge number of satellites of the LEO constellation requires new ODTS technique be suitable for the autonomous operation of LEO satellites, which includes: short-term and long-term orbit prediction method considering low thrust perturbation, clock predictions of chip level atomic clocks etc.

(3) **Constellation Monitoring**

The collision risks of the LEO navigation constellation satellites is much higher than current systems, the anti-collision monitoring database has to be established based on long-term orbit prediction.

5 Conclusion

As the Chinese contribution to the global satellite system, BDS applies many innovative approaches in the GCC segment to ensure its high accuracy, high availability and high reliability. Parts of these innovative approaches and their performance are introduced in this paper introduces.

We further discuss some key concepts of the ongoing comprehensive PNT system, which is regarded to be able to enhance the maintenance ability of satellite systems and to innovate the future PNT modes. The ongoing study of BDS performance improvement and ODTS technique of LEO/BDS constellation will contribute to set the technical basis for the construction scheme of future satellite navigation system.

References

1. Yang Y, Mao Y, Sun B (2020) Basic performance and future developments of BeiDou global navigation satellite system. Satell Navig 1:1
2. Chen J, Zhang Y, Zhou J et al (2018) Zone correction: a SBAS differential correction model for BDS decimeter-level positioning [J]. Acta Geod Cartogr Sin 47(9):1161–1170
3. Chen J, Zhang Y, Wang A et al (2022) Models and performance of SBAS and PPP of BDS[J]. Satell Navig 3:4
4. Chen J, Wang A, Zhang Y et al (2020) BDS satellite-based augmentation service correction parameters and performance assessment[J]. Remote Sens 12(5):766
5. Chen J, Yang S, Zhou J et al (2017) A Pseudo-range and phase combined SBAS differential correction model [J]. Acta Geod Cartogr Sin 46(5):537–546
6. Zhou J, Chen J, Zhang J (2019) Service performance enhancement technologies for BeiDou navigation satellite system along the belt and road[J]. Strat Study Chin Acad Eng 21(5):69–75
7. Chen Q, Chen J, Wu S et al (2020) Post-maneuver orbit determination based on the predicted clocks[J]. Acta Geod Cartogr Sin 49(1):24–33
8. Tang C, Hu X, Zhou S et al (2016) Improvement of orbit determination accuracy for Beidou navigation satellite system with two-way satellite time frequency transfer. Adv Space Res 58(7):1390–1400
9. Tang C, Hu X, Zhou S et al (2018) Initial results of centralized autonomous orbit determination of the new-generation BDS satellites with inter-satellite link measurements. J Geodesy 92(10):1155–1169
10. Montenbruck O, Steigenberger P, Hauschild A (2020) Comparing the 'Big 4'-a user's view on GNSS performance[J]. In: 2020 IEEE/ION position, location and navigation symposium (PLANS). pp 407–418
11. Tan S (2017) Innovative development and forecast of BeiDou system[J]. Acta Geod Cartogr Sin 46(10):1284–1289
12. Chen J, Zhang Y, Yu C et al (2022) Processing algorithms and performance evaluation of BDS RDSS location reporting service[J]. Acta Geod Cartogr Sin 51(4):511–521

Junping Chen is a professor at Shanghai Astronomical Observatory (SHAO), Chinese Academy of Sciences. He is the deputy director of the Center for Astro-geodynamics, SHAO and is the head of the GNSS data analysis center of SHAO. His research area covers precise orbit determination, clock analysis, SBAS, PPP, PPP-RTK, LEO augmentation, reference frame and Geodynamics etc. He authors more than 170 peer-reviewed paper, 3 monographs and 13 patents.

Novel Coronavirus National Science and Technology Resource Service System

Juncai Ma, Linhuan Wu, and Jianyuan Zhang

Abstract The Novel Coronavirus National Science and Technology Resource Service System which jointly constructed by Institute of Microbiology, Chinese Academy of Sciences and Chinese Center for Disease Control and Prevention was officially launched on Jan 24, 2020. The system authoritatively released scientific data of COVID-19 immediately and without any delay including the first electron micrograph of the virus, nucleic acid sequence information and primer design suggestions which provide important data support for global pneumonia epidemic prevention, control and scientific research work. On the basis of the management and release of scientific data resources, the service system further utilizes the advantages of long-term data accumulation and analysis model development experience in the field of microbiology, and strengthens support for microbiological data analysis and mining. Focused on genomics and structural biology, the Novel Coronavirus National Science and Technology Resource Service System 2.0–New Coronavirus Variation Evaluation and Early Warning System (VarEPS) was officially released. It is the world's first system for multi-dimensional risk assessment and early warning of known and virtual variants in the SARS-CoV-2 genome. This system has received widespread attention from the domestic and international community. The system supports Chinese scientists to publish articles in internationally renowned journals such as *Nature*, *Science* and *Lancet* by publishing genome sequence data and protein crystal structure data. On November 23, 2020, the system was selected as one of the 15 world's leading scientific and technological achievements of Wuzhen World Internet Conference.

Keywords New coronavirus · Data · Assessment and early warning

Text

The COVID-19 epidemic continues to erupt worldwide and spread rapidly, which is triggering an epidemic crisis that is raging around the world. New and recurring viruses are a global challenge to public health. The World Health Organization

J. Ma (✉) · L. Wu · J. Zhang
Institute of Microbiology, Chinese Academy of Sciences, Beijing, China
e-mail: ma@im.ac.cn

China's e-Science Blue Book 2023, https://doi.org/10.1007/978-981-99-8270-7_19

announced on March 11 that the new crown pneumonia epidemic has the characteristics of a "pandemic". After the outbreak, China's public health and scientific research institutions "raced at full speed" with the virus. Since January 3, China has regularly notified the WHO and relevant countries and regions of the epidemic situation and shared scientific research data.

The virus knows no borders, and the epidemic has no distinction of race. Global cooperation is the only good medicine to resolve this epidemic crisis. Beijing Institute of Genomics, Chinese Academy of Sciences (China National Center for Bioinformation) has established the Resource for Coronavirus 2019 (RCoV19). Wuhan Science Library, Chinese Academy of Sciences and National Science Library Chinese Academy of Sciences have jointly established a COVID-19 scientific research dynamic monitoring platform.

The Novel Coronavirus National Science And Technology Resource Service System jointly constructed by the Institute of Microbiology of the Chinese Academy of Sciences and the Chinese Center for Disease Control and Prevention was officially launched in January 24, 2020. The system released the world's first electron microscope photos of COVID-19 and established a scientific data release and sharing platform for the first time. Global pneumonia epidemic prevention and control and scientific research work provide important support.

1 Authoritative Data Release and Sharing to Help the Global Fight Against Epidemic

Novel Coronavirus National Science and Technology Resource Service System was released on January 24, 2020. The aim of the service system is to authoritatively release the publicly available virus strain resources and scientific data related to the epidemic, including the collection of virus strain resources (National Pathogenic Microorganism Resource Bank), electron microscopy photos, detection methods, genomes, scientific literature and other comprehensive information. It provides an important reference for the design and manufacture of nucleic acid detection reagents (Fig. 1).

2 Provide a Global Data Release Platform to Help Vaccine and Antibody Research and Development

Fighting the epidemic is a race between human society and the virus, and scientific data is an important "weapon" to help. It is an important measure to exert the value of scientific data and to gather global scientific research forces to jointly fight the virus by utilizing the national scientific data platform to gather internationally open data, and then publishing and sharing them in a timely manner.

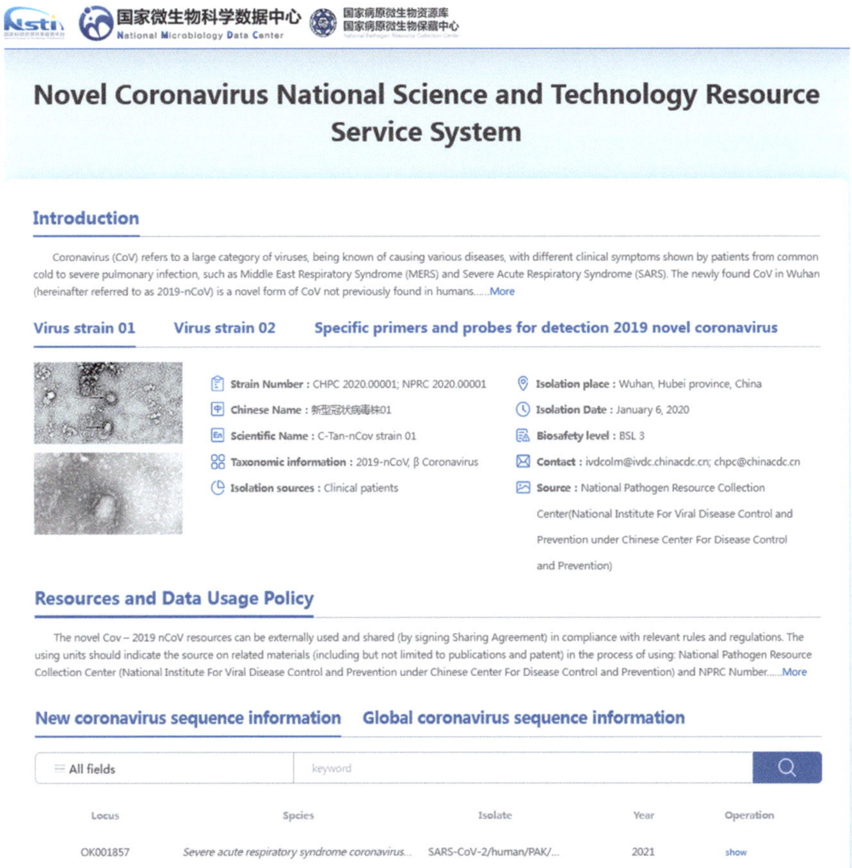

Fig. 1 The system released the the first electron micrograph of COVID-19

Since then, the service system cooperated with scientists to establish a fast track, and released a number of important data such as nucleic acid sequences and crystal structures at the first time, which supported the international community to use these data in a timely manner for joint research, and played an important role in the global COVID-19 traceability research, virus infection mechanism analysis, vaccine and drug development. It provides important support for publication of relevant research achievements in international high-level journals such as *Nature*, *Science and Lancet*, and comprehensive analysis and sharing of domestic and foreign coronavirus data. This move fully embodies Chinese scientists' responsibility for the global data open and sharing and plays an important role in the global new coronavirus traceability research, virus infection mechanism analysis, vaccine and drug development (Fig. 2).

On February 20th, the Chinese team completed the analysis of the structure of COVID-19 S protein, and uploaded the structure data to the National Microbial Science Data Center immediately before the article was published. Six hours later,

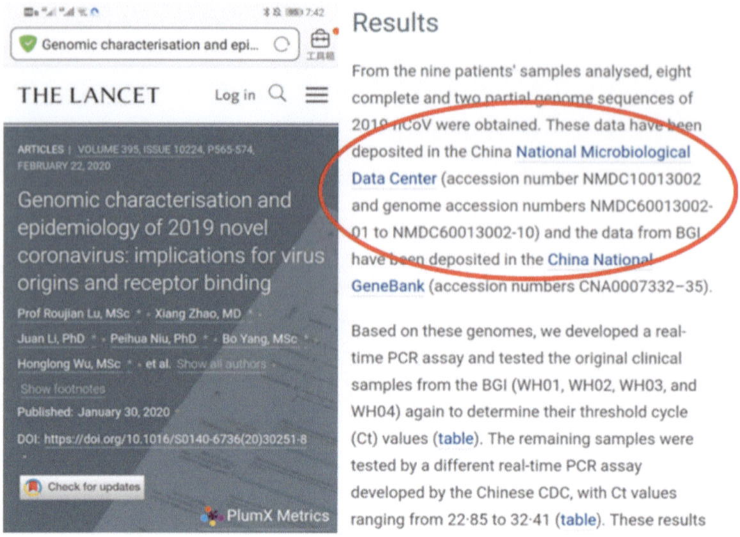
Genomic characterisation and epi...

THE LANCET Log in

ARTICLES | VOLUME 395, ISSUE 10224, P565-574, FEBRUARY 22, 2020

Genomic characterisation and epidemiology of 2019 novel coronavirus: implications for virus origins and receptor binding

Prof Roujian Lu, MSc • Xiang Zhao, MD • Juan Li, PhD • Peihua Niu, PhD • Bo Yang, MSc • Honglong Wu, MSc • et al. Show all authors

Show footnotes

Published: January 30, 2020

DOI: https://doi.org/10.1016/S0140-6736(20)30251-8

Check for updates

PlumX Metrics

Results

From the nine patients' samples analysed, eight complete and two partial genome sequences of 2019-nCoV were obtained. These data have been deposited in the China National Microbiological Data Center (accession number NMDC10013002 and genome accession numbers NMDC60013002-01 to NMDC60013002-10) and the data from BGI have been deposited in the China National GeneBank (accession numbers CNA0007332–35).

Based on these genomes, we developed a real-time PCR assay and tested the original clinical samples from the BGI (WH01, WH02, WH03, and WH04) again to determine their threshold cycle (Ct) values (table). The remaining samples were tested by a different real-time PCR assay developed by the Chinese CDC, with Ct values ranging from 22·85 to 32·41 (table). These results

Fig. 2 The system supports Chinese scientists to publish articles in internationally renowned journals *Lancet*

the data was online and available for download by scientists worldwide, and the international PDB database did not publish the structure data until March 18. On April 9, the internationally renowned journal "Cell" officially published related articles. Chinese scientists quickly shared the data to the world through the national data center in the first time, which fully reflects the responsibility and contribution of Chinese scientists. The rapid release of data is of great value for drug research related to the fight against the epidemic (Fig. 3).

The genome of COVID-19 continues to mutate during the epidemic. With the deepening of the research on the mutation, the functional impact of the mutation has gradually become the focus of attention. The "New Coronavirus Variation Evaluation and Early Warning System (VarEPS)" (www.nmdc.cn/ncovn) was officially released on October 15, 2021 by National Microbiology Data Center and published online on Nucleic Acids Research. It is the world's first system for multi-dimensional risk assessment and early warning of known and virtual variants in the SARS-CoV-2 genome [1].

The VarEPS system is designed to provide users with the latest status of global SARS-CoV-2 genome sequence and variation analysis, as well as different levels of variation such as translation efficiency, secondary structure, ACE2 binding capacity, neutralizing antibody binding capacity, and RT-PCR primer efficiency. The system combines online analysis tools as a navigation and recommendation tool for global virus mutation monitoring. It will help design stable vaccines and neutralizing monoclonal antibodies in the future. The risk level assessment based on virtual mutations

Fig. 3 The structure of COVID-19 S protein

provides key information for the design of preventive antibodies and vaccines against mutations with higher risk levels.

As of January 2022, the number of confirmed COVID- 19 patients worldwide reached 366 million with > 5.6 million deaths. Over 70 vaccines are currently under development and 4 billion vaccine doses have already been administered [2]. Rapid diagnosis and vaccination are still the most effective methods for controlling the pandemic. As a result, it remains crucial to understand whether SARS-CoV-2 variants impact the affinity of current neutralizing antibodies under development or the performance of current diagnostic methods. It is also critical to pay close attention to variants that may escape from protective immune responses induced by population-level immunity. The VarEPS system presented here allows close monitoring and evaluation of the current global status of genetic variations of SARS-CoV-2.

In November, 2021, the number of confirmed cases of COVID-19 in South Africa surged. On November 25th, 2021, experts from the Ministry of Health of South Africa announced the discovery of the COVID-19 mutant and called it a "major threat." On November 26th, 2021, the WHO held an emergency meeting and listed it as a "Variant of Concern" and named it Omicron [3]. Based on the genome data of the virus strain, the VarEPS system provided early warning analysis of the Omicron virus strain through the artificial intelligence model. The VarEPS system evaluates the variation in multiple dimensions from the difficulty of nucleotide variation, the difficulty of amino acid substitution, the effect of variation on protein secondary structure, the change of binding free energy of ACE2 and neutralizing antibodies caused by single amino acid mutations. (Fig. 4).

The prediction of the VarEPS system shows that Omicron may become more infectious through the enhancement of its binding to ACE2, which will affect some

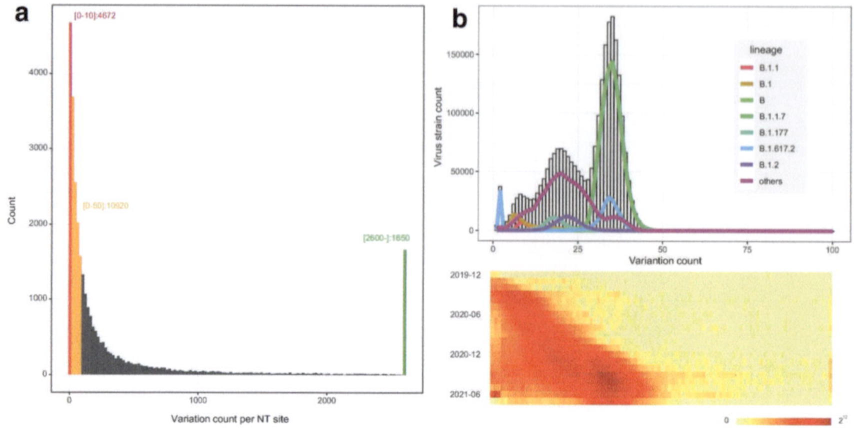

Fig. 4 Statistics of nucleotide mutation numbers in SARS-CoV-2 genomes. **a** Histogram of the mutation count at all nucleotide positions. Red, orange and green bars refer to the frequency of mutation count below 10, below 50 and above 2600, respectively. **b** Histogram of total mutation count in one strain. The heatmap shows the distribution of total mutation count in each month. Mutation counts are accumulating over time and coordinate with lineages

of the current antibodies. The genomic analysis of the Omicron variant strains showed that the mutation sites of the Omicron variant strains were mainly concentrated in the highly variable region of the S protein gene, not in the target regions of the primers and probes of nucleic acid detection reagents published in my country. The site does not affect the sensitivity and specificity of mainstream nucleic acid detection reagents in China.

3 Provide Scientific Data Analysis for Epidemic Decision-Making and Support Precise Prevention and Control

Relying on National Microbiology Science Data Center to integrate relevant domestic and foreign data, a large-scale reference database was quickly established for virus traceability and mutation research, which has formed a rapid response mechanism. After the outbreak in Beijing, using this data platform, researchers quickly and accurately carried out the traceability study of the virus, which provided an important scientific basis for the precise prevention and control of the epidemic, and provided important support for the prevention and control of the epidemic and the resumption of production.

On the basis of the management and release of scientific data resources, the service system further utilizes the advantages of long-term data accumulation and analysis model development experience in the field of microbiology, and strengthens support for microbiological data analysis and mining. It provides a one-stop, standardized data analysis environment for COVID-19 and even the global coronavirus research.It has overall realized an integrated analysis service method from data to results to help scientists around the world to quickly and conveniently conduct research on virus mutation, traceability, and evolution.

4 Help the Global Fight Against COVID-19 with Broad International Influence

Since the outbreak of the epidemic, China has always adhered to the concept of a community with a shared future for mankind by releasing epidemic information in a timely manner, sharing prevention and treatment experience with the World Health Organization and the international community without reservation, and strengthening cooperation in scientific research. The release of information from Novel Coronavirus National Science and Technology Resource Service System has aroused great attention at home and abroad.

As of January 2022, it has provided 21.35 million times of data browsing and searching for 584,000 users from 177 countries and regions around the world, of which 140,000 international users in 176 countries have accessed 4.398 million times, downloaded data 29 million times and files about 50 TB. This system has received widespread attention from the domestic and international community. On November 23, 2020, the system was selected as one of the 15 world leading internet scientific and technological achievements of Wuzhen World Internet Conference (Fig. 5).

5 Prospect for Nest Step

We will carry out scientific and technological anti-epidemic cooperation with the BRICS countries, by providing services for the data management and scientific and technological support of the COVID-19 on the basis of a big data platform, and then form a knowledge base with mutations as the core through comprehensive analysis of public data and self-produced data. And through the research on the functional effects caused by polymorphisms in key regions, a new rapid early warning system will be developed, and by combining imported and local new case epidemiological data to monitor the virus sequence characteristics of imported cases in real time, study and judge the world epidemic situation, and provide support data for epidemic prevention decision-making.

Fig. 5 The system was selected as one of the 15 world leading internet scientific and technological achievements

References

1. Sun Q, Shu C, Shi W, Wu L, Ma J (2021) VerEPS:an evaluation and prewarning system of known and virtual variation of SARS-CoV-2 genomes. Nucl Acids Res, gkab921
2. McCallum M, Bassi J, Marco DA, Chen A, Walls AC, Iulio JD, Tortorici MA, Navarro MJ, SilacciFregni C, Saliba C et al (2021) SARS-CoV-2 immune evasion by the B.1.427/B.1.429 variant of concern. Science 373:648–654
3. Chen JH, Wang R, Wang ML, Wei GW (2020) Mutations strengthened SARS-CoV-2 infectivity. J Mol Biol 432:5212–5226

Juncai Ma Ph.D., senior engineer, is currently the director of the Center for Microbial Resource and Big Data, Institute of Microbiology, Chinese Academy of Sciences, director of National Micorbiology Data Center, director of WFCC-MIRCEN World Data Centre for Microorganisms, director of the Biotechnology and Bioindustry Information Center of Chinese Society of Biotechnology, member of Executive Committee of WFCC, chairman of the data management working group of Asian Network of Research Resource Centers (ANRRC) , and Co-chairman of the Working Group of BOLD Mirror, iBOL.

Data Intelligence-Driven Prevention and Control of COVID-19

Gang Li, Yanlin Gao, Jing Du, Ping Wang, Yuan Li, and Jun Yan

Abstract China has entered a period of ongoing COVID-19 prevention and control. However, imported cases, domestic resurgence and frequent mutations still place intense pressure on the country's COVID-19 prevention and control system. This paper introduces the wide application of data intelligence in the prevention and control of COVID-19, illustrating how data intelligence has contributed to efficient, targeted and scientific COVID-19 prevention and control, including early detection and warning systems, source control, transmission chain disruption and protection for vulnerable populations. This paper provides an insightful analysis of the short-comings in the application of data intelligence and offers an approach to building a data intelligence-driven COVID-19 prevention and control platform and using data intelligence to enhance the capabilities of public health emergency response systems.

Keywords COVID-19 · Data intelligence · COVID-19 prevention and control · Public health · Emergency response system

1 Introduction

COVID-19 has been a major public health emergency for the past several years, with the rapid development of the pandemic, a large number of people were infected worldwide, making it difficult for countries to prevent and control the spread of the disease. It has been a tough test for the world. On March 11, 2020, the World Health Organization declared the COVID-19 outbreak a global pandemic. By September 27, 2022, cumulative confirmed cases and deaths had reached 613.9 million and 6.53 million worldwide and 1.29 million and 5,211 in China, respectively. There have been multiple strains of the virus that causes COVID-19, from Alpha, Beta, Gamma,

G. Li (✉) · Y. Gao · J. Du · P. Wang · Y. Li
Beijing Center for Disease Prevention and Control, Beijing, China
e-mail: ligang@bjcdc.org

J. Yan
Yidu Cloud (Beijing) Technology Co. Ltd., Beijing, China

China's e-Science Blue Book 2023, https://doi.org/10.1007/978-981-99-8270-7_20

Delta to Omicron, putting constant pressure on global prevention and control systems to respond [1–3]. Since the outbreak of COVID-19, the government of China has always put people's lives and health first, making COVID-19 prevention and control a top priority [4–7]. In the face of COVID-19, nationwide preventative and control measures were taken, effectively stopping the spread of COVID-19 and protecting people from contracting the virus. Despite the stringent measures taken, there are still sporadic cases across the country, from Wuhan, Beijing, Ruili, Shenzhen, Yangzhou, Nanjing, to Hulunbuir and Xi'an. The Delta variant has spread to many provinces and cities, and the risk from the Omicron variant is increasing.

As the core driving force of the latest scientific and technological revolution, artificial intelligence (AI) technology is now being applied at a rapid rate in medical care, transportation, education, life, services and other fields. AI is also playing an active role in COVID-19 prevention and control, ensuring people's safety and livelihoods as they return to work.

AI has become the backbone of technology-driven COVID-19 prevention and control, contributing to scientific decision making and targeted measures. However, factors such as data, algorithm innovation and technology integration limit AI application in some complex scenarios and make it difficult to fully exploit the potential of AI. Making practical recommendations based on the impact of and problems with AI application is an important basis for improving the application of AI to public health emergency response systems. With the continuous integration of big data and AI in recent years, the concept of "data intelligence" has gradually emerged. Data intelligence refers to the processing, analyzing, and mining of massive data with the help of big data engines and machine learning technologies; these technologies make data "intelligent" by extracting valuable information, producing solutions to problems and making predictions by building models [8]. It is imperative to shift to data intelligence-driven COVID-19 prevention and control.

2 Data Intelligence-Powered COVID-19 Prevention and Control

Traditional epidemic prevention and control rely on human inputs, including the past experiences of U.S. Centers for Disease Control and Prevention (CDC) staff [9]. Staff members manually fill out many forms, analyze a large amount of information collected, identify the source of diseases and assess the situation based on past experience and cases. The classification of at-risk places and people relies on traditional epidemiological models and experience, making it difficult to evaluate the response to an epidemic [10]. Data intelligence can produce structured information through semantic recognition and information extraction, facilitating analysis and retrieval for professionals. This leads to a more efficient identification of the source of infection [11, 12], at-risk places and people, risk control prioritization (places and people), and evaluation of the epidemic response (Fig. 1).

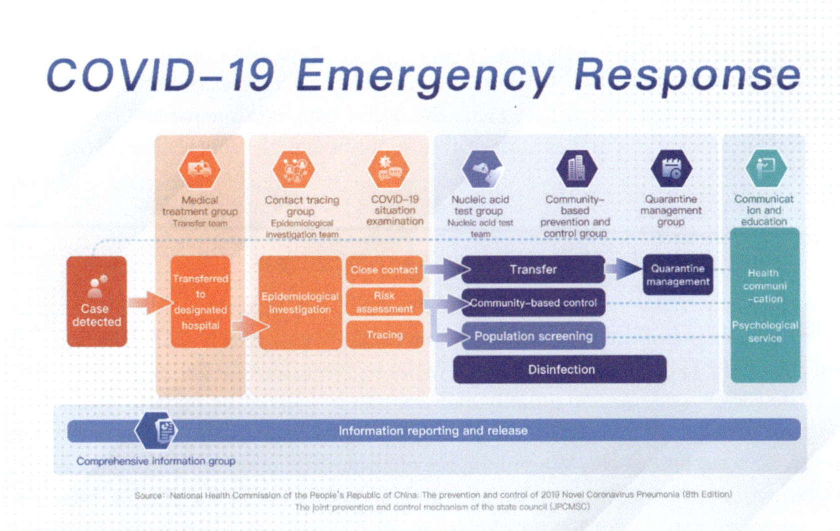

Fig. 1 Higher work efficiency and data accuracy

2.1 *More Intelligent Public Health Data Due to Technological Advances*

In today's world, major public health emergencies have become a considerable threat to human life and health. Uncertain, contagious, systemic and complicated in nature, these emergencies pose serious challenges to traditional governance models. With big data technology, massive, multidimensional, heterogeneous, dynamic data can be collected, but these data are just "big". The key is how to extract valuable information through data mining to turn big data into intelligent data.

AI algorithms and cloud computing can improve the predictions, early warnings and responses to major public health emergencies, optimize risk governance structures and reduce risk governance costs. Core data intelligence technologies can be broadly divided into platform, collation, analysis, interaction and visualization technologies.

2.2 *Data Intelligence-Powered COVID-19 Prevention and Response*

COVID-19 prevention and control are very challenging and involve the use of unprecedented resources and technologies. The application of AI and big data has provided technological support, laying the foundation for targeted and efficient prevention and control policies [13]. Meanwhile, real-time updates and the timely release of COVID-19 information have prevented the spread of rumors and public panic [14]. In particular, governments at all levels have used big data and AI technology to review the COVID-19 situation and develop policies on the resumption of work and production accordingly [15]. The use of data intelligence will be a key tool for targeted prevention and control of future public health emergencies [16].

2.3 The Upgrade of Intelligent Thermometers

A new generation of intelligent thermometers has been used for COVID-19 prevention and control [17]. Thermal cameras, facial recognition, and thermal body temperature detection systems can quickly and accurately collect the body temperature of a person within 2 m and match the person's identity with his or her body temperature to create a data sheet. When a higher-than-normal temperature is detected, the system will automatically issue an alert to help staff identify the person. Intelligent thermometers can detect the temperatures of 200–300 people moving in a single direction within one minute. Upgraded facial recognition systems can recognize masked faces. These kinds of devices are efficient, safe and reliable, reducing human resource needs and the risk of infection transmission among staff and the public.

2.4 Health Code-Based Dynamic Management

Since the outbreak of COVID-19, China has used big data and AI technology to track cases of infection and trace contacts and relied on facial recognition cameras to track people with a history of travel to afflicted areas. Hangzhou, Zhejiang Province, was the first city to introduce a health code to dynamically manage residents and those who visit Hangzhou through green code, red code, and yellow code [18]. The green, red and yellow codes are based on real data, and a color is generated and assigned to individuals based on a review of the information they provide online. These health codes determine travel in Hangzhou. Individuals apply for a health code only once, and it can be used across cities. The introduction of health codes has enabled a targeted, scientific and orderly resumption of work and production [19]. The combination of big data and AI technology makes fast self-tests possible for close contacts during quarantine [20].

2.5 AI Model-Powered Accurate Predictions

In collaboration with several big data and AI technology companies, Zhong Nanshan's team developed an AI model through machine learning based on the classic SEIR model to derive infection curves for Hubei and the country; the model accurately predicted the inflection point and the number of infections in China [21] and demonstrated that the measures taken by the Chinese government, such as delayed resumption of work and production and return to school, played a key role in COVID-19 prevention and control [22]. The team also collaborated with Tencent to develop a model that can predict the progression of COVID-19 illness based on deep learning and big data analysis. With this model, doctors are able to calculate the risk coefficients for severe disease within 5, 10, and 30 days based on 10 characteristics, including patient age, history of malignant tumor, and shortness of breath [23], which helped the treatment of patients at high risk of severe symptoms. The study was published in *Nature Communications*, and the model codes were open source to assist in fighting the pandemic worldwide [24].

3 Whole-Process COVID-19 Prevention and Control Driven by Data Intelligence

The transmission of an infectious disease requires three conditions—source of transmission, transmission chain, and susceptible people—to form the biological basis for transmission [25]. Transmission is also influenced by natural and social factors, which result in different intensities and natures throughout the process. Big data and AI technology cover the whole process of COVID-19 prevention and control, including routine surveillance, case detection, epidemiological investigation, establishment of the source of infection, risk analysis and management of key locations and areas, population risk analysis, at-risk population management, and policy adjustment and evaluation (Fig. 2).

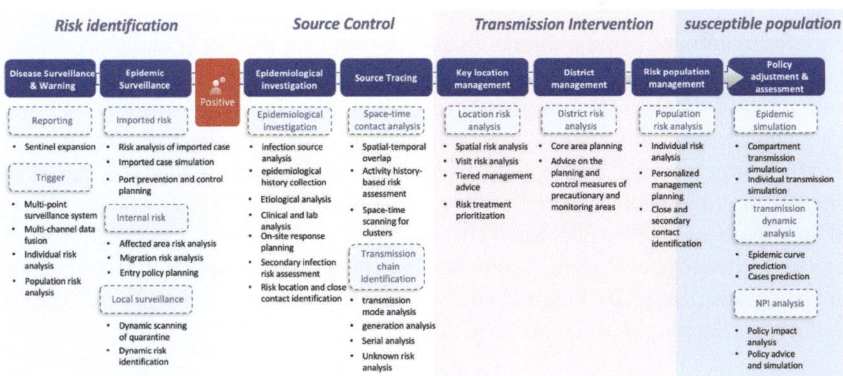

Fig. 2 Whole-process COVID-19 prevention and control

3.1 COVID-19 Surveillance and Warning Systems

The first requirement of a surveillance and warning system is early risk monitoring. Before cases are detected, public health analysts need to continuously monitor potential risks from outside the country and from afflicted areas within the country [26]. Usually, there is both active and passive surveillance [27], and active surveillance is the main tool. The improvement of active surveillance relies on the continuous extension of surveillance points and surveillance workers' strong capability to identify and report risks. Timely detection of cases through active surveillance needs the improvement of the disease surveillance network, including better use of channels and touch points within the network. The second is the risk monitoring of secondary cases [28]. During ongoing COVID-19 prevention and control, public health analysts need to closely follow domestic and foreign COVID-19 situations, analyze risks, and provide global COVID-19-related information in real time to automatically generate risk reports. In addition, other factors, such as the scale of the movements of people across areas, interaction patterns and characteristics, should be considered. These factors can be used to identify the potential local impact of COVID-19 and guide the strategies of ports (Fig. 3).

3.2 Source of Infection Control

3.2.1 Problems

After the outbreak of COVID-19, finding and tracing the movements of infected individuals and their close contacts were problematic [29]. First, the information collected by epidemiological investigations was not accurate because epidemiological investigators were unfamiliar with the geography of an area or patients provided inaccurate information. Second, extracting information, such as time and location from an epidemiological report, is time consuming. Third, it is difficult to integrate and analyze information. It is challenging to identify distance, time of visit, activity pattern or the overlapping movements of infected individuals based on whether multiple people visited a place of the same name or a neighboring place epidemiological investigators know. Big data and AI technology can produce structured epidemiological reports, automatically extract time and space information, and create a movement map of individuals, which supports the use of space and time to identify overlapping movements [30].

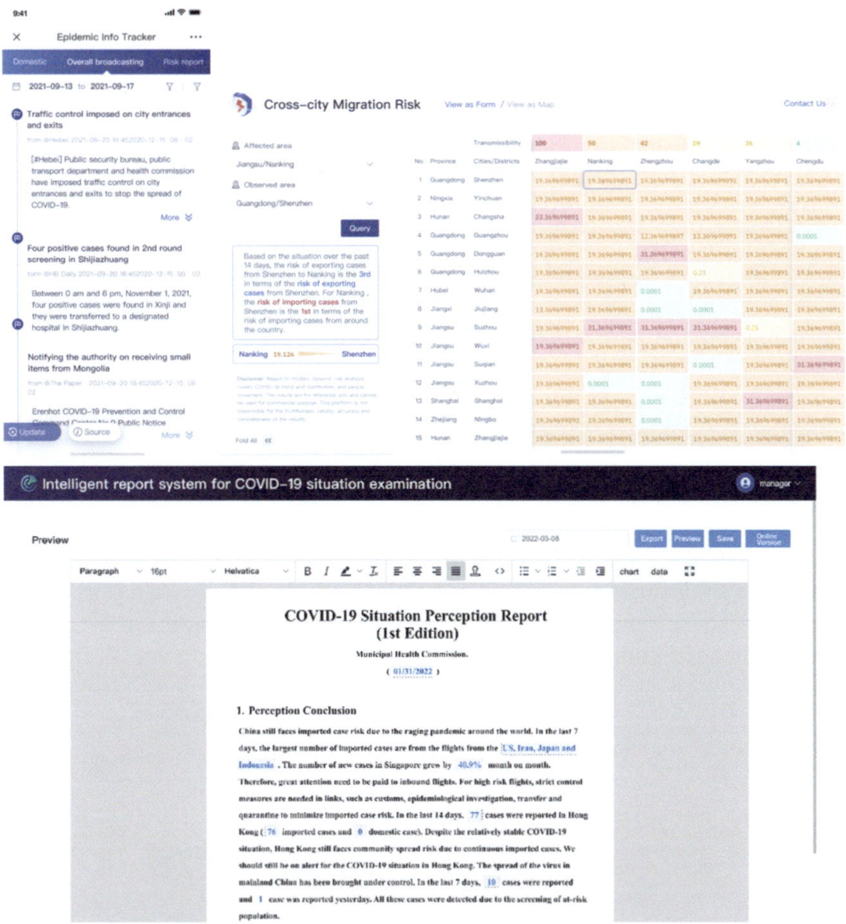

Fig. 3 Domestic COVID-19 situation analysis and customized and intelligent risk evaluation

3.2.2 Intelligent Epidemiological Investigation Information Collection

AI technology can be used to collect epidemiological investigation information from audio recordings, pictures, texts, forms, etc. Nature language processing (NLP) technology can turn audio and pictures into text to identify and extract key information. By building a standard public health terminology system and applying machine learning to text materials, accurate key information and identification modules can be established [31]. AI systems can accurately identify key fields, including basic information (name, age, sex, date of birth, residential address, company name, etc.), onset of disease and consultation information (symptoms, signs, time of onset, complications, routine blood test results, imaging features of pneumonia, consultation records, etc.), risk factors and exposure history (occupation, past medical history, travel history to risk area, history of being part of cluster cases, etc.), epidemiological information (recent 14-day travel history, contact with other people, etc.) and laboratory test results (nucleic acid test results, viral load, antibody levels, etc.). The structured information processed with NLP technology is automatically entered on an epidemiological investigation form, enhancing the flexibility, ease of use and fault tolerance of information collection [32].

3.2.3 Intelligent Epidemiological Investigation Information Processing

The spatiotemporal information visualization system for epidemiological investigation can link the home address, company name and movements of an individual collected from epidemiological investigation with geographic information, including latitude and longitude, to spatially visualize an afflicted location. Spatial and temporal information are used together to define high-risk movements and analyze the time of movements based on pathogenic findings. These are used to identify movement patterns and create a relationship diagram. Based on close contact information, the tasks—the epidemiological investigation of close contacts and the screening of key places—are automatically generated. Finally, based on epidemiological investigation results and the templates of epidemiological investigation reports from CDCs at all levels, the information is automatically entered into a report in doc format. This report can be used for reviewing, publishing and archiving. (Diagram 4) (Fig. 4).

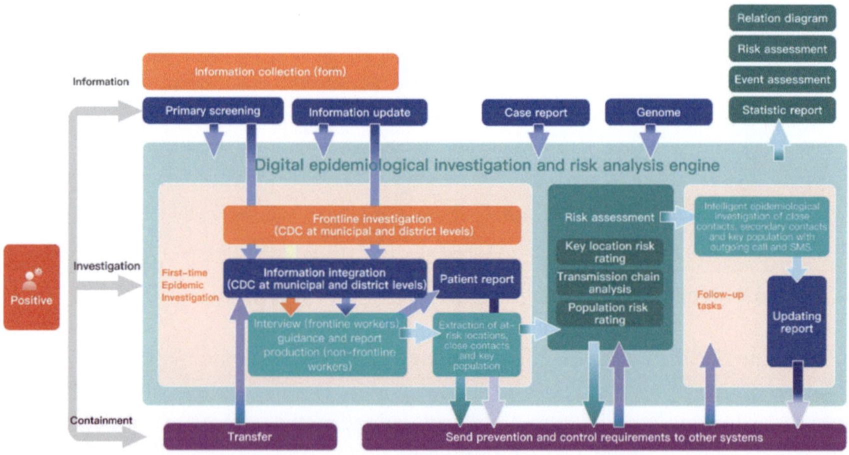

Fig. 4 Intelligent epidemiological investigation

3.2.4 Intelligent Case Investigation

Space–time contact is an important basis for case investigation. AI technology can dynamically adjust risk factors: time, space and transmissibility (whether staying in the same space with a confirmed COVID-19 patient, within an area 800 m in length and 800 m in width, for over 10 min four days before the appearance of symptoms) to determine risk threshold and conduct space and time overlap analysis (using big data technology to find close contacts based on overlapping time and space). Points with cross-infection risk are quickly identified and marked in space and time dimensions based on space–time overlap analysis. Transmission mode, the number of generations of spread, time of infection and future unknown risks are analyzed to determine the transmission chain and identify the cross infection risk [33, 34]. Models such as the virus spread inference diagram—a diagram that uses the relationships among confirmed cases of the disease to determine the order of the spread—can calculate the probabilities of virus spread among individuals. This model proved successful in identifying 95% of the related cases in the COVID-19 waves in Shunyi District, Beijing and Guang Zhou (May 21, 2021).

AI technology can automatically extract case-related information from epidemiological investigation reports and create a relationship diagram with name, event, location, etc. The diagram supports the demonstration of relationships among infected individuals and relationship description, providing a basis for the analysis of the source of infection in cluster cases and individual cases. The points in the relationship diagram represent infected individuals, and the lines among the points are their relationships. Each relationship is described in epidemiological investigation reports. The relationship descriptions in epidemiological investigation reports and the whole report can be accessed online (Figs. 5, 6 and 7).

3.3 Cutting off the Transmission Chain and Protecting Susceptible People

3.3.1 Dynamic Risk Tracking of Key Locations and Areas

Based on the epidemiological investigation of an infected person, all the locations visited by the infected person since he or she theoretically became infectious, including indoor and outdoor locations, should be identified. The locations should be sampled and disinfected, and people at risk of infection should be quarantined and managed. For major infectious disease events, in the event of resource shortage, the screening of risk locations should be prioritized. In cross-sectional analysis, public health analysts need to make an overall risk assessment based on the movements of all cases. When it has become clear that the transmission chain cannot be cut off, they need to advise local governments on control measures based on differentiated risk ratings, the locations covered by the measures, and how to implement the measures

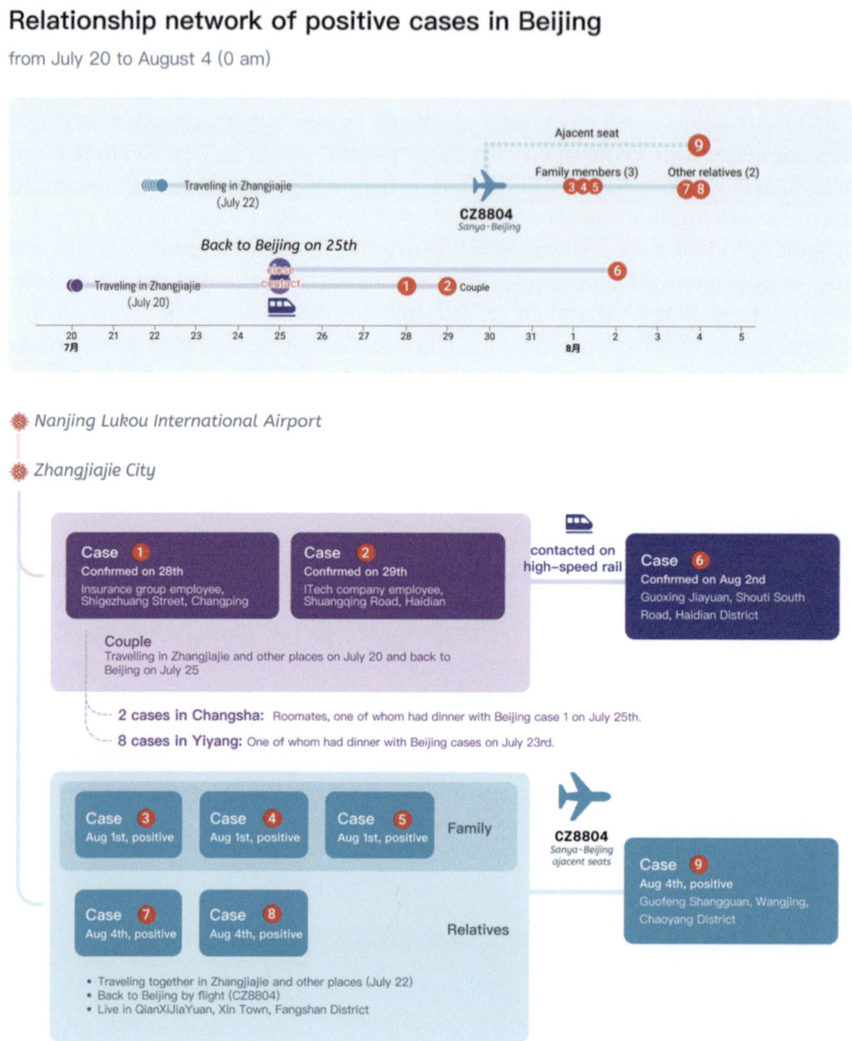

Fig. 5 Transmission chain

(e.g., duration, sampling, disinfection, etc.). The use of machine learning to identify and calculate risk factors can make this process significantly less subjective by relying less on the experience of public health analysts to rate the risks of related locations [35].

The first step in dynamic risk tracking of key locations and areas is judging area risks. AI and big data technology can be used to define the boundaries of areas with different risk levels (high, moderate and low) with coordinates in a two-dimensional

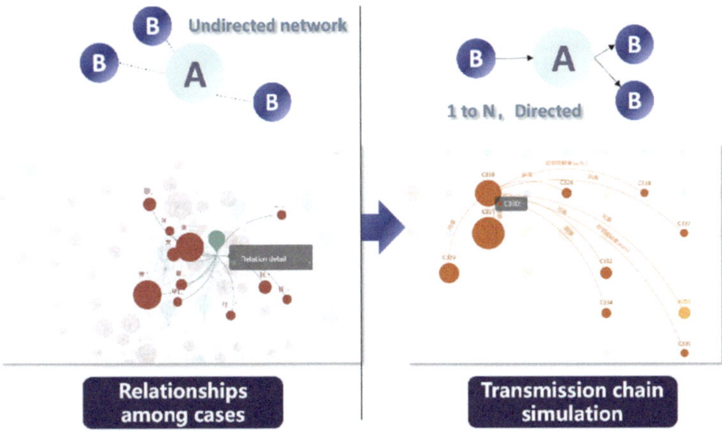

Fig. 6 Space and time overlap and relationship identification

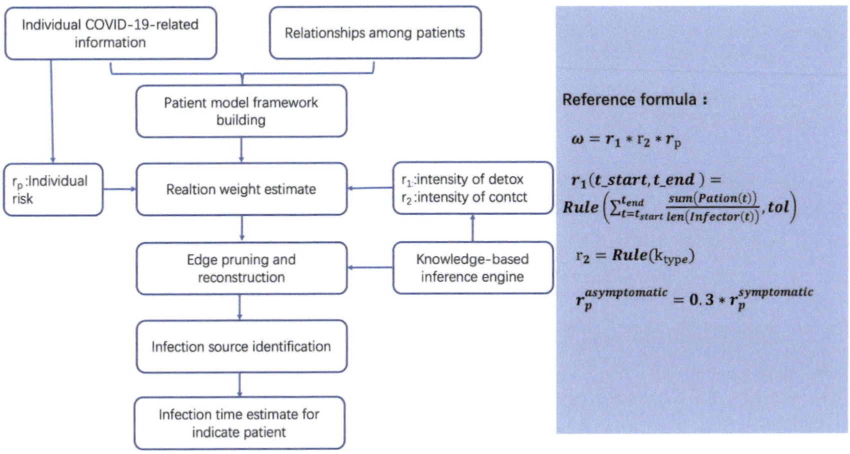

Fig. 7 Inference figure-based probabilities of virus spread among cases

space and assign different colors to different risk levels. Then, based on the movements of and locations frequently visited by the infected individual, high-risk locations and susceptible people can be identified to provide a basis for policy. The second step is to analyze location risk. Based on experience, rules and research findings, some factors, including the time of the individual's visit to the location, the health status of the individual during the visit and the social and physical characteristics of the location, are factors in disease transmissibility and key to judging the degree of risk. At the same time, based on secondary cases and the health status of the individual's close contacts, a dynamic risk adjustment model is used to rate the risks of the locations visited by the infected individual, including space risk, visit

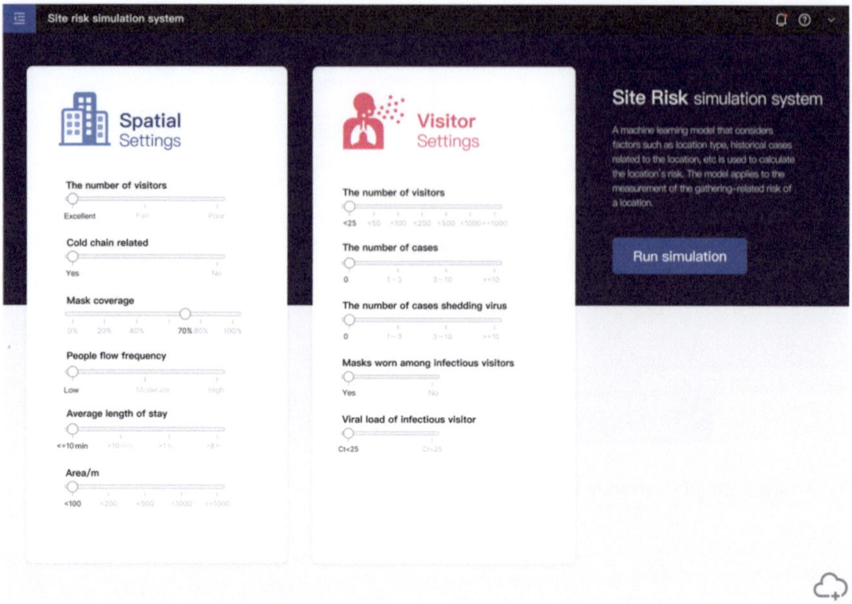

Fig. 8 Case-visited location risk calculation (the spread of infection)

risk, dynamic risk and overall risk. Omission analysis refers to the use of AI-based visualized analysis to show all locations visited by the individual to ensure that they are all covered by control measures (Fig. 8).

3.3.2 Identification and Management of At-Risk Individuals

Larger cities with greater population densities are more likely to face greater risks and find it more difficult to identify and manage at-risk people. Based on the investigation of an infected individual's interactions with others (the distance between the individual and others, interaction duration and interaction method) after the individual theoretically became infectious, close contacts and secondary contacts should be identified. Personalized control measures are applied to the two kinds of contacts [36]. For highly transmissible diseases with a short incubation period, further investigation should be carried out on close contacts, and control measures should also cover close contacts to stop the spread of the disease [37]. A multisource and multidimensional data fusion model featuring transmission risk analysis and space and time data analysis can identify the movements of the individual and people exposed to him or her, slowing the spread of the disease at the source and improving the management of at-risk people.

The first step in the identification and management of at-risk individuals is the risk analysis of close contacts. Based on the data of past COVID-19 close contacts, machine learning can be used to dynamically learn the characteristics of close contacts who become confirmed cases and build a corresponding risk probability model. The model can automatically assess the risks of close contacts and calculate the likelihood of a close contact becoming a confirmed case. Public health workers can rate the risks of close contacts based on this likelihood and prioritize control measures for close contacts. The system can show the current residential addresses of all close contacts on a map, assess the risks of areas according to close contact distribution and filter close contacts by administrative division and risks. The second step is to identify the relationship between an individual and others. Data intelligence systems can create a relationship diagram by automatically extracting information from epidemiological reports, such as name, event and location. The system supports the visual display of the relationship between the infected individual and others with a relationship description, thereby assisting with the analyses of both cluster and individual cases. The third step is the identification of the transmission chain of the cluster case. The transmission chain identification model can identify the direction and intensity of the spread of the disease through accumulated molecular epidemiology knowledge and the real data from field epidemiology. The model automatically generates patient zero, shows the transmission chain (cluster case) and indicate the existence of sporadic cases based on the individual's relationship network, leading to scientific and efficient epidemiological investigation.

3.3.3 Dynamic Risk Monitoring During Disease Control

Risk monitoring and identification have always been the preconditions for risk mitigation [38]. On the basis of routine monitoring, dynamic risk monitoring is enabled for targeted control measures. Even if known risks can be controlled, continuous risk monitoring and perception are still needed for the implementation of control measures, quarantine locations, etc. Data analysis and reporting of abnormalities can be used to identify risks, such as quarantine location-related exposure to infection and in-hospital infection due to mismanagement. The building of an intelligent emergency response system enables data analysis, simulation and forecasting. A unified interdepartmental platform for data fusion and efficient coordination that integrates data from different sources (public health, healthcare and population) pushes data sharing across departments, industries and levels, supports the intelligent analysis of big data, including space and time analysis and forecasting, data visualization and dynamic risk monitoring and provides data support for policy.

3.3.4 Policy and Response Evaluation

Control measures and the implementation of these steps play a decisive role in controlling virus spread. Public health analysts need to conduct a detailed analysis of detected cases and the response (what went well and what went wrong) to provide important advice for process optimization, organizational structure adjustment and team capability building related to major infectious disease response. AI can be used to improve the dynamic learning ability of models to improve their adaptability and scalability to enable intelligent forecasting and warning and policy simulation. An agent-based model [39] can be used for local transmission simulation (a method of using physical and mathematical models to analogize and simulate the real situation and its evolution to identify patterns). The model, which considers the impact of individual health status and activity on disease spread, is applicable to gatherings in a specific space such as conferences and sporting events. A compartmental model [40] can be used for massive simulations. The model, which considers the outbreak and development of COVID-19, is applicable to spread across populations and areas.

Since the outbreak of COVID-19, countries have taken various measures to stop the spread of the virus. However, it is a struggle to control the spread due to the continuous emergence of new variants. Intelligent models can be used for policy simulation, which predicts spread speed and population infection rate based on factors such as control measures targeting schools and people, public transportation restrictions, vaccination rate and vaccine effectiveness. These models can provide a basis for governments' policy and response evaluations by showing the impact of different policies on virus spread (Fig. 9).

Fig. 9 Policy simulation and forecasting based on variant and vaccination coverage

4 Problems Identified in the COVID-19 Response

4.1 Lack of Intelligent Epidemiological Investigation and Efficient Information Collection

Currently, epidemiological investigators interview people face to face to obtain information to identify close contacts and secondary contacts. Then, they create epidemiological reports and a task list. The whole process requires 6–12 h. Significant human resources are needed to respond to variants or complicated individual cases [40]. Epidemiological investigations often contain asymmetric and incomplete information, which means that more work needs to be done, creating an extra burden [29]. An efficient epidemiological investigation is key to COVID-19 prevention and control. Data are now calculated based on the confirmation dates of symptomatic and asymptomatic cases and do not include data about people who tested positive after the first test [41]. It takes a long time to match close contacts with data such as their cell phone numbers, and information does not flow smoothly (1–2 days in key links). In addition, letters for interprovincial cooperation on the COVID-19 response need to be sent manually.

4.2 Lack of IT-Based Prevention and Control and Interconnected Multisource Data

Due to the lack of a unified data platform for prevention and control, all kinds of data, including infected people, close contacts, secondary contacts, key population management, environmental monitoring, fever clinics, pharmacy drug purchases, nucleic acid tests, etc., are isolated information islands and need to be manually integrated, which results in the need for considerable human resources, inefficiency and various sources of statistical data. This also leads to mismanagement in some closed loop work, such as the identification, confirmation, transfer and management of close contacts [42]. The Ministry of Industry and Information Technology can only provide activity history at the level of prefecture-level cities. Provinces need to collect the activity history at the county-level cities before the information can be used. In practice, data are from different sources (health authority, public security, etc.) and the same data are sent to public health workers at the grassroots level many times, hence the low efficiency of the work [43]. The data of close contacts are also problematic.

4.3 Intelligent Situation Examination and Data Processing and Analysis Capability to Be Improved

A large amount of information of different kinds is needed. However, it is not efficient to collect the information, and the information is of varying quality, making the examination of the situation difficult [37]. As COVID-19-related data are multimodal and heterogeneous, classification, cleaning, integration, reduction and desensitization are needed to improve the usability of data. For data to be used in highly accurate analyses, it is necessary to label the data with domain expertise and human engineering, turn raw data into data suitable for machine learning and deep learning algorithms, and establish a domain knowledge base, training resource base, and evaluation sample base to meet the application requirements of AI [36]. Situation examination parallels prevention and control. The heavy workload of data analysis and inefficient situation examination affect decision making.

4.4 Ineffective Simulation Exercise and Insufficient Data Intelligence Scenarios

COVID-19 is a highly transmissible infectious disease with insidious onset, posing a new challenge to health workers. Simulation exercises are an important part of preparatory work and a main way to improve emergency response, covering emergency plans, emergency preparation, emergency teams and emergency mechanisms. China lacks a standardized simulation exercise focusing on early warning, mobilization, situation handling, material management, alert and control, on-site control and recovery. Therefore, the evaluation of simulation exercises is subjective. COVID-19 has revealed many problems in emergency response systems, and simulation exercises need improvement [44, 45].

4.5 Bottlenecks in Underlying Algorithms and Application Algorithms and Insufficient AI Technology Integration Capability

In the face of algorithm bottlenecks, China's research institutes and AI enterprises should work on innovating underlying algorithms and application algorithms and promote the research and development of common key AI technologies. They should also push the integration and application of a new generation of technologies with AI as the core, guided by the "new infrastructure" strategy (including 5G base stations, ultrahigh voltage, intercity high-speed railroads and urban rail transport, new energy vehicle charging piles, big data centers, AI and industrial internet). Studies on convolutional neural networks and recurrent neural networks (original core model, algorithm and framework) should also be undertaken for improvement. Work should be done to explore the application of deep reinforcement learning, transfer learning, meta-learning and generative adversarial networks in the analysis of unlabeled data sets, small sample data and semistructured and unstructured data to enable machines to better automatically identify rules, modes, connections and trends in large amounts of data and to optimize the whole process, from mathematical modeling and algorithm design to training. With the defects in the interpretability of black box AI's analysis of results, efforts should be made to improve modeling theory for vaccine manufacturing, drug research and development, etc., to advance the integration and application of a new generation of information technologies [34, 36].

5 Improvement and Innovation Directions

5.1 "New Infrastructure" Synergy-Propelled Prevention and Control

Synergy from "new infrastructure" should be created to meet the diverse needs of COVID-19 prevention and control [35]. The first step would be to build an intelligent cloud platform by integrating cloud computing with AI. The second step would be to build an Internet of Intelligence by integrating the Internet of Things with AI to enable the interconnection of intelligent entities (human, machine, and object) at the semantic level to meet key needs: real-time monitoring of residents' health status, medical material dispatching and tracking, targeted management of people under quarantine, intelligent production and intelligent community service to lend strong support for large-scale social collaboration and intelligent service. The third step would be to reshape the trust mechanism by integrating blockchain with AI to enable the transparency of COVID-19-related information and secure use of data among medical institutions at different levels. On this basis, a blockchain-driven automatic data synchronization network at the national, provincial, and city levels should be established to better respond to public health emergencies. The fourth step would be to build digital infrastructure for COVID-19 prevention and control by integrating 5G with AI to accelerate the validation and demonstration of high-definition video, virtual reality, autonomous driving and robotics to enable high-definition cloud broadcasting, remote diagnostics and unmanned operations in high-risk scenarios during ongoing COVID-19 prevention and control.

5.2 *Shifting Toward Data Intelligence-Driven Prevention and Control*

In terms of COVID-19 prevention and control, data intelligence should be applied to more scenarios. Achieving full scenario coverage from dynamic warning, diagnosis, treatment, and social governance to the resumption of work and production is an important means to better leverage AI to respond to major public health emergencies. It is essential to move toward an intelligent prevention and control mode with data at the core: intelligent warning, intelligent diagnosis and treatment, intelligent social governance and intelligent resumption of work and production. Epidemiological investigation should be based on big data-enabled government service platforms, and it is imperative to break down data silos to integrate data from different sources: medical data (diagnosis, treatment, medical resources, research progress, etc.), transportation data (road monitoring, traffic, etc.), public security data (population, law enforcement, etc.) and community management data (user-provided information, volunteer information, etc.). The integration channels of government service data, internet user behavior data and operator data should be expanded with open partnerships to provide sufficient data for the building and application of AI algorithms covering clinical diagnosis, drug research and development, situation examination, traffic control, resource allocation and the communication of public opinion to present a full picture of the situation for scientific and accurate decisions [34].

5.3 *Optimizing Public Health Service Resources to Reduce Costs for Society*

Data intelligence technology should play a larger role in prevention and control, situation examination, transmission chain analysis, health monitoring of the floating population and targeted policy. The data channel should remain smooth to create more efficient COVID-19 prevention and control and reduce costs for society. A big data command center should be established to build an efficient mechanism for data communication among provinces, cities and counties; to collect, integrate, analyze and share data from various sources; to guide local COVID-19 responses; and to achieve more efficient allocation of public health service resources.

6 Summary and Outlook

We are in the era of big data and digital transformation, and data are ubiquitous. The need to be data driven has become a consensus, and the value of data has been demonstrated in scientific research and other areas. However, without the extraction of knowledge and information from data and their effective use, data alone cannot lead to the success of digital transformation. Data intelligence is the key to maximizing the value of data.

Data intelligence is an interdisciplinary research field that combines large-scale data processing, data mining, and machine learning with human–computer interaction, visualization and other technologies. It can discover, extract and obtain revealing and actionable information from data to provide intelligent support for data-based decisions or performing tasks. Data intelligence produces value with data analysis, turning raw data into information and knowledge that can support decisions or actions. It has become a key technology for digital transformation, a technology that is increasingly important and that has experienced rapid development in recent years. With the implementation and development of data intelligence, new applications, scenarios and challenges will sustain its momentum and push it to a higher level. Going forward, data intelligence will become more automatic, more intelligent, more reliable, more widely applicable and more efficient.

At the same time, we should be keenly aware that in the era of broad AI applications, the development of AI involves not only technological advances but also ethical and legal issues for AI tech companies, China and the world. AI legislation has a direct impact on its application. In fact, making AI laws is a major issue that needs to be thoroughly investigated and may be the most difficult legislation in human history. AI technological development and AI legislation should go hand in hand because boundaries are required to avoid the abuse of AI and to ensure that it can truly benefit humanity. General Secretary Xi Jinping noted that it is a priority to enhance the early monitoring and warning capability of the public health system, including improving the monitoring system of infectious disease and public health emergencies and the monitoring mechanism for diseases of unknown cause and abnormal health events and establishing an intelligent early warning mechanism with multiple trigger points. During the prevention and control of COVID-19, the big data-driven response became a buzzword in the news media. Big data analysis can support situation examination and control measures and help monitor the condition of the floating population. The prevention and control of major infectious diseases is a long-term, complex and systemic project and an important part of modern national governance. Since the fight against COVID-19 began, many local governments have worked with high-tech companies through big data platforms, leading to more efficient responses and valuable experiences. A still raging COVID-19 pandemic means greater difficulties and challenges ahead for the world. In the future, data intelligence-driven technologies, such as big data and AI, will likely play a more prominent role in the prevention and control of COVID-19.

References

1. Xuefeng H (2021) Reflections on the prevention and control of COVID-19 in Wuhan [J]. Sociol Rev China 8(2):8–12

2. Liu NQ, Zhang F, Wei C et al (2020) Prevalence and predictors of PTSS during COVID-19 outbreak in China hardest-hit areas: gender differences matter [J]. Psychiatry Res 287:112921
3. Smith JA, Judd J (2020) COVID-19: vulnerability and the power of privilege in a pandemic [J]. Health Promot J Austr 31(2):158–160
4. Sisi K (2020) China's COVID-19 response as a responsible major power [J]. Hongqi Wengao 4:19–20
5. Xin S (2020) China's contribution to the world's fight against COVID-19 as a responsible member [J]. Lei Feng Magazine 4:2
6. Liu S (2020) World situation under COVID-19 (IV): China's choice as a responsible member [J]. 2020(7):2
7. Huiming L (2021) The complex ecological relations between China and the world and China's missions under profound changes unseen in a century [J]. Teach Res 9:13
8. Xiaofeng M (2021) Scientific data intelligence: AI for scientific discovery [J]. Bull Natl Nat Sci Found China 35(03):419–425
9. Zhang H, Hanlei S, Lun X (2020) The capability building of COVID -19 prevention and control from big data perspective [J]. Acta Math Appl Sin 43(2):468–481
10. Wang S (2021) Big data-based monitoring, warning and response system of major infectious diseases [J]. 2021:72–75
11. He SB, Peng YX, Sun KH (2020) SEIR modeling of the COVID-19 and its dynamics [J]. Nonlinear Dyn 18(3):1–14
12. Yin S, Zhang N (2021) Prevention schemes for future pandemic cases: mathematical model and experience of interurban multi-agent COVID-19 epidemic prevention [J]. Nonlinear Dyn 27:1–36
13. Yaoting C, Jinping S (2020) The inspiration of artificial intelligence application in the prevention and control of COVID-19 virus [J]. J Res Nurs 34(7):1117–1118
14. Xiaoxiang H (2020) On the warning and information release mechanism of infectious diseases [J]. J Nanjing Med Univ (Soc Sci) 1:1–4
15. Yifei C, Wei Z (2020) Thoughts and suggestions on post COVID-19 resumption of work and production [J]. Decis & Inf 5:21–27
16. Xu H, Peng K, Wei N (2021) The inspiration of COVID-19: prediction-empowered decision—from traditional models to big data solutions [J]. J Public Manag 18(4):116–125+173
17. Whitelaw S, Mamas MA, Topol E et al (2020) Applications of digital technology in COVID-19 pandemic planning and response [J]. Lancet Digit Health 2(8):435–440
18. Yunfei Z (2020) Health code: automatic rating of individual risks [J]. Zhejiang Acad J 3:8
19. Zhen Y (2020) The important role of telecommunication technology in the fight against COVID-19 (part 2 of 2) [J]. Forum Adv 5:6
20. Mashamba-Thompson TP, Crayton ED (2020) Blockchain and artificial intelligence technology for novel coronavirus disease 2019 self-testing [M], vol 10, no 4. Multidisciplinary Digital Publishing Institute, p 198
21. Xiao S, Wang P (2020) Academician Zhong Nanshan's latest view on scientific approach to fighting COVID-19. 12(3):14–15
22. Li D, Yu J (2021) The application of AI in the fight against COVID-19 [J]. Sci Technol & Innov 2021(7):136–137+139
23. Zhong Nanshan team and Tencent's cooperation on using a deep learning model to assess a patient's risk of critical illness from COVID-19 [J]. Shanghai Med & Pharm J 41(17):13. (2020)
24. Liang WH, Yao JH, Chen A et al (2020) Early triage of critically ill COVID-19 patients using deep learning [J]. Nat Commun 11(1):1–7
25. Siyan Z (2012) Epidemiology [M], 7th edn. People's Medical Publishing House, Beijing
26. Zijian F, Keli L, Lianmei J et al (2008) Study on the occupational distribution of people who discover and report public health events through internet-based surveillance system [J]. Chin J Epidemiol 1:1–4
27. Jian C, Yaxu Z, Dechuan K et al (2020) Practice and thinking of acute respiratory infection surveillance for the response of emerging respiratory diseases in Shanghai [J]. Chin J Epidemiol 41(12):1994–1998

28. Liu B, Xiao G, Chao S (2021) Research on the construction of risk assessment and decision-making model of major public health emergency [J]. J Inf Resour Manag 11(5):17–26+37
29. Guangjun C, Jijiang S, Yunxi L et al (2020) Preliminary study on experience in epidemiological survey of COVID-19 epidemic [J]. Chin J Nosocomiology 30(8):1147–1151
30. Sun P, Wang L, Zhang J (2020) Design and implementation of aided decision-making system for epidemiological investigation of suspected COVID-19 cases [J]. Chin Med Equip J 41(6):5–10+31
31. Ting L, Mengxuan W (2020) Design and implementation of a NLP-based system for the examination of COVID-19 situation [J]. Telecommun Inf 6:21–25
32. Fei X, Jiang L, Chen P et al (2020) Exploration of the change trend analysis of the epidemiological history of COVID-19 patients based on natural language processing [J]. China Digit Med 15(5):76–78+106
33. Yang W, Lan Y, Lyu Wei et al (2020) Establishment of multi-point trigger mechanism and multi-channel surveillance mechanism for intelligent early warning of infectious diseases in China [J]. Chin J Epidemiol 41(11):1753–1757
34. Zhao Y, Cao W (2020) The inspiration of artificial intelligence application in the prevention and control of COVID-19 virus [J]. J Inf Resour Manag 2020(6):20−27+37
35. Allam Z, Jones DS (2020) On the coronavirus (COVID-19) outbreak and the smart city network: universal data sharing standards coupled with artificial intelligence (AI) to benefit urban health monitoring and management [J], Healthcare (Basel) 8(1):46
36. Alimadadi A, Aryal S, Manandhar I et al (2020) Artificial intelligence and machine learning to fight COVID-19 [J]. Physiol Genomics 52(4):200–202
37. Qu S, Yang D (2020) Intelligent response to public health emergencies: theoretical retrospection and trend analysis [J]. Reform 3(14):14–21
38. Hongshan Y (2020) Grassroot level experience and thoughts on improving major risk identification capability [J]. Governance 27:21–24
39. Cuevas E (2020) An agent-based model to evaluate the COVID-19 transmission risks in facilities [J]. Comput Biol Med 121:103827
40. Yuan L, Yanlin G, Gang L (2020) The construction and application of the field epidemiological investigation system for COVID-19 [J]. Chin J Health Inform Manag 17(5):627–631
41. Chuan T, Ruonan L (2020) Analysis and discussion on the application of information technology in COVID-19 prevention and control [J]. World Sci-Tech R & D 42(4):426–438
42. Wenjing X (2020) Data governance from the perspective of major epidemic prevention and control: main value, realistic dilemma and optimization path [J]. Shanxi Archives 251(3):22–28
43. Hui W, Chengyun Z, Junhua C et al (2020) Characteristics of trans-provincial close contacts of patients with coronavirus disease 2019 in Sichuan province [J]. Chin J Public Health 36(11):1579–1581
44. Hongyan L, Pengju W, Tao Z et al (2021) Application of emergency drill in emergency management of COVID-19 [J]. Chin Nurs Manag 21(4):555–558
45. Libo L, Minglei S, Dandan Z et al (2020) Improvement of public health emergency preparedness system in China: in view of coronavirus disease 2019 epidemic [J]. Chin J Public Health 36(12):1693–1696

Gang Li has a medical doctorate degree. He is a chief physician, a postgraduate supervisor, the director of the Information Statistics Center of the Beijing Center for Disease Prevention and Control, the vice chairman of the Committee on Health Big Data and Intelligent Application of Chinese Preventive Medicine Association, the chairman of the Committee on Health Statistics and Big Data Application of Beijing Preventive Medicine Association and on the editorial board of Capital Journal of Public Health. His researches focus on the application of statistical methods, epidemiology and etiology.

Construction and Current Status of the International Authoritative PIWI-Interacting RNA Database Platform

Shunmin He, Jiajia Wang, and Peng Zhang

Abstract piRNA (PIWI-interacting RNA) is a kind of small noncoding RNA mainly expressed in germ cells and bound to PIWI proteins. It plays an important role in germ cells and gene regulation. Due to the huge amount and complexity of related data resources and the lack of systematic integration, a large number of data have not been fully mined and utilized, which limits the further study of piRNA. In order to interpret the massive data more effectively and assist in its functional studies, Center for Big Data Research in Health, Institute of Biophysics, Chinese Academy of Sciences has built a comprehensive database platform piRBase (piRNA database) that integrates a variety of piRNA data and targets information. Now the platform has collected more than 180 million non-redundant piRNA sequences, covering 440 datasets of 44 species. These data resources are integrated, analyzed and annotated from many aspects, which provides important support for the functional study of piRNA. This paper introduces the piRBase database platform from the aspects of platform composition, data collected, multidimensional information annotation, and platform application services, and briefly summarizes outcomes based on the platform and current status.

Keywords PIWI protein · piRNA · Database platform · Functional study · High-throughput sequencing

1 Introduction

PIWI-interacting RNA (piRNA) is a class of small noncoding RNA interacting with PIWI proteins. piRNAs are usually 24–31 nucleotides (nt) in length, and mainly expressed in animal germ line [1, 2]. The discovery of piRNA has opened up a new field for noncoding small RNA. Due to its large number, piRNA has attracted extensive attention once discovered, and was rated as one of the top 10 Breakthrough

S. He (✉) · J. Wang · P. Zhang
Institute of Biophysics, Chinese Academy of Sciences, Beijing, China
e-mail: heshunmin@ibp.ac.cn

China's e-Science Blue Book 2023, https://doi.org/10.1007/978-981-99-8270-7_21

of the Year in 2006 by *Science* [3]. As an important class of noncoding RNA, piRNA plays key roles in transposon silencing and post-transcriptional gene regulation, as well as in stem cell differentiation, development, chromosome integrity, epigenetic regulation, heterochromatin formation and sex determination [4–14]. Also, there are increasing studies indicating that piRNAs are involved in many diseases [15–19].

There are also more and more piRNA data resources released. Since 2006, there are as many as 1600 related papers and 3500 related datasets of various types. However, the lack of systematic integration for the huge amount of data hindered the full use of them and limited further studies on piRNAs. For better interpreting these data and assisting piRNA functional studies, it is necessary to integrate different types of piRNA data and related information. A number of piRNA databases emerged since 2008. However, information included in these databases are relatively simple. For example, piRTarBase [20] only collected target genes in worm, piRNAclusterDB [21] only included piRNA cluster related data, piRDisease [22] only collected disease related piRNAs, and is currently unavailable. piRNABank [23] and piRNAdb have limited species, with no follow-up updates. It is necessary to build a comprehensive database with wide coverage, full annotation, more data types, user-friendly interface and continuous updates.

piRBase is a database platform that integrated different types of piRNA data and target information to assist piRNA functional study. piRBase collected huge amount of records and related functional studies, especially piRNA target genes. piRBase was published for the first time in 2014, and updated twice up to now. The platform annotates piRNAs in various aspects, provides gold standard piRNA sets, collects high quality piRNA clusters and sequence variants, and provides data basis for different levels of piRNA studies. In addition, many useful applications are provided, such as online tools, batch download, and data visualization for piRNA and target regulatory networks, showing their relationships more intuitively. piRBase provides researchers with a database resource with more species, comprehensive annotation, complete data types, better user experience, and frequent updates. The platform is a basis and support for studying piRNA functions and mechanisms.

2 Composition of piRBase Platform

piRBase is a comprehensive resource platform based on piRNA related big data, which unifies data collection, storage, analysis, integration and application.

Based on efficient, professional, safe and stable software and hardware facilities (infrastructure layer and platform software layer), the platform imports a large amount of data collected in literature and relevant databases into the database (data support) after cleaning, processing, analysis and integration, and then uses various means of application service layer to interact with scientific researchers through interface display layer according to the data content. The detailed composition of the platform is shown in the figure below (Fig. 1).

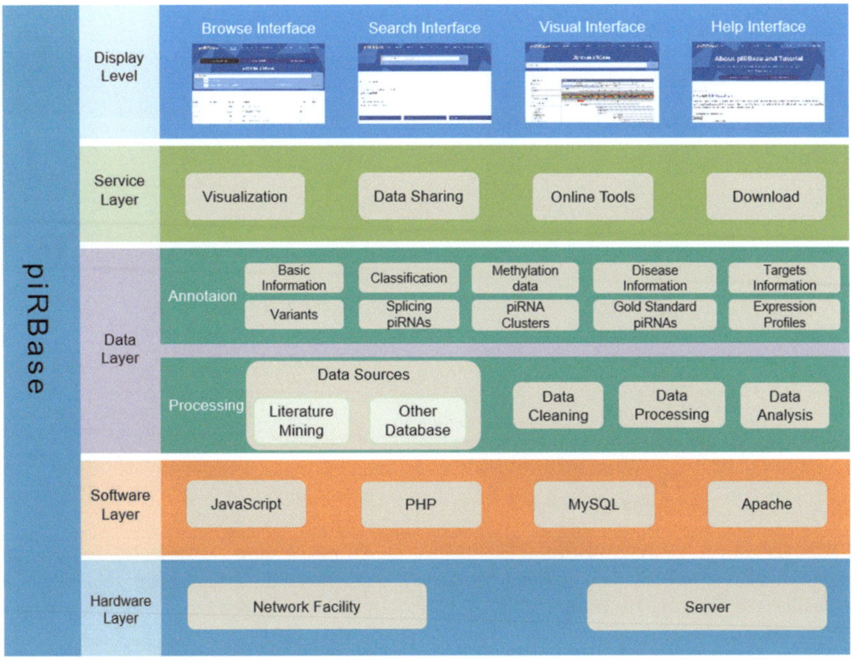

Fig. 1 Composition of piRBase platform

3 Data in piRBase

3.1 Amount of Data in piRBase

The number of non-redundant piRNA sequences in piRBase reached more than 180 million, covering 440 datasets of 44 species (Fig. 2), surpassing other relevant databases in the world (Table 1), such as piRNABank [23] (0.18 million piRNAs in 5 species), piRNAQuest [24] (0.99 million piRNAs in 3 species), and IsopiRBank [25] (60 million piRNAs in 4 species). piRBase is the piRNA database with the largest amount of data at present, and it is updated periodically to ensure the timeliness and cutting-edge of data contents.

3.2 Multidimensional Annotation in piRBase

In view of the functional diversity of piRNAs, in order to facilitate studies, piRBase not only collects the basic information of piRNA, such as sequence, length, source datasets, genome location, and expression level, but also annotates from many other

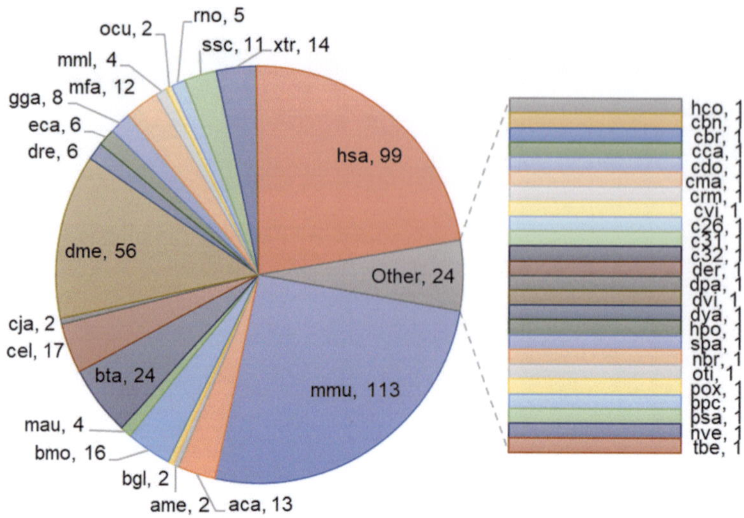

Fig. 2 Datasets in piRBase platform. hsa: human, mmu: mouse, dme: *D. melanogaster*, xtr: *X. tropicalis*, ssc: pig, rno: rat, ocu: rabbit, mml: rhesus, mfa: crab-eating macaque, gga: chicken, eca: horse, dre: zebrafish, cja: marmoset, cel: *C. elegans*, bta: cow, mau: golden hamster, bmo: silkworm, bgl: *B. glabrata*, ame: giant panda, aca: sea hare, hco: barber pole worm, cbn: *C. brenneri*, cbr: *C. briggsae*, cca: *C. castelli*, cdo: *C. doughertyi*, cma: *C. macrosperma*, crm: *C. remanei*, cvi: *C. virilis*, c26: *Caenorhabditis* sp26, c31: *Caenorhabditis* sp31, c32: *Caenorhabditis* sp32, der: *D. erecta*, dpa: *D. pachys*, dvi: *D. virilis*, dya: *D. yakuba*, hpo: *H. polygyrus*, spa: mud crab, nbr: *N. brasiliensis*, oti: *O. tipulae*, pox: *P. oxycercus*, ppc: *P. pacificus*, psa: *P. sambesii*, nve: starlet sea anemone, tbe: tree shrew

Table 1 Comparison of data in piRNA databases

	piRBase	piRNABank	piRNAQuest	IsopiRBank	piRNAdb
Human (hsa)	8,592,949	35,356	41,749	4,564,080	27,700
Mouse (mmu)	68,542,499	55,359	890,078	38,093,003	54,865
Rat (rno)	4,081,625	46,444	66,758		54,475
D. melanogaster (dme)	41,950,613	44,417[a]		15,058,093	21,570
C. elegans (cel)	30,036				15,913
Zebrafish (dre)	1,330,692			6,367,811	
Platypus		51[b]			
Cricetulus griseus					25,600[b]
Other species	57,088,926				

[a] The number of unique piRNAs may be less than shown. [b] piRNA sequences were not provided in the literature, and were not included in piRBase

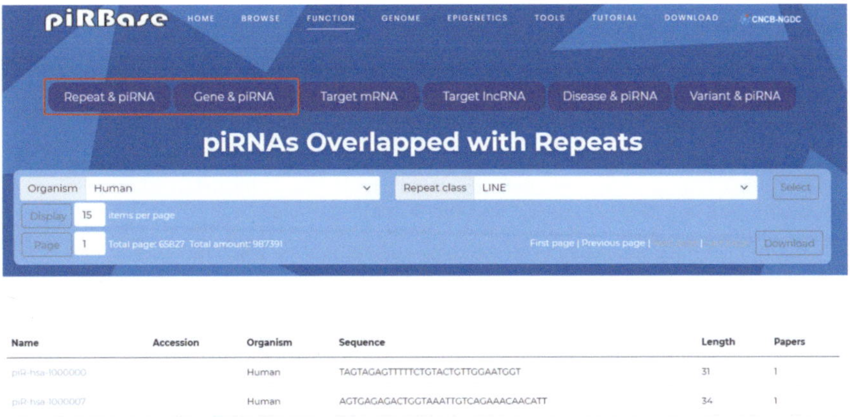

Fig. 3 Repeat derived and gene derived piRNA sequences

aspects, so as to expand the direction of studies and provide more candidate interpretations for the functional abnormalities caused by piRNA, promoting piRNA function related researches from more aspects.

3.2.1 Source of piRNA Sequences

Bowtie software [26] was used to align piRNA sequences to the corresponding genome to obtain the location information. According to the genome location, piRBase divides piRNAs into two categories: repeat-derived and gene-derived (Fig. 3) according to piRNA sequence sources.

3.2.2 Gold Standard piRNA Set

Given the large number of piRNAs and many detection methods, in order to obtain a more representative piRNA set and make studies more effective, the platform introduces the concept of gold standard set based on two typical characteristics of piRNA (one is interaction with PIWI protein, and the other is 2′-O-methylation modification at the 3′ end). According to these two characteristics, piRNAs can be enriched by immunoprecipitation of PIWI protein and oxidation treatment. Firstly, the datasets obtained from the above two types of experiments and background control small RNA sequencing were collected. Then the collected data sets were analyzed using binomial distribution model to obtain significantly enriched piRNAs. Finally, the gold standard set is annotated in piRBase (Table 2).

Table 2 Statistics of gold standard piRNA set

Organism	No. of sequences
Human	77,242
Mouse	85,563
Rat	2578
D. melanogaster	68,707
Caw	5640
Crab-eating macaque	26,309

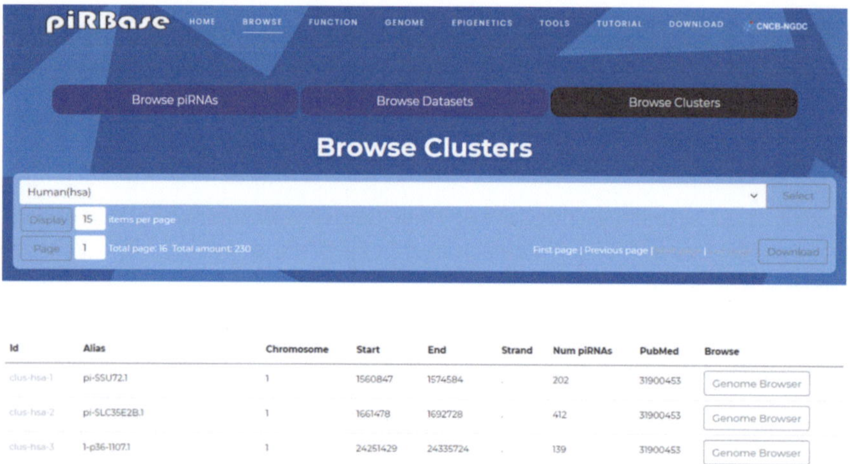

Fig. 4 Information of piRNA clusters in piRBase platform

3.2.3 piRNA Clusters

Many piRNAs originate from large genomic regions, which are called piRNA clusters. piRNA clusters can be transcribed to long single-stranded precursors and further processed into mature piRNA sequences. piRBase collects and organizes high-quality piRNA cluster-related information (Fig. 4, Table 3), such as the genomic location, length, and literature sources. Users can access the detail page by clicking the name of the piRNA cluster to obtain a list of piRNAs originated from the cluster. The inclusion of piRNA cluster-related information provides support for various levels of piRNA studies.

3.2.4 Variants in piRNA

Single nucleotide polymorphism (SNP) is a common type of genomic variation, and studies have shown that many SNPs are closely associated with phenotypes and

Table 3 Statistics of piRNA clusters

Organism	No. of piRNA clusters
Human	230
Mouse	620
Rat	123
Rhesus	168
Golden hamster	2246

Table 4 Statistics of piRNA variants in piRBase

Organism	No. of variants	Organism	No. of variants
Human	1151	Silkworm	313
Mouse	53,473	*B. glabrata*	119
Cow	15,657	*D. virilis*	103
D. melanogaster	12,504	*D. erecta*	90
Crab-eating macaque	3445	*D. yakuba*	81
Rabbit	2315	Rhesus	49
Horse	2222	Tree shrew	21
Marmoset	2044	Giant panda	10
Chicken	1430	Zebrafish	8
Rat	1086	Mud crab	2
X. tropicalis	868	Golden hamster	2
Pig	528		

diseases. The genome alignment results showed that not all piRNAs are perfectly matched in the genome, and mismatches between piRNA and the reference genomic sequences may be caused by genomic variants. In order to obtain the variant sites in piRNA sequences, we selected the alignment records with only one mismatch from the Bowtie results of uniquely mapped sequences, and counted the number of piRNAs with the same mismatch sites. The variants on piRNA sequences were identified if the mismatch sites were supported by more than 10 sequences. piRBase recorded variant loci with genomic position, type of locus variation and the number of piRNAs supporting the variation (Table 4). For human piRNA variants, we also annotated with dbSNP database and provided links to other SNP functional annotation databases (Fig. 3, Variant & piRNA).

3.2.5 Splicing-Junction piRNAs

It was reported that some piRNA precursors contain introns, while some piRNAs cannot be localized to the genome by Bowtie's full-length alignment strategy. The

Table 5 Number of piRNA target information in piRBase

Organism	mRNA	lncRNA
Mouse	3240	1199
D. melanogaster	57	
C. elegans	27,040	
Silkworm	1	

alignment strategy of STAR [27] was added to the analysis pipeline to mine some potential piRNA sequences of splice site origin. The information about the splicing-junction has been added to the piRNA detail page. These piRNAs were also annotated based on the positional relationships with repetitive sequences and genes.

3.2.6 piRNA Targets

Based on datasets collected by piRBase, our team found that mammalian piRNAs have post-transcriptional cleavage effect on mRNAs (messenger RNAs). By integrated analysis of piRNA and 5' RACE data, and combining with experimental validation, the phenomenon of piRNA guided cleavage of mRNA target was demonstrated in mouse [14]. Based on these findings, a method was developed to predict piRNA target genes based on high-throughput sequencing data [28]. In addition to the piRNA target genes collected from the literature, the target lncRNAs (long non-coding RNAs) were also predicted and displayed based on the prediction method and are available for query download, providing reliable candidate target genes for relevant researchers (Fig. 3, Target mRNA and Target lncRNA). Currently piRBase contains mRNA target genes for four species (mouse, *D. melanogaster*, *C. elegans* and silkworm) and lncRNA target genes for one species (mouse) (Table 5).

3.2.7 piRNA/piRNA-like sRNAs and Disease

More and more studies show that there is a correlation between piRNA and diseases. In order to assist further studies on related diseases, piRBase has added a module on diseases. At present, more than 300 reported piRNA records in 13 types of cancer (breast, bladder, pancreatic, gastric, liver, kidney, myeloma, colorectal, thyroid, lung, prostate, malignant glioma, ovarian), cardiovascular disease, stroke, Parkinson's and Alzheimer's disease are included (Table 6), and the information recorded includes disease type, piRNA name, expression (up/down regulation and variation), functional description, and literature sources. These information can be queried and displayed on the disease module and piRNA information page, which can be used as an important basis for precision medicine research.

Table 6 Number of records in disease related piRNAs

Disease	No. of records	Disease	No. of records
Bladder cancer	198	Lung cancer	3
Stroke	24	Cardiovascular diseases	3
Parkinson	20	Ovarian cancer	2
Breast cancer	16	Prostate cancer	2
Colorectal cancer	12	Liver cancer	1
Kidney cancer	6	Myeloma	1
Gastric cancer	5	Thyroid cancer	1
Glioblastoma	4	Pancreas cancer	1
Alzheimer's disease	3		

4 Application Services in piRBase Platform

Since the release of piRBase, our team has been continuously developing and updating the application services of the platform (Fig. 5) to provide users with more convenient operation and more intuitive presentation. In addition to data browsing and data searching provided in other piRNA databases (e.g., piRNAQuest [24], piRNABank [23], etc.), piRBase also provides visualization, quick query, online tools, data sharing, and batch download of piRNA-related data.

4.1 Data Visualization

piRBase uses JBrowse Genome Browser to visualize piRNA loci and other annotation, including variant loci, target gene, related epigenetic data, etc. In addition, the expression in different datasets and the regulatory relationship between piRNAs and target genes (Fig. 6) are also visualized to provide users with a more intuitive display.

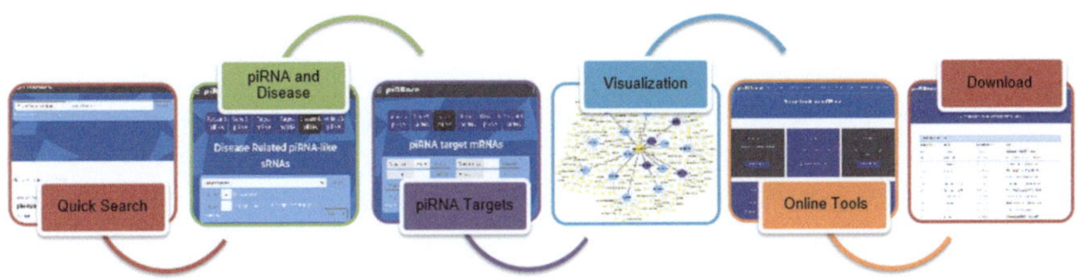

Fig. 5 Application service and display in piRBase

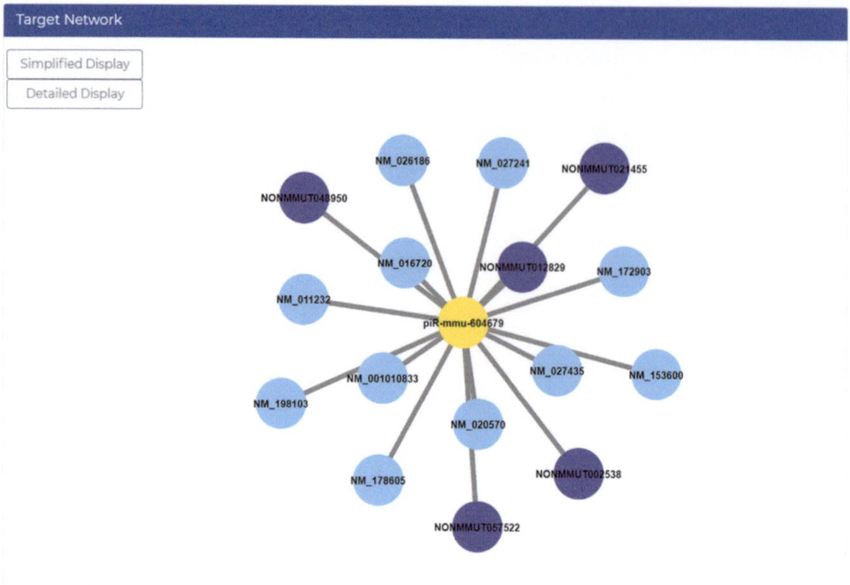

Fig. 6 Visualization of piRNA target regulatory relationships

4.2 Quick Query

The platform home page provides a quick search tool (Fig. 7), which allows users to query for relevant information by entering the piRBase ID. The search tool provides two search modes: piRNA full collection search and gold standard set search, which allows users to conduct more accurate searches according to their needs.

4.3 Online Tools

For better user experience, an online tools module is also provided in piRBase (Fig. 8). This module provides researchers with some quick start data analysis tools, including ID conversion, sequence to name, sequence reverse complement, online Bowtie, hypergeometric test, etc. Among them, the ID conversion tool supports the conversion of IDs from multiple database sources to piRBase IDs for data standardization, which facilitates further querying of piRNA-related information.

Fig. 7 Quick search tool in piRBase home page

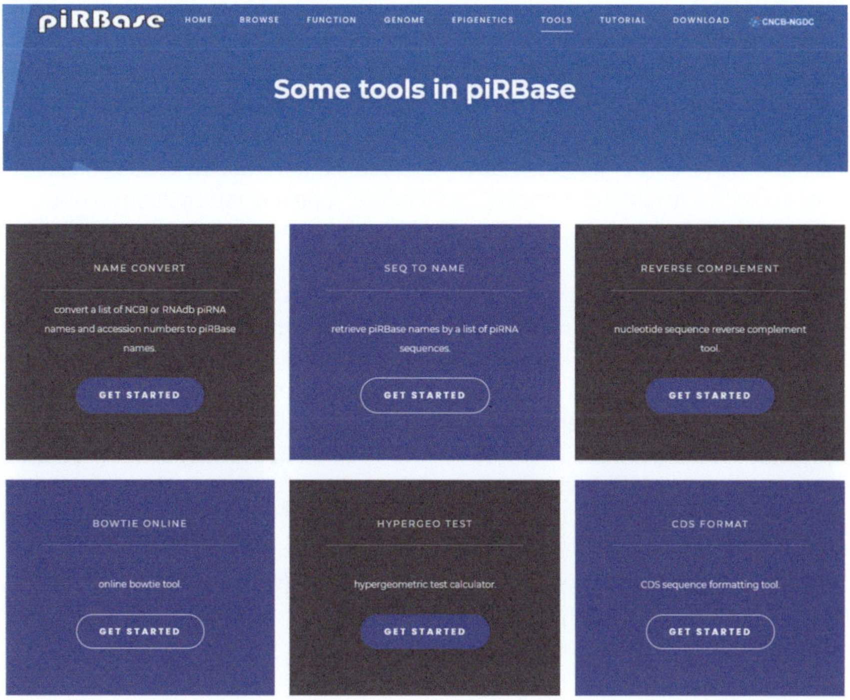

Fig. 8 Online tools in piRBase

4.4 Data Sharing and Batch Download

piRBase does not require registration, and relevant data information can be accessed, browsed, searched and downloaded directly. The download page provides the corresponding piRNA data files according to species, and users can directly download the compressed files in batch. In addition, the platform also provides customized data download. Users can search and get the corresponding result list according to their needs in the browse module, and download the relevant results directly through the result page.

5 Outcomes and Current Status of piRBase Platform

5.1 piRBase Related Outcomes

In 2014, piRBase was officially released. After that, our team systematically studied piRNAs in nematodes and the regulatory relationships between these piRNAs and other small RNAs. In 2015, we systematically discovered the cleavage and regulatory effects of piRNAs on coding genes [14], and developed the target gene prediction algorithm [28]. Based on these results, piRBase was upgraded to greatly improve the comprehensiveness. In addition, we periodically updates and expands the data content to ensure the timeliness and cutting-edge content. As related studies are going on, multi-dimensional information of piRNAs is mined, new modules and contents are added according to research progress, and the corresponding information is collected, organized, included and displayed to present a more comprehensive annotation of piRNAs.

Based on piRBase, a series of scientific researches have been published. Through studying the related data, it was found that piRNAs in mice recognize mRNA target genes through incomplete base pairing and cut mRNA targets through piRNA-binding MIWI protein, thus promoting sperm development and maturation. If this regulatory mechanism is abnormal, spermatogenesis will be affected, leading to male infertility. Similarly, it has been shown that in human, mutations in the piRNA-interacting protein HIWI can lead to abnormalities in the regulatory pathway, blocking spermatogenesis and leading to male infertility. piRBase and relevant scientific publications are shown in the following figure (Fig. 9).

piRBase is currently the only expert database on piRNA included in RNAcentral, the international RNA database consortium. piRBase joined the Chinese National Genomics Data Center (https://ngdc.cncb.ac.cn/) in 2019.

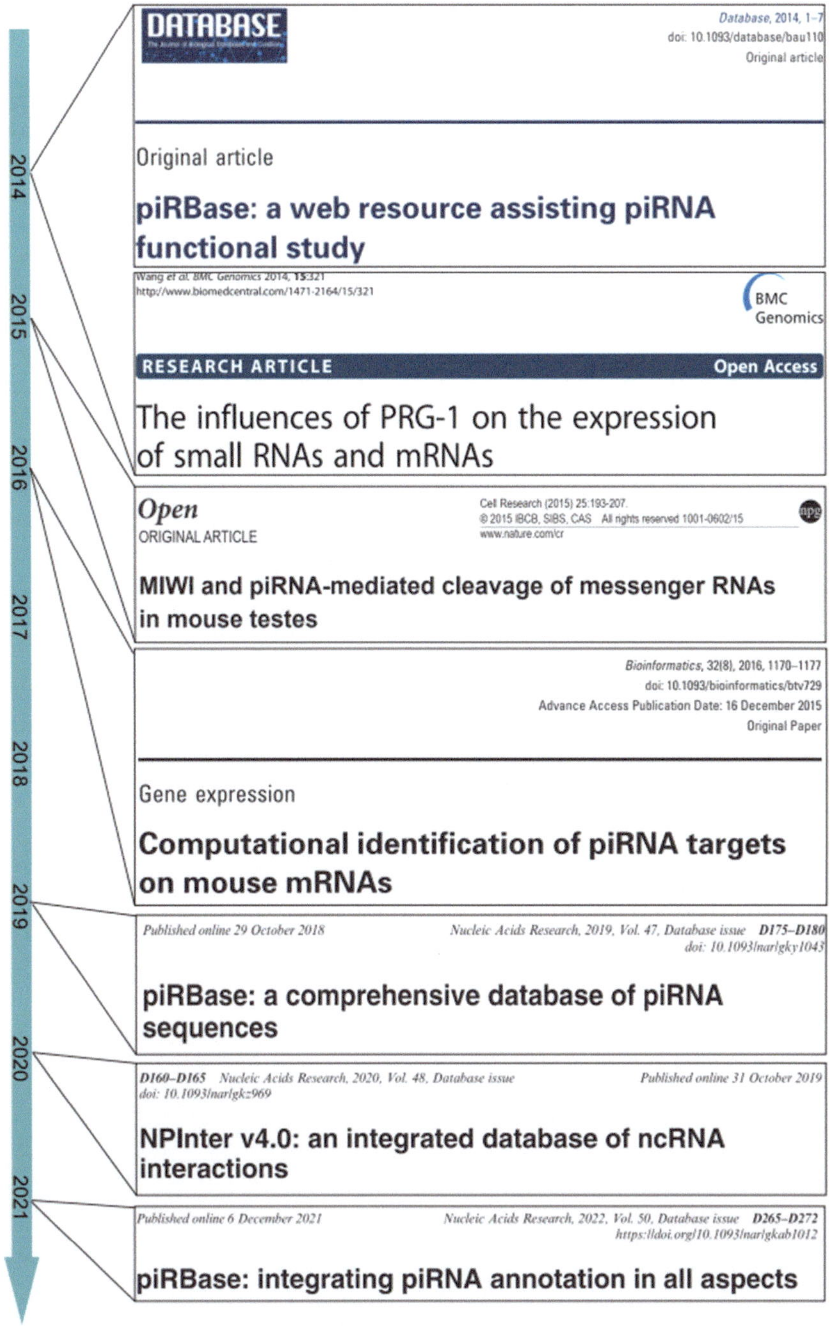

Fig. 9 Publications related to piRBase platform

5.2 *Current Status and Impact of piRBase Platform*

The piRBase was published in 2014, and has updates twice since then. The citations of piRBase related scientific publications have increased year by year, and the total citations have reached more than 300 (Fig. 10), including citations from top journals like Nature and Cell (Fig. 11). About two-thirds of the citations are from foreign researchers in related fields (Fig. 12). For example, based on the genomic positions of human piRNAs provided by piRBase, Martinez, V.D. et al. found that 15 piRNAs highly expressed in human placenta are localized in the DLK1-DIO3 (14q32.31) imprinted region, suggesting their potential roles in placental function [29]. All these reflect the strong support role of the platform for scientific research. Since the official release of the third version in November 2021, by the end of February 2022, the website have accumulated more than 60,000 total visits, with 1,588 unique access IPs.

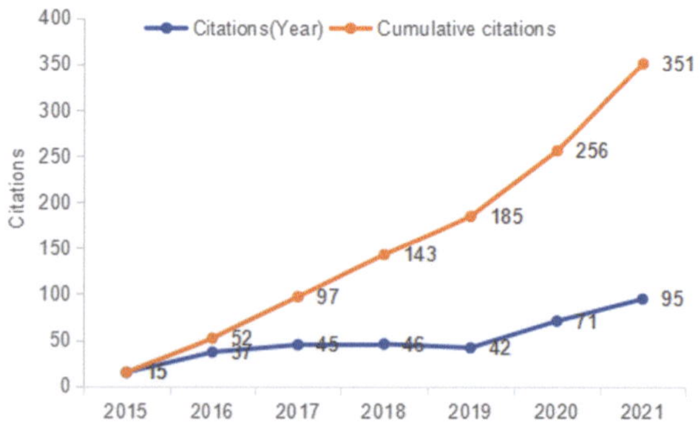

Fig. 10 Citations of piRBase platform. X axis: year; Y axis: number of citations

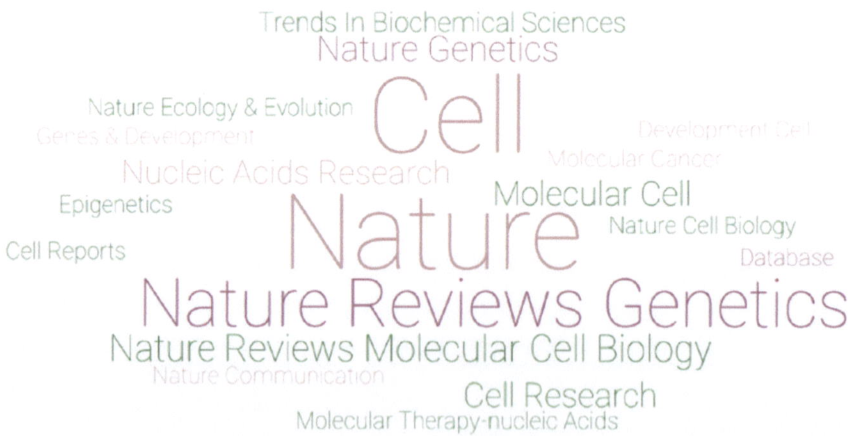

Fig. 11 Citation profile of piRBase platform

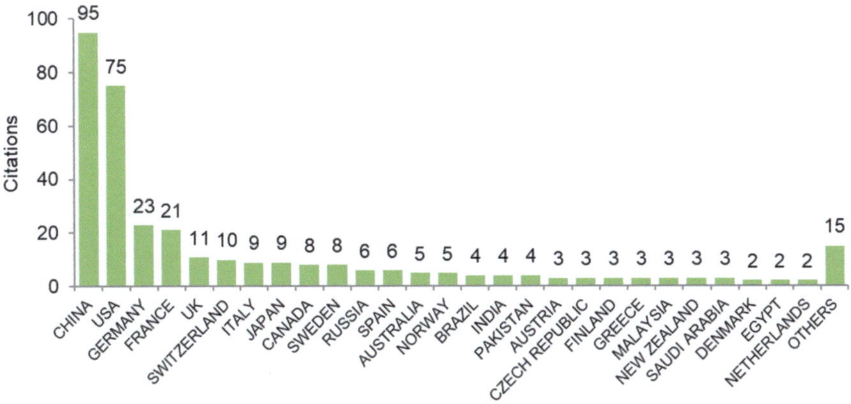

Fig. 12 Number of citations in different countries

The RNAcentral is a non-coding RNA database resource developed by EBI, integrating non-coding RNA information from several expert databases such as Ensembl, miRBase, Rfam, etc., aiming to provide a unified reference for non-coding RNA research. piRBase is included as an expert database by RNAcentral, which is the only piRNA expert repository currently included. Through data sharing with RNAcentral by March 2021 (Fig. 13), the international influence was expanded to contribute to further studies in the field of piRNA.

Since the release of piRBase, timely updates have been made based on user feedback to make piRBase continuously improved and provide better user experience. Through academic communications with researchers in related fields in China and abroad, the content of piRBase platform has been enriched to further expand the influence of the platform in the field of piRNA study.

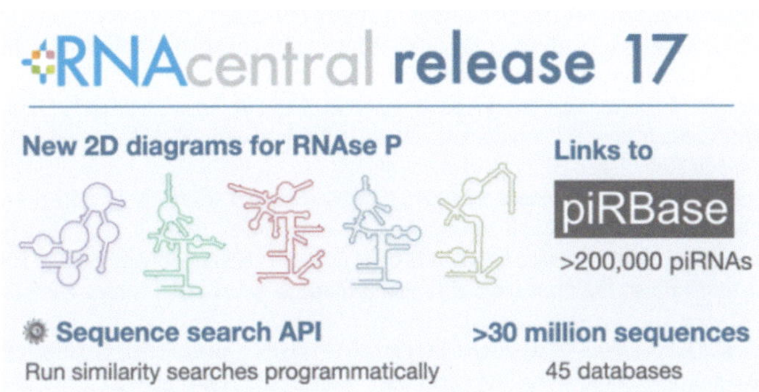

Fig. 13 Data sharing with RNAcentral

6 Conclusion

The database platform of piRBase aims to integrate a variety of piRNA data and target information for supporting piRNA functional studies, and is currently the largest piRNA database in the world. Integration and informatization of piRNA data provide convenience for researchers to utilize published omics data. In order to better utilize the informatization technology as well as to ensure the timeliness and cutting-edge of data content, piRBase will be comprehensively updated and improved periodically.

piRBase platform provides featured application services, such as quick query, online tools, batch download, etc. In addition, the platform also provides customized result download, users can search according to their needs and download relevant results on the result page. Based on the feedbacks from worldwide users, piRBase is constantly improved, and good user experience helps generate greater social value. Through data sharing with RNAcentral, piRBase expands its international influence and further promotes the development of related research fields. piRBase conducts academic communications actively with international researchers to promoting international scientific cooperation and further realizing the value of the piRBase platform.

In this paper, piRBase is introduced from the composition, the data included, annotation of piRNAs and application services. piRBase provides an important support for the functional study of piRNAs and good user experience. The platform will surely make valuable contributions to the progress of piRNA field.

References

1. Girard A, Sachidanandam R, Hannon GJ et al (2006) A germline-specific class of small RNAs binds mammalian Piwi proteins. Nature 442(7099):199–202
2. Aravin A, Gaidatzis D, Pfeffer S et al (2006) A novel class of small RNAs bind to MILI protein in mouse testes. Nature 442(7099):203–207
3. Breakthrough of the year (2006) Areas to watch in 2007. Science 314(5807):1854–1855
4. Ozata DM, Gainetdinov I, Zoch A et al (2019) PIWI-interacting RNAs: small RNAs with big functions. Nat Rev Genet 20(2):89–108
5. Sienski G, Donertas D, Brennecke J (2012) Transcriptional silencing of transposons by Piwi and maelstrom and its impact on chromatin state and gene expression. Cell 151(5):964–980
6. Pezic D, Manakov SA, Sachidanandam R et al (2014) piRNA pathway targets active LINE1 elements to establish the repressive H3K9me3 mark in germ cells. Genes Dev 28(13):1410–1428
7. Teixeira FK, Okuniewska M, Malone CD et al (2017) piRNA-mediated regulation of transposon alternative splicing in the soma and germ line. Nature 552(7684):268–272
8. Czech B, Munafo M, Ciabrelli F et al (2018) piRNA-guided genome defense: from biogenesis to silencing. Ann Rev Genet 52:131–157
9. Houwing S, Kamminga LM, Berezikov E et al (2007) A role for Piwi and piRNAs in germ cell maintenance and transposon silencing in Zebrafish. Cell 129(1):69–82
10. Kiuchi T, Koga H, Kawamoto M et al (2014) A single female-specific piRNA is the primary determiner of sex in the silkworm. Nature 509(7502):633–636

11. Tang W, Seth M, Tu S et al (2018) A sex chromosome piRNA promotes robust dosage compensation and sex determination in C. elegans. Dev Cell 44(6):762–770.e3
12. Watanabe T, Cheng EC, Zhong M et al (2015) Retrotransposons and pseudogenes regulate mRNAs and lncRNAs via the piRNA pathway in the germline. Genome Res 25(3):368–380
13. Zhang D, Tu S, Stubna M et al (2018) The piRNA targeting rules and the resistance to piRNA silencing in endogenous genes. Science 359(6375):587–592
14. Zhang P, Kang JY, Gou LT et al (2015) MIWI and piRNA-mediated cleavage of messenger RNAs in mouse testes. Cell Res 25(2):193–207
15. Muller S, Raulefs S, Bruns P et al (2015) Next-generation sequencing reveals novel differentially regulated mRNAs, lncRNAs, miRNAs, sdRNAs and a piRNA in pancreatic cancer. Mol Cancer 14:94
16. Liu Y, Dou M, Song X et al (2019) The emerging role of the piRNA/piwi complex in cancer. Mol Cancer 18(1):123
17. Mao Q, Fan L, Wang X et al (2019) Transcriptome-wide piRNA profiling in human brains for aging genetic factors. Jacobs J Genet 4(1):014
18. Qiu W, Guo X, Lin X et al (2017) Transcriptome-wide piRNA profiling in human brains of Alzheimer's disease. Neurobiol Aging 57:170–177
19. Li M, Yang Y, Wang Z et al (2021) Piwi-interacting RNAs (piRNAs) as potential biomarkers and therapeutic targets for cardiovascular diseases. Angiogenesis 24(1):19–34
20. Wu WS, Brown JS, Chen TT et al (2019) piRTarBase: a database of piRNA targeting sites and their roles in gene regulation. Nucl Acids Res 47(D1):D181–D187
21. Rosenkranz D, Zischler H, Gebert D (2022) piRNAclusterDB 2.0: update and expansion of the piRNA cluster database. Nucl Acids Res 50(D1):D259–D264
22. Muhammad A, Waheed R, Khan NA et al (2019) piRDisease v1.0: a manually curated database for piRNA associated diseases. Database (Oxford), 2019:baz052
23. Sai Lakshmi S, Agrawal S (2008) piRNABank: a web resource on classified and clustered Piwi-interacting RNAs. Nucl Acids Res 36(Database issue):D173–177
24. Sarkar A, Maji RK, Saha S et al (2014) piRNAQuest: searching the piRNAome for silencers. BMC Genom 15:555
25. Zhang H, Ali A, Gao J et al (2018) IsopiRBank: a research resource for tracking piRNA isoforms. Database (Oxford), 2018:bay059
26. Langmead B, Trapnell C, Pop M et al (2009) Ultrafast and memory-efficient alignment of short DNA sequences to the human genome. Genome Biol 10(3):R25
27. Dobin A, Davis CA, Schlesinger F et al (2013) STAR: ultrafast universal RNA-seq aligner. Bioinformatics 29(1):15–21
28. Yuan J, Zhang P, Cui Y et al (2016) Computational identification of piRNA targets on mouse mRNAs. Bioinformatics 32(8):1170–1177
29. Martinez VD, Sage AP, Minatel BC et al (2021) Human placental piwi-interacting RNA transcriptome is characterized by expression from the DLK1-DIO3 imprinted region. Sci Rep 11(1):14981

Shunmin He Professor, Executive Deputy Director, Center for Big Data Research in Health, Chinese Academy of Sciences Key Laboratory of Nucleic Acid Biology, Institute of Biophysics, Chinese Academy of Sciences. He is mainly engaged in the bioinformatics analysis of biological big data, and has developed more than 20 related algorithms, softwares and databases, and published more than 30 SCI papers. He has experience and ability in processing and analyzing petabytes of big data; specializes in analysis of genomic data and interpretation of variant loci, especially for non-coding regions that take up 97% of the genome; specializes in integrative analysis of data from different sources and multi-omics data.

Opportunities, Challenges, and Development of Building Tumor Big Data Platforms in China

Qiang Guo

Abstract Globally, malignant tumor prevention and treatment are major challenges. Utilizing big data to study the pathogenic factors, main causes, and development trends of tumors can not only provide scientific basis for cancer prevention and treatment for Chinese people but also provide references for cancer prevention and treatment worldwide. In this article, we analyze the opportunities and challenges involved in the construction of tumor big data platforms and offer constructive suggestions for their future development. The project is expected to contribute to the overall construction level of China's tumor big data platform.

Keyword Tumor big data · Top-level design · Data quality · Data security

In today's world, medicine is an information science, and health and medical data are exploding at an unprecedented rate thanks to emerging technology. Advanced data and computing technology, represented by big data, artificial intelligence and cloud computing, has become a huge driving force for scientific and technological innovation, which is promoting the rapid development of medical services, new drug research, scientific research, hospital management and other fields. Due to the complex pathogenic factors, high morbidity and mortality of malignant tumors, the prevention and treatment of malignant tumors is a difficult problem in the world. With the help of big data and artificial intelligence technology, the correlation between tumors and multiple factors can be more effectively analyzed, and the occurrence and development of tumors can be monitored earlier, so as to make prevention and treatment easier and more effective. However, we should also understand that due to the multi-source heterogeneity, structure, standardization and other issues of tumor diagnosis and treatment data, there are still many challenges in sharing, governance,

Q. Guo (✉)
National Clinical Research Center for Cancer, National Cancer Center/National Clinical Research Center for Cancer/Cancer Hospital, Chinese Academy of Medical Sciences and Peking Union Medical College, Beijing 100021, China
e-mail: 64917027@qq.com

China's e-Science Blue Book 2023, https://doi.org/10.1007/978-981-99-8270-7_22

analysis and application. It can be said that the construction and application of tumor big data coexist with risks and opportunities, and challenges and development.

1 The Development Opportunities of Tumor Big Data Platform

1.1 The Policy System is Becoming More Perfect

In September 2015, the State Council of China reviewed and released the “Action Plan for Promoting the Development of Big Data.” This plan outlined the significance of big data, provided top-level planning and deployment for its development, and elevated the big data industry to a national strategic level. In June 2016, the State Council issued the “Guiding Opinions on Promoting and Regulating the Development of Health and Medical Big Data Applications,” which required the consolidation of the foundation for health and medical big data application and its comprehensive deepening. Further in April 2018, the State Council published the “Opinions on Promoting the Development of ‘Internet Plus Medical and Health’,” which proposed the acceleration of medical and health information interoperability and sharing, and the establishment and improvement of the “Internet Plus Medical and Health” standard system. In July of the same year, the National Health Commission issued a “Notice on Printing and Distributing the Management Measures for National Health and Medical Big Data Standards, Security, and Service (Trial),” regulating health and medical big data from standard, security, service, and supervision management aspects. Following these significant steps, the National Health Commission has successively issued a series of documents such as the “Internet Diagnosis and Treatment Management Measures (Trial),” “Opinions on Strengthening the Construction of National Health Information Standardization System,” and “Notice on Deeply Promoting the ‘Internet Plus Medical and Health’ ‘Five-in-One’ Service Action.” These documents have guided the promotion of the standardization work mechanism and encouraged medical institutions to use information technology like the internet and big data to expand their service space and contents. 2021 is the beginning year of the “14th Five-Year Plan,” and a series of heavyweight documents related to the development of medical and health big data have been successively released: in March, the “14th Five-Year Plan for National Economic and Social Development and the Long-term Goals for 2035” was released; in June, the National Development and Reform Commission and other four departments announced the “Implementation Plan for Building a High-quality and Efficient Health Service System during the 14th Five-Year Plan”; in December, the Central Committee for Cybersecurity and Informatization issued the “National Informatization Plan for the 14th Five-Year Plan.” Promoting information interconnection, mutual recognition and sharing, data standardization and integration, building high-level smart hospitals, and constructing major disease data centers are all important tasks in the above-mentioned documents.

Top-level planning aligned with industry-specific and professional policies, enabling the big data industry to rapidly develop in terms of time and space. In 2016, the construction of the first batch of national pilot health and medical big data centers began in Fujian and Jiangsu. In 2017, the National Development and Reform Commission approved 13 national big data engineering laboratories. From 2018 to 2019, governments in various regions established 109 big data parks, and most provinces and cities have already established district level or above big data parks. In 2019, the total number of open datasets in China was 62,801, which is about 7 times that of 2017 [1]. In February 2022, the National Development and Reform Commission and other four ministries and commissions jointly issued a notice agreeing to start the construction of national computing hub nodes in eight places, including Beijing Tianjin Hebei, Yangtze River Delta, Guangdong Hong Kong Macao Bay Area, Chengdu Chongqing, Inner Mongolia, Guizhou, Gansu and Ningxia, and planned 10 national data center clusters. The overall layout design of China's integrated big data center system was completed.

1.2 Flourishing Development of New Technologies

The emergence of the Hadoop project marked the beginning of the era of big data technology, while the commercialization of AWS (Amazon Web Services) signaled the exponential growth of cloud computing, accelerating the pace of the information age [2]. Since then, mature commercial products have emerged in areas such as large-scale cluster computing, servers, processor chips, basic software, etc. In recent years, new computing engines represented by Spark and Flink have replaced the MapReduce framework, better meeting the real-time processing needs of data. The Greenplum framework provides powerful parallel data computing performance and massive data management capabilities, while tools such as Impala and Spark SQL solve structured data problems. With factors such as computing power, data, and algorithms gradually becoming available, advanced algorithmic structures continue to emerge. The Transformer model has not only achieved great success in the field of natural language processing but has also performed well in computer vision, becoming an important network architecture in the field of computer vision. Prompt Tuning has become a new training paradigm for pre-training language models in the field of natural language processing. The emergence of the GPT-3 model has inspired researchers to explore larger-scale and more amazing performance pre-training models, and multi-modal pre-training models have become the next major focus of development [3]. With various technologies constantly developing and research results emerging one after another, the big data technology architecture system continues to be improved and developed.

Next-generation sequencing technology has enabled the realization of precision medicine. Since the introduction of second-generation sequencing, the cost of gene sequencing has continuously decreased at a rate that surpasses Moore's Law. According to research data from the US National Human Genome Research Institute

(NHGRI), the cost of sequencing a single human genome was close to $100 million in 2001 but had dropped to around $1,000 by 2015. This number has been further reduced to around $500 in 2021. With the continuous improvement of sequencing technology and the decreasing cost of detection, the integration of clinical medical data and omics data has become an unstoppable trend, and biomedical research has entered the era of big data.

1.3 There is a Wealth of Data Resources, and the Advantages in Scientific Research are Obvious

China, the world's most populous country, boasts the most abundant clinical research resources. The total number of diagnosis and treatment visits to medical and health institutions surpasses 7 billion annually, reaching 8.47 billion in 2021, with an average of 6 visits per capita [4]. According to "Healthy China (2019–2030)," chronic disease patients' population, represented by cardiovascular diseases, tumors, diabetes, and respiratory diseases, exceeds 500 million (without deduplication), and nearly 180 million elderly people suffer from chronic diseases. Taking malignant tumors as an example, the number of new cases in China reached 4.064 million in 2016, with a standardized incidence rate of 186.46/100,000, and 2.414 million deaths, with a mortality rate of 105.19/100,000 [5]. With the aging population, environmental factors, social and economic development, and changes in people's lifestyles in China, the incidence and mortality of malignant tumors will continue to rise. The vast patient population and different epidemiological and genomic characteristics, relying on the integration and management of the tumor big data platform, will undoubtedly provide abundant and high-quality data support for medical research and new drug development.

2 The Current Status of the Construction of Tumor Big Data Platforms Both Domestically and Internationally

2.1 The Current Status of the Construction of Tumor Big Data Platforms in International

Developed countries like Europe and America have been building and utilizing tumor data platforms and biobanks since the 20th century, providing important support for the rational formulation of tumor prevention and control strategies and diagnostic and treatment norms. The United States has a complete medical health database, including 12 regional electronic medical record data centers, 9 medical knowledge centers, and 8 medical imaging and bioinformatics data centers covering the country. SEER database is one of the most representative large-scale cancer registry databases in North America, covering 28% of the US population, collecting a large amount of evidence-based medicine data, and open to all researchers, making positive contributions to reducing the incidence of cancer and improving the quality of life of the population in the United States. The CANCER-LINQ, pioneered by the American Society of Clinical Oncology (ASCO), is composed of an intelligent computer network. This network collects and analyzes millions of patient visit data and expert guidance information, and feeds analyzed information back to data providers, providing support for their health care services. The All of Us Research Program of the US National Institutes of Health (NIH) is expected to collect data from about one million people, using genomics to manage and analyze large datasets, accelerating biomedical discovery, improving individual health, and revealing pathways to achieve precision medicine. The UK National Health Service (NHS) has a large and comprehensive medical data, including patient health records, disease data, etc., which can be used to create more value for public health services, medical research, etc. The UK Biobank, a biobank of biological samples, currently has the world's largest genomic data, which is open to global researchers for application.

2.2 The Current Status of the Construction of Tumor Big Data Platforms in China

China has not yet established an open multi-center research network based on clinical tumor data. However, with policy support, top hospitals, and well-known data enterprises in China have made rapid progress in health and medical big data, as well as biological big data.

The National Health Medical Big Data Center (Fuzhou) was launched in April 2017. As of July 2020, it had completed the data aggregation of medical insurance users in 14 provincial hospitals, 37 public hospitals at or above the second level, and 174 primary medical and health institutions in Fuzhou. The center has stored over 4 billion pieces of structured data, totaling over 180TB. More than 440 leading digital economy enterprises such as China Electronics, Microsoft, and Tencent, have been attracted to settle down with a registered capital of 38.6 billion yuan, including 122 health and medical big data enterprises with a total registered capital of over 5.306 billion yuan [6].

The West China Biomedical Big Data Center was established in 2016, and its scientific research big data platform currently has a high-performance computing (HPC) cluster, cloud platform, and AI computing cluster, with a total capacity of 103 computing nodes, over 7,200 CPU cores, 68 V100 GPU cards, and 14 PB of storage space, with a total computing power of 1.16 PFLOPS. It has produced breast cancer, COVID-19, colon cancer, depression, kidney disease, and esophageal cancer specialty databases and supports data traceability and security level management. It has supported 42 research teams and over 100 researchers, hosting more than 70 research projects.

The Medical Big Data Research Center was established at the PLA General Hospital in 2019. Its medical big data platform includes three basic databases: specialty disease databases, medical image databases, and electronic medical record text databases. The center has aggregated and integrated all clinical data from the past 25 years, established a clinical basic data resource pool, realized online access to historical images throughout the hospital, and integrated all text data into the database, supporting free data retrieval. The center focuses on clinical research and has established more than 30 specialty disease databases, including heart failure databases, emergency case databases, and lung cancer databases. Based on this, it has further expanded the scope of data utilization by organizing various themed data competitions and publishing topic-specific data sets [7].

In 2019, the National Cancer Center established the "National Clinical Application Monitoring Network for Anti-Tumor Drugs" (referred to as the Monitoring Network). The network aims to comprehensively understand the clinical application of anti-tumor drugs in China, strengthen standardized diagnosis and treatment of tumors, and provide timely and reliable data for health administrative decision-making. The monitoring network covers more than 1,400 hospitals in 31 provinces, municipalities, and autonomous regions across the country. It mainly monitors the procurement and use of anti-tumor drugs and the standardization of diagnosis and treatment of cancer patients. After three years of data governance, the monitoring network has begun conducting in-depth research on the rationality and accessibility analysis of clinical application of anti-tumor drugs, standardized diagnosis and treatment, disease-specific quality control, and more.

3 Challenges Faced by the Construction of Cancer Big Data Platform

3.1 Data Sharing

Open sharing is essential for maximizing data efficiency, but there are still many challenges to truly achieving sharing. Ownership, usage, and legal consequences of violating personal privacy data are unclear [8]. Data barriers exist widely between different regions, industries, and departments, and mechanisms for data openness, flow, and trading require improvement. If all data belongs to individuals, it may not benefit technological innovation and data circulation. If all data belongs to companies, data monopolies and user privacy breaches may occur. If all data belongs to the state, collection and storage costs may be unaffordable. The non-exclusivity and different values of data in different hands or scenarios make it difficult to evaluate and distribute data profits. In China, individual users struggle to obtain additional benefits beyond free services [9].

In 2021, the "Data Security Law" and "Personal Information Protection Law" were issued and implemented. The law clarified the relationship between data security and development, and personal privacy was protected by law. However, how to balance the three aspects of personal privacy protection, data development and utilization and public interests also needed further dynamic research and judgment and definition by various industry authorities, government agencies and enterprises.

3.2 Data Security

The cybersecurity situation in the healthcare industry is troubling. On one hand, new technologies and formats like big data, "Internet + healthcare," and intelligent health devices present new challenges to the existing data security system. On the other hand, the healthcare industry still lags behind in security protection levels for medical information systems. According to the "2018–2019 China Hospital Informatization Survey Report" by the China Hospital Information Management Association (CHIMA), only 43.95% of the 839 hospitals surveyed passed the level protection evaluation, and 75% of hospitals below Level 3 did not conduct level protection evaluations. In 2018, Tencent released a special report on ransomware viruses in the healthcare industry, which showed that since July of that year, 247 hospitals in China's top three hospitals had detected ransomware viruses, and there were almost 3–4 major medical data leaks every month. In the first half of 2021, traditional industries, healthcare, and government agencies with high-value data were targeted in ransomware virus incidents, accounting for a total proportion of up to 69%. Improving network security capabilities in the healthcare industry and balancing efficiency and risk are challenges that need to be addressed in the big data platform construction process.

3.3 Data Standardization

During the 13th Five-Year Plan period, the National Health Commission issued a total of 597 health and health standards. However, due to the late start and low starting point of health information standards, the implementation of standards lacks mandatory and effective measures, and there are still many problems in the application of standards [10]. There are many medical manufacturers, and the information system designs of various medical manufacturers lack standards. The data quality of each medical unit is generally incomplete, non-standard, inaccurate, and lacking correlation, making it difficult to achieve the requirements of data interconnection, data sharing [11, 12]. According to the "2018–2019 China Hospital Informatization Survey Report," 60.35% of the surveyed hospitals believed that there was a lack of standards and difficulty in integrating medical information products. Only 39.71% of hospitals participated in the application level grading evaluation of electronic medical record systems, and only 14.61% participated in the standardization maturity evaluation of interconnection. In 2022, the National Health Commission issued the "14th Five-Year Plan for Health and Health Standardization Work Plan," proposing to enhance the timeliness, pertinence, and effectiveness of standards, strengthen the evaluation of the application effects of health information standards, and promote information sharing and mutual recognition. It can be foreseen that to establish a sound health and health standard system and further drive the coordinated development of the industry chain, a gradual and persevering process is still needed.

3.4 Data Governance and Knowledge Transformation

Medical data face many quality issues, hindering the conversion of data resources into value, and it is urgent to explore suitable data governance techniques. In clinical practice, approximately 80% of data is unstructured and freely entered text, such as admission and discharge records, diagnostic imaging reports, pathology reports, and surgical records, which contain a wealth of high-value information. However, clinical text context information often has multimodal and multi-source characteristics, and is further compounded by Chinese character ambiguity and incomplete original information structure. Existing information extraction, knowledge fusion, and knowledge processing techniques are still unable to meet the demands of big data applications, lacking efficient algorithm implementation for "finding a needle in a haystack". In recent years, although BERT and ERNIE have made significant progress in the field of natural language processing and have repeatedly topped the charts, they still cannot handle common sense and reasoning issues well in the medical field. The scarcity of annotated data and the complexity of the annotation process are another important aspect that hinders the development of AI technology. Medical concepts are complex, and data involving privacy leads to difficulties in data acquisition. The data annotation process is long and costly. According to a report from Cognilytica, about 80% of AI project time is spent on aggregating, cleaning, labeling, and expanding data for machine learning models, and only 20% of the time in AI projects is spent on algorithm development, model training and adjustment, and machine learning operations.

Converting high-quality data into useful information and knowledge still requires a gradual, step-by-step process. In the construction of knowledge graphs and knowledge bases, there is no widely adopted universal terminology standard in China. Inconsistencies and conflicts exist in the content of medical literature, clinical guidelines, evidence-based medicine, and real-world clinical evidence. New medical concepts, new drugs, and treatments are constantly emerging, and the challenge lies in how to standardize terminology, how to construct ontologies, and how to maintain and update them continuously to ensure the correctness and timeliness of knowledge. These are all enormous challenges.

4 Prospects for the Construction of Tumor Big Data Platforms

Big data in the internet industry has achieved great success in fields such as mobile payments, online credit reporting, and e-commerce. The construction of tumor big data platforms should not only fully draw from the successful experiences of big data in the internet industry, but also combine the characteristics of their own industry, change their mindset, plan for the long-term, and provide important platform support for the digitalization and intelligent transformation of medical services.

In October 2016, the Communist Party of China Central Committee and the State Council issued the "Healthy China 2030 Planning Outline," which established the "big health" and "big healthcare" perspectives centered on promoting health. If "continuing to promote the transformation from disease-centered to people-centered" is the future focus of national healthcare reform, then building an efficient, unified, and collaborative multi-center and even national tumor big data platform, achieving the standardization and scientific management of the three-level prevention and diagnosis and treatment processes, will become a digital focus in the future of tumor prevention and treatment. The planning and design of the tumor big data platform should closely revolve around this focus and follow the 1234 principle, which services one center transformation, focuses on two dimensions, achieves three interconnections, lays a solid foundation for four links, and implements gradually in stages (see Fig. 1) [13].

One center transformation: The development of the tumor field in the future will shift from a center of treatment to a center of comprehensive health and tumor prevention, from "treating diseases" to "health," emphasizing early diagnosis, early treatment and early recovery, which will affect the scope, channels and methods of data acquisition.

Fig. 1 Principle 124 of the top-level design and planning for the tumor big data platform

Two development dimensions: The future big data platform will face the wide-ranging management and research needs of different groups, including government, universities, medical institutions, pharmaceutical companies, and insurance companies, with data integrity, confidentiality, usability, and veracity becoming the key considerations for platform value; the diversified development of technology, such as 5G, cloud computing, big data, internet of things, and blockchain will create a ubiquitous intelligent ecosystem, and the diagnosis and treatment mode, research mode, and management mode will undergo disruptive changes.

Three interconnections: The tumor big data platform should be able to exchange data with various medical institutions, connect with regional health information platform business, and share results in key technology research fields. The standardized construction of the tumor big data platform is both the foundation and the soul (see Fig. 2) [13].

Four links: In the data collection link, full-dimensional automated collection is achieved. With the standardization of data and the standardization of the data collection process, passive data collection methods such as manual filling or file uploading will be eliminated, and automated collection technology that supports multi-party data aggregation will become mainstream; in the data governance link, fragmented data governance will evolve into process-oriented, automated governance, and the platform will have a more flexible metadata configuration management function,

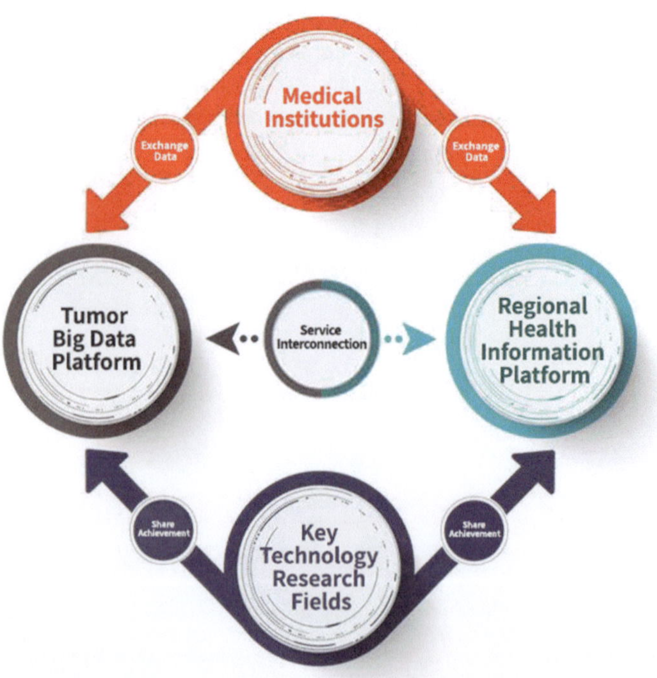

Fig. 2 Principle 3 of the top-level design and planning for the tumor big data platform

using highly reliable, interpretable governance technology and tools to achieve structured and standardized data; in the data empowerment link, a knowledge graph and knowledge base are constructed to convert data into information and knowledge, forming business intelligence; in the data linkage link, openness, sharing, interactivity and collaboration are achieved, and the medical, medical insurance, medical platform, and tumor big data platform achieve business and project linkage, improving overall level of tumor prevention and treatment.

5 Conclusion

Over the past few years, the government has played a critical role in providing macro-power, policies, and regulations to advance the development of big data platforms, while advanced data and computing technologies such as big data, artificial intelligence, and cloud computing have offered technical support to facilitate the construction of big data platforms. Simultaneously, improvements in the information standards of the medical and health industry, along with increased interconnectivity among medical institutions, have created a fertile environment for the construction of a tumor big data platform. Despite some challenges, the construction of the platform provides ample opportunities, and known factors outweigh unknown ones. Building a tumor big data platform not only requires seizing on policy and industry opportunities, but enhancing the big data technology system based on its unique characteristics, planning strategically for the long-term, continually adapting to changing times, and becoming an important support platform for the digital and intelligent transformation of the medical and health industry.

References

1. Chen J, Duan M, Hongxing Wu (2019) Report on the development of big data applications in China (2019). Social Sciences Academic Press, Beijing
2. Liu W, Sun Y (2019) Big data 3.0—the core technology of big data in the post-Hadoop era. Data Comput Dev Front 1(1):94–104
3. Beijing Zhigen Artificial Intelligence Research Institute. Zhigen artificial intelligence frontier report for 2021–2022 [EB/OL]. https://www.shangyexinzhi.com/article/4812076.html. 26 October 2022
4. National Health Commission of the People's Republic of China. Statistical bulletin on the development of China's health and health care in 2021 [EB/OL]. http://www.gov.cn/xinwen/2022-07/12/content_5700670.html. 26 October 2022
5. Zheng R, Zhang S, Zeng H et al (2022) Cancer incidence and mortality in China 2016. J Natl Cancer Cent 2(1):1–9
6. Liu Y (2022) Fuzhou health commission: successful trial of national health and medical big data fuzhou pilot project [EB/OL]. http://fj.people.com.cn/n2/2020/1006/c181466-34334945.html. 26 October 2022

7. Shi J (2021) Construction and research utilization of medical big data platform in the Chinese people's liberation army general hospital. In: The 5th world intelligence conference intelligent health and medical summit forum, Tianjin
8. GuoQ, Wang L, Mu J et al (2020) Research on medical data information security policies. Med Inf 41(1):22–25
9. Chen M (2022) Some thoughts on data ownership [EB/OL]. https://www.zhonghongwang.com/show-278-241412-1.html. 26 October 2022
10. Jin X (2018) Health and medical big data. People's Medical Publishing House, Beijing
11. Fu H, Xu F, Fan M (2019) On the governance and system construction of hospital health and medical big data. China J Tradit Chin Med Pharm Inf 43(3):1–5
12. Ruan T, Qiu J, Zhang Z et al (2019) Medical data governance: building the intelligent analytical data foundation for high-quality medical big data. Big Data 1:12–24
13. Guo Q, Wang C, Heng F (2021) Opportunities, challenges, and development of medical big data platform construction. Med Inf 42(1):2–8

Qiang Guo Director of Big Data Office, National Cancer Center/ Cancer Hospital Chinese Academy of Medical Sciences; Vice Director of Information Center, Project Management Specialist, and Senior Engineer. He is responsible for the construction and management of several projects, including the National Cancer Big Data Platform, the National Cancer Prevention and Control Information Management Platform, and the National Anti-tumor Drug Surveillance System. His main social appointments include: Vice Secretary-General of the Expert Committee on the National Anti-tumor Drug Surveillance System; Director of Intelligent Medical Research Institute of China Research Hospital Association; Standing Committee member of CSCO on Tumor Big Data Expert Committee.

Research and Exploration of Endoscopic Image Big Data Technology Serving Precision Medicine

Jingang Zhang, Liang Zong, Min Wei, Wenqi Ren, Xiongzhi Wang, Yanbo Dong, Kun Liu, Liangfa Liu, and Shiming Yang

Abstract According to the outline of the national precision medicine development plan, our research team is focused on promoting the application of big data technology to clinical diagnosis. We aim to explore a new paradigm of big data from endoscopic images for Smart HealthCare and build a strong data foundation to assist in endoscopic diagnosis and treatment. Our ultimate goal is to improve the accuracy of diagnosis and patient comfort to meet the growing demand for high-quality medical services. In this paper, we explain in detail the new endoscopic imaging technology, intelligent labeling method, and multi-modal fusion technology. We systematically review the development status of endoscopic image big data and introduce the current research situation and key technologies of the endoscopic image service platform. Finally, we propose the prospect of endoscopic image big data technology in the field of precision medicine.

Keywords Endoscope image · Multimodal fusion · Super-resolution reconstruction · Intelligent labeling · Aided diagnosis

J. Zhang · W. Ren · X. Wang
Center for Intelligent Imaging, University of Chinese Academy of Sciences, Beijing, China

J. Zhang
Industrial Research Institute of Huairou Science City, Beijing, China

L. Zong · K. Liu · S. Yang (✉)
Department of Otorhinolaryngology-Head and Neck Surgery, Chinese People's Liberation Army (PLA) General Hospital, Beijing, China

National Clinical Research Center of Otolaryngology Diseases, Moscow, Russia

M. Wei
Department of Orthopedics, Chinese People's Liberation Army (PLA) General Hospital, Beijing, China

Y. Dong · L. Liu (✉)
Beijing Friendship Hospital Affiliated With Capital Medical University, Beijing, China

China's e-Science Blue Book 2023, https://doi.org/10.1007/978-981-99-8270-7_23

1 Introduction

In October 2016, the State Council issued the "Healthy China 2030" Planning Outline, which proposed to strengthen key technological breakthroughs such as precision medicine and smart medicine. The plan aims to integrate modern scientific and technological means with traditional medical methods and explore a new paradigm of medical services [1]. Endoscopic diagnostic systems, such as electronic nasopharyngeal endoscopy and electronic gastrointestinal endoscopy, can obtain more intuitive lesion image data in real-time, assisting doctors in quickly drawing diagnoses and treatment plans. Therefore, it has significant clinical application value and has become one of the key development technologies in the field of precision medicine.

The endoscopic image big data service platform is a prime example of the combination of big data and artificial intelligence technology. It typically consists of three parts: an endoscopy image data acquisition system, a pathological data comprehensive processing system, and an auxiliary diagnosis application system. The endoscopic image data acquisition system collects and stores massive endoscopic image data, patient medical record data, and expert knowledge through the Hospital Information System (HIS). The pathological data comprehensive processing system uses the stored big data samples to preprocess and automatically label data such as endoscopic images and train the lesion identification model. The auxiliary diagnosis application system uses the trained model to provide visual diagnosis results and can conduct friendly human-computer interaction with doctors. Currently, endoscopic images have been successfully applied in the fields of cystoscopy and gastrointestinal endoscopy. However, intelligent medical diagnosis requires deep learning-based models and massive samples, and it requires intelligent analysis with artificial intelligence algorithms. This will face new challenges such as collecting enough image data, labeling it fast enough, and accurately integrating multi-dimensional information.

The development of precision medicine has shown that endoscopy and microscopic biopsy are the gold standards for early detection and screening of lesions, particularly for early cancer prevention of respiratory and digestive diseases. However, due to the limitations of cavity space and differences in image information characteristics of each part of the endoscopy method, there have always been challenges such as difficult acquisition, low resolution, and difficulty in multimodal fusion analysis. Consequently, the precision diagnosis and treatment plan based on artificial intelligence-assisted diagnosis lacks sufficient basic data support, and therefore, the intelligent development and application of precision medicine are relatively slow. To address these challenges, the research team proposed to build an endoscopic image big data service platform. This platform can promote the top-level design of precision medical solutions, provide powerful data resources and platform support, and accelerate the development and application of intelligence in the field of endoscopic diagnosis and treatment.

2 Current Status and Challenges of Endoscopic Image Big Data Techniques

2.1 Domestic and International Status

(1) Endoscopic image capture has stepped into the digital era

Endoscopy is the primary step in endoscopic image big data techniques and plays a crucial role in acquiring high-quality image data. With the rapid development of technologies in optics, mechanics, electronics, semiconductor, computer science, communication, robotics, and more, integrated medical endoscopes have been widely adopted in clinical applications, providing a significant means for the massive acquisition of endoscopic image data. However, conventional endoscopes suffer from undiversified imaging modes and low information dimensionality. As a result, the acquired endoscopic images can only play an auxiliary role in diagnosis.

Endoscope is a medical instrument to observe the condition of internal organs of the human body. Its development history can be coarsely divided into three phases [2–4], rigid endoscope, fiber endoscope, and digital endoscope. The rigid endoscope is relatively bulky, potentially risky to the human cavity, and not able to detect deep or complicated organs. UK scientists Hopkins et al. introduced fiber to the design of endoscopes, and invented the fiber endoscopy, which breaks through the limitation of observing the complicated organs [5]. A US company, Welch Allyn, developed and applied charge coupled device (CCD) to replace the fiber bundles in fiber endoscopes, which proclaimed the birth of digital endoscope [6]. Digital endoscopes feature compact form factors, and releases a tremendous space at the end tip of the instrument to realize other functionalities. In the meantime, it is capable of transmitting images and videos in real-time to the monitor for display to multiple audience. It can also record the image and video information to a hard disk. Digital endoscope has been a widely applied medical instrument so far in multidisciplinary clinical scientific fields. It has been an indispensable tool in the inspection and diagnosis of in vivo mucosal lesions [7].

Currently, the high-end endoscope market is largely dominated by Japanese, European, and American companies. Their products are characterized by wide field-of-view (better than 80°), high definition (better than 1920×1200), high speed (60 Hz), and automated high-quality imaging. In China's high-end digital endoscope market, Japanese vendor Olympus and German vendor XION hold a combined market share of 87.5%, while other foreign vendors such as Pentax from Japan and STORZ from Germany cover 11.5%. Domestic digital endoscope products are limited by several factors, including a lack of techniques, funding, brand recognition, talent, and experience. As a result, there is still a significant gap between domestic products and world-leading levels in terms of instrument resolution, fine manufacturing of optical systems, instrument compactness, and integration. Independent domestic endoscope enterprises are still at the mid-to-low levels, with only a few large-scale enterprises achieving breakthroughs in high-end products. These enterprises provide a preliminary localization solution for the acquisition of endoscopic image big data [8].

(2) **Continuous development and refinement of endoscopic image datasets have been a significant basic support for diagnosis**

The endoscopic image dataset is a carrier for the data of patient samples, which records the multi-dimensional image data of the patient lesion sites, and the doctor's diagnosis information. It would serve as the basis of future training of lesion recognition models and auxiliary diagnosis. The construction of endoscopic image datasets has developed from simple relational datasets to distributed and multi-mode datasets. In 1980, I. I. Rosen developed a computer dataset system (DAISTU) to store and query selectively the results of medical diagnosis research [9]. In 2008, Jing Zhu conducted research on the image datasets of a wireless endoscopic system, with SQL Server2000 as the back-end technique [10]. In 2015, Hongtao Ji et al. constructed a breast ultrasonic workstation based on the medical image information management system (PACS), and collected 27,432 samples of breast ultrasonic inspection results for their ultrasonic dataset [11]. In 2016, Hui Wu et al. proposed a distributed storage framework based the prevailing data storage and mining techniques of medial image big data. Meanwhile, A multidisciplinary team of pathologists and engineers in Spain constructed a hyperspectral dataset of in vitro human brain tissues [12]. They discovered an automated in vitro way to distinguish healthy brain tissues from the tumor tissues in samples using spectral information [13]. In 2019, Himar Fabelo et al. applied hyperspectral imaging in neurosurgery for real-time delineation of brain tumors, and introduced the first in vivo hyperspectral image dataset of human brain issues. With the development of a wide variety of endoscopic image datasets [14], including gastroenterology and respiratory clinics, the storage capabilities are maturing. However, there are still challenges in low query efficiency and low intelligent management. In particular, an efficient intelligent storage framework is still lacking for the auxiliary diagnosis with hyperspectral and multi-modal image fusion.

(3) **Endoscopic image big data serves precision medicine and gradually matures**

Artificial intelligence possesses powerful capabilities for data processing. It can learn contextual information, such as the features of endoscopic images, and establish correlations between endoscopic images and disease symptoms. It can imitate doctors' cognitive judgment skills, provide high-quality real-time endoscopic images, assist in detecting the lesion area, shorten operation time, and help doctors achieve fast and accurate diagnoses. Artificial intelligence technology is gradually being applied to various medical fields, with research on its application in gastrointestinal endoscopy being particularly prominent.

In gastroscopic detection, Hirasawa et al. [15]. developed an artificial intelligent system to detect gastric cancer from gastroscopic images with the aid of convolutional neural network. The system uses a collection of broadband gastroscopic images as training dataset, and can detect thousands of gastroscopic images in one minute. The overall sensitivity for identifying gastric cancer is 92.2%, with a positive predictive rate of 30.6%. Zhu et al. [16]. adopted a pre-trained convolutional neural network ResNet50 as the network framework, and trained the model with 790 gastric cancer images. The accuracy and specificity of the system has been improved. The sensitivity is 76.47%, the specificity is 95.56%, and the overall accuracy is 89.16%. In colon polyp detection, an artificial intelligence model based on convolutional neural network (CNN) has a sensitivity higher than 90% in colonoscopy video detection, a specificity of 63–95%, and can process 25–98 images per second [17–19]. In esophageal cancer detection, the sensitivity of CNN comprehensive diagnosis is 98%. The diagnostic accuracy of superficial cancer is 99%, and the diagnostic accuracy of advanced cancer is 92%. The diagnosis accuracy rates of esophageal squamous cell carcinoma and esophageal adenocarcinoma are 99 and 90%, respectively. Tencent Miying takes less than 4 s to screen an endoscopy, and has a detection accuracy of 90% for early esophageal cancer [20].

The above application studies demonstrate that artificial intelligence can significantly enhance the efficiency and accuracy of diagnoses, thereby assisting endoscopists and reducing the workload of doctors. Therefore, it is crucial to introduce artificial intelligence technology in the field of precision medicine.

2.2 *Challenges in the Construction and Development of Data Service Platforms*

Intelligent diagnosis of endoscopic images generally involves technologies such as whole-chain data acquisition, high-precision extraction and labeling, multi-type data fusion, and intelligent case analysis. To enhance the coverage, information volume, and analysis efficiency of endoscopic image data, it is essential to conduct in-depth research from various aspects of hardware design, software processing, and application integration.

(1) **Higher precision and multi-modal requirements in the design and implementation of endoscopic image acquisition equipment**

The primary function of an endoscope is to extend human vision to the inside of organs and overcome the problem of limited visibility. From rigid tubes to optical fibers and digital endoscopes, hardware design has continuously adapted to the conditions of narrow and tortuous cavities. The miniaturization of endoscopes has been a constant theme in development to overcome the space limitations of existing endoscopes. In recent years, emerging technologies have offered new ideas for the miniaturization of endoscopes [21, 22], such as meta-lenses with ultra-thin structures [23], lensless computational imaging that removes optical lenses, and single-pixel imaging that reduces imaging detection devices [24]. It is anticipated that endoscopes will be able to provide vision at the sub-millimeter level. However, the development of ultra-miniature endoscopes that are suitable for narrow and complex cavity spaces remains a major challenge in endoscope system design.

Second, the introduction of digital image processing technology into image sensors is solving the problem of "seeing clearly". It all depends on the selection of datasets whether the neural network model can be well trained, and whether the neural network model can achieve excellent results. Both low-resolution endoscopic images and high-resolution endoscopic images are required to support the development of super-resolution models for endoscopic images [25]. Currently, there is a shortage of publicly available endoscopic image datasets, and it is crucial to maximize the detection capabilities of existing hardware to achieve a high-definition standard by increasing image resolution by up to 4 times or more. Another significant challenge is obtaining a dataset that includes low-resolution and high-resolution endoscopic images of various human body parts.

Finally, the use of new endoscope hardware to acquire multi-modal image data can address the issue of accurately identifying pathological tissues in the early stages of diseases. At this stage, the differences in morphology and color between normal and diseased tissues are not always apparent. Traditional methods of simulating human vision can only collect basic plane geometry and color information, which may not be sufficient to quickly and accurately reveal the biochemical characteristics of pathological tissues. The material property differences of organs and tissues can be expressed in high-dimensional multimodal forms such as narrow-band spectroscopy [26], hyperspectral imagery [27], and polarization endoscopy [28]. Near-infrared narrow-band spectroscopy can effectively distinguish the distribution of blood vessels and other tissues, while hyperspectral information can aid in segmenting and identifying early diseased tissue regions. Additionally, polarization information can provide valuable insights into the pathological changes of cell tissue. However, the development of new endoscopic acquisition equipment is currently facing several challenges, including the selection of available optical modalities, the integration of multi-modal imaging systems, and the implementation of snapshot computational imaging acquisition technologies.

(2) **Automated processing, format conversion, and multi-dimensional fusion problems in clinical lesion data labeling**

The development and implementation of artificial intelligence technology for intelligent medical diagnosis heavily relies on a large amount of labeled data. However, the annotation of multimodal endoscopic image data requires significant human resources, time, and economic costs. The quantity and quality of annotated data still require improvement, which significantly limits the development and promotion of artificial intelligence models. Currently, clinical lesion data labeling faces three main challenges. To upgrade the manual labeling to semi-automatic and fully automatic labeling for endoscopic image data [29], it is necessary to study semi-automatic or fully automatic intelligent labeling that simulates professional endoscopists, reduces the burden on doctors, improves the ability to build large-scale labeling data, and solves the source problem of clinical lesion data labeling. To standardize the labeling format and tools, it is important to research on the standardization of a universal labeling format, and develop a unified annotation tool suitable for multi-disciplinary and multi-omics to solve the problem of a wide variety of labeling tools and inconsistent labeling formats [30]. In terms of the fusion of annotation of multi-modal annotation data, it is necessary to aim at the optical properties and biochemical differences of multi-modal endoscopic images. Research is necessary to carry out to analyze the appearance, intensity and numerical distribution of image data; to study data dimensionality reduction, separation, extraction, and fusion technologies for high-dimensional information; and to propose corresponding methods, standards, and mechanisms to solve the quality of labeled data [31].

(3) **Modal selection, fusion strategy, and privacy protection issues in multimodal data fusion**

The application of endoscopic image big data fusion differs significantly from traditional diagnosis methods. Multimodal, non-structural, and cross-domain data must be fused, and the fusion mode should be selected based on the pathological personality to ensure the accuracy and effectiveness of diagnosis. Additionally, data security and privacy must be considered, which has led to new requirements for data storage and operating environments. Multimodal image fusion generates one or several images from different types of images through a specific algorithm, extracting increasingly accurate information to adapt to human eyes/machine vision or image processing tasks. The images collected by the endoscope include multi-modal data such as ultrasound images, fluorescence images, hyperspectral images, and white light images. Different modal data combination schemes are selected based on the needs of information compensation and improvement to avoid performance degradation due to incorrect combinations. This is the primary challenge for modal combination. The modal fusion strategy includes three types of fusion: input level, hierarchical level, and decision level. Input level fusion integrates the original information of different modalities, while hierarchical level fusion fuses the extracted features. Decision fusion fuses the final prediction result in the output stage. Ensuring that the adopted fusion strategy can make full use of the information of each modality is another challenge. Training neural network model algorithms requires a large amount of data, including sensitive data involving personal information. Strengthening data protection is another challenge that must be addressed.

3 Construction and Application Practice of Endoscopic Image Big Data Service Platform

With the rise of artificial intelligence and big data technology, high-quality database construction has become an important guarantee for the development of smart medical care. In order to meet the huge demand for intelligent endoscopic diagnosis and treatment, the endoscopic image big data service platform came into being. For example, the United Kingdom has been developing an automated National Endoscopy Database (NED) [32] since 2013 to collect endoscopic image data collected during endoscopic procedures in the United Kingdom. The acquisition and analysis of multimodal big data are gradually used for endoscopic navigation in surgery, etc. [33]. However, the existing big data platforms are mainly based on the acquisition and management of image data, which lacks the connection with the intelligent diagnosis system. Furthermore, for the endoscopic image big data service platform covering the whole process, it is difficult to realize the function of intelligent diagnosis.

The Intelligent Imaging Center of the University of Chinese Academy of Sciences has taken on the important task of building an "endoscopic hyperspectral database and intelligent diagnosis system" as part of the "Network Security and Informatization Project of the Chinese Academy of Sciences". Collaborating with the National Clinical Research Center of Otolaryngology, the General Hospital of the People's Liberation Army, and Beijing Friendship Hospital affiliated with Capital Medical University, the center has explored the development path of the research, trial, and promotion of the endoscopic image big data service platform for precision medicine. The general goal of the endoscopic hyperspectral database and intelligent diagnosis system is to establish a large endoscopic atlas database (including MRI, CT, and multi-omics data), introduce endoscopic hyperspectral data, and combine machine learning algorithms to achieve fast, real-time, and efficient online accurate diagnosis.

The project team has successfully completed the system development according to the implementation plan. The system utilized state-of-the-art endoscopic imaging equipment to collect, label, and fuse image data for lesions in over 200 patients. Thanks to the multi-angle and multi-method shooting of the lesion area, the total number of endoscopic images collected has reached thousands, and a basic database of nasopharyngeal endoscopic images has been established. With this preliminary dataset, the team can train an endoscopic image super-resolution convolutional neural network, which can automatically convert subsequently collected low-resolution endoscopic images into high-resolution ones through the network model. In the future, a vast database of high-definition endoscopic images can be constructed, and with the addition of more samples, it can be extended to cover the digestive tract, ear canal, and other fields.

The establishment of this system and the pathological research it contains holds immense significance for the development of precision medicine. Firstly, the non-invasive and real-time nature of the data collection process minimizes patient discomfort and improves the speed of on-site diagnosis by doctors. Secondly, the large database constructed can provide a wealth of case reference materials for front-line doctors in primary hospitals, which can enhance the accuracy of disease diagnosis and provide doctors with intelligent auxiliary diagnoses and treatment plans, thereby reducing their workload. Finally, the massive sample data of labeled cases provide a solid foundation for the study of pathological diagnosis models, which can improve the accuracy of lesion diagnosis models. Therefore, the establishment of endoscopic hyperspectral big data will comprehensively promote the transformation of traditional empirical medicine into a modern concept and paradigm of precision medicine aimed at minimizing iatrogenic damage and maximizing health benefits.

3.1 The Overall Technology

The project plans to establish a big data service platform for endoscopic images. Aiming at the multi-modality of endoscopic images, insufficient resolution, and difficulty in labeling sample data, it develops intelligent diagnostic technologies for multimodal endoscopic imaging, super-resolution endoscopic images, and multimodal endoscopic image fusion. Ultimately, the storage of massive endoscopic data and real-time intelligent diagnosis of conditions will be realized.

First, a high-definition endoscopic image module was developed using novel endoscopic imaging techniques, including miniaturized, multimodal acquisition. It includes the degradation prediction network model and the illumination enhancement-super-resolution model. It needs to acquire pairs of low-light-low-resolution and normal-light-high-resolution image pairs to establish a multi-dimensional endoscopic image dataset as a training set for the image deep learning network model.

The next step is to use the network model based on Retinex theory and frequency domain to realize the enhancement and super-resolution of low-light, low-resolution endoscopic images. Thereby, a high-definition multimodal endoscopic image dataset is established. Then, the big data intelligent labeling algorithm is used to obtain the data set with the lesion label, which is used as the training set of the artificial intelligence real-time intelligent diagnosis system.

Based on the above datasets, an artificial intelligence real-time intelligent diagnosis system is developed. It puts multiple modal endoscopic data into a deep feature extraction network and performs hierarchical fusion in the network training process. Through the efficient use of various data, the identification and segmentation of the lesion area can be finally realized, and the diagnosis pre-selection plan is provided for doctors. The overall technical route of the system is shown in the figure below (Fig. 1).

3.2 Novel Endoscopic Imaging Techniques

(1) **Miniaturized Endoscopic Imaging Techniques**

Three major technical routes are available for endoscopic imaging: meta-lens imaging, lensless imaging, and single-pixel imaging. Meta-lens employs 2D planar subwavelength structures as building blocks to achieve full modulation of light, including amplitude, phase, and polarization. This technique opens up a new way to greatly reduce the volume and weight of lenses, making it possible for large field-of-view and larger aperture flat optics imaging. The working spectral bands of this technique are gradually moving from infrared to visible, and from single wavelength to full spectrum, and even multispectral and hyperspectral imaging. The requirement on polarization states also accommodates unpolarized, linearly polarized, and circularly polarized light, and even full Stokes and Muller matrix imaging. The meta-lens

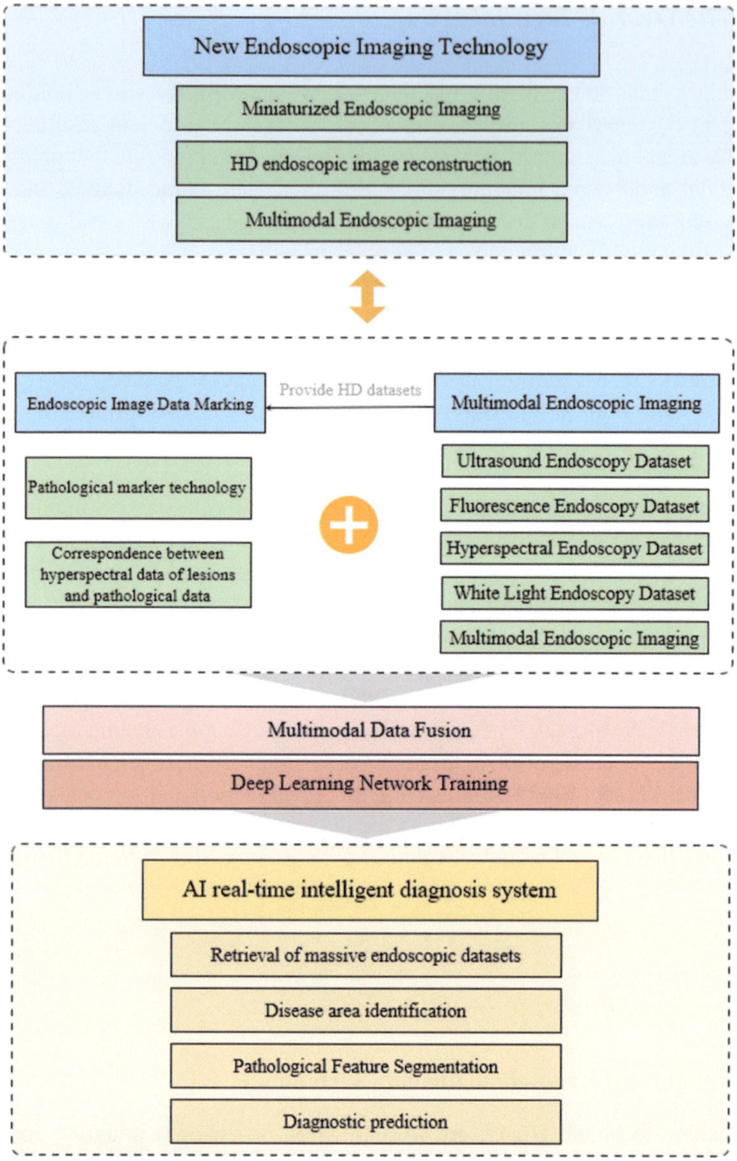

Fig. 1 Overall technology roadmap

technique has set a new record for the smallest size of miniaturized optical imaging systems. It also features great flexibility in light modulation, making it promising in broad applications. The lensless imaging technique makes use of amplitude modulated or phase modulated components to encode optical information by designing the topology of optical structures. In the software part, image reconstruction algorithms are designed to solve an inverse problem mathematically to achieve high image quality. Lensless imaging components are flexible to design and can be used along with spectral filters to achieve snapshot lensless spectral imaging with a single image. They can also be used with a polarization sensor to recover linear polarization information from a single image as lensless polarization imaging. Therefore, it is promising to be a new snapshot multi-modal endoscopic imaging method. Single-pixel imaging roots in the compressive sensing principle and can avoid the capture of redundant information in hardware design. A single photosensitive unit is used to collect signals, and computational reconstruction algorithms are adopted to achieve 2D imaging. The size of endoscopes is limited by the dimension of the imaging detector itself. Although the pixel pitch is approaching 1 μm, it is still mandatory to use pixel arrays to acquire full spatial images in conventional endoscopes. Single-pixel imaging can either sample the signals in the spatial domain or in the Fourier domain. With the aid of structural light illumination, it offers an effective solution to solving the bottleneck of sensor size, paving the way for novel miniature endoscopic imaging instruments.

(2) **High-definition endoscopic image reconstruction technology**

We have designed a light enhancement and super-resolution algorithm based on Retinex theory and frequency domain to address the common issue of low or uneven illumination in endoscopic images. This algorithm enables real-time automatic conversion of endoscopic images into high-quality ones, taking advantage of the complementary promotion relationship between endoscopic image illumination and super-resolution.

The high-definition endoscopic image reconstruction technology mainly includes two modules. The degradation prediction module models the illumination, blur, and noise degradation of real high-resolution endoscopic images, and obtains real low-resolution images paired with real high-resolution images. The super-resolution-light enhancement module divides the image into two branches: high-frequency and low-frequency. The low-frequency branch corresponds to the illuminance map in Retinex theory and uses a bilateral grid for super-resolution and enhancement, which can speed up the process and enable the module to run in real-time. The high-frequency branch corresponds to the reflection map in the Retinex theory and uses the octave convolution to extract the high-frequency information of the image as a guide, making the super-resolution effect of the reflection map more accurate. The low-frequency and high-frequency branches have mutual guidance and communication, which can make the generated images have high-definition details and textures and good lighting.

(3) **Multimodal endoscopy**

Multimodal endoscopy technology is a new type of endoscopic imaging system designed by hardware encoding and software decoding. It is used to obtain high-dimensional multimodal information from two-dimensional raw data. Therefore, the multimodal endoscope system features the joint optimization of software and hardware. At present, matured multimodal endoscopes mainly include narrow-band and multispectral endoscopic imaging technology, polarization endoscopic imaging technology and ultrasonic endoscopic imaging technology.

Narrow-band and multi-spectral endoscopic imaging technologies usually employ active light illumination, including free-form surface large field of view, large-aperture front-end imaging module, continuously adjustable narrow-spectrum illumination light source module, and spectral image reconstruction algorithm. The imaging module adopts "single-wavelength design, wide-spectrum application", combined with the advanced free-form and multi-field direct optical design method, using free-form surface lenses, diffractive optical elements, sub-wavelength meta-surface structures and other optical elements of different scales. The component is used as the realization unit to transfer the aberrations that are difficult to optically correct in the optical hardware design to the subsequent image reconstruction algorithm. Transferring the pressure of the optical component design helps minimizing the complexity of the optical system. The active lighting unit provides two working modes: visible light illumination and narrow-band illumination. In the spectral band, four separate optical fiber channels are used to combine narrow-band light sources with different brightness and spectral range to provide corresponding illumination for continuous narrow-band hyperspectral imaging. Image reconstruction utilizes computational imaging techniques and considers spectral aberration priors in hardware design to build an end-to-end model from optical imaging to image reconstruction, including the matching function of pseudo-color spectra from RGB or RGB-IR images to multispectral images. It utilizes the fixed and unique PSMF coefficient matrix of the detector to flexibly convert between RGB images and multispectral images, which can be inverted from color images to obtain multispectral images [34]. Polarization endoscopic imaging technology introduces the polarization information of light into the design of the imaging system. A metlens unit can be designed to realize the modulation of linear polarization, circular polarization, and non-polarization state, which generates an optical point spread function that depends on the polarization states. Multi-dimensional endoscopic imaging of polarization-related information, such as Stokes vector and Mueller matrix, can be realized by modulating and transforming the polarization states of the illumination light source and using a polarization focal plane device. With the principle of lensless imaging, the lens components in an imaging system can be replaced by a phase-type diffractive optical element, which greatly reduces the size of the optical system and realizes polarization-dependent point spread functions. Polarization information is encoded by convolution into the detectable intensity signals. Prior knowledge of the image is then applied to solve the inverse problem and reconstruct the polarization information through an efficient optimization algorithm or artificial neural network. With a focal plane polarization detector, it is possible to further optimize the electronic components and simplify the optics. In the polarization image reconstruction, a deep learning algorithm is used to recover the polarization-related information, such as degree-of-polarization, and angle-of-polarization [35, 36]. The ultra-sonic endoscopic technique is built by the principle of acoustic attenuation by human organs. The human organs and tissues, including pathological tissues, have specific acoustic impedance and attenuation characteristics. As a result, different reflections and attenuations are generated when ultrasonic waves enter the body. These different reflections and attenuations form ultrasound images. Currently, multi-frequency transduction technology is the mainstream research direction of ultrasound endoscopic imaging. Multimodal endoscopic imaging requires the integration of various modalities to achieve a device that supports imaging of various modalities at the same time.

3.3 Intelligent Data Labeling Method for Cross-Modal Fusion

Intelligent labeling of cross-modal fusion data is a new trend in addressing the challenge of mass sample labeling required for precision medicine. This method labels visible, near-infrared, narrow-band, and multi-spectral images to quantitatively extract features, mine correlations between images of different modalities, and form fusion information of lesion images, enabling automatic marking of abnormal lesions in images.

Intelligent labeling of cross-modality fusion data enables the acquisition of images of multiple modalities (visible, near-infrared, narrow-band, multi-spectral) through a special electronic endoscopic optical system. Different types of images have different characterization effects on the same lesion, such as the narrow-band mode being able to easily detect early cancer of the mucosal surface. This provides a data basis for the intelligent labeling of lesion images. Once pathological features are found in one image sample of the same lesion, other image types of the same lesion can be automatically correlated. Additionally, with the use of CNN feature extraction network and cost-sensitive vector machine, auxiliary labeling of abnormal areas of other images can be achieved. Finally, professional medical staff conducts proofreading and approval based on pre-intelligent labeling, resulting in high-quality lesion sample data.

Through the fusion of multi-modal images and automatic marking of abnormal points based on deep learning, lesions and normal tissue boundaries can be better identified. We have carried out relevant basic research work on the analysis of the characteristics of throat lesions. The collected samples of throat cancer tissue were analyzed by hyperspectral image information, and the consistency rate with the clinician's marking reached 96%. The identification algorithm is being further optimized.

3.4 Multimodal Data Fusion Technology

The multi-modal data fusion technology draws on the application experience of CT image and MRI image fusion, and fuses the unique multi-modal data such as ultrasound, fluorescence, hyperspectral, and white light collected by endoscopy equipment.

Ultrasound images can display both the anatomical structure of internal organs and the physiological functions of organs. Fluorescence endoscopy images have the advantage of wide-spectrum imaging, and the visibility of lesions and pre-lesion areas is high, which can be used to observe lesions or pre-lesion areas that cannot be effectively captured by real images. Hyperspectral images can distinguish the boundaries of diseased tissue from normal tissue. White light endoscopic images are the most common endoscopic images, which can clearly display the condition of various body tissues.

Based on this, multimodal fusion technology uses a feature-level fusion strategy to extract features from each modal data. These feature information is the feature extraction of the lesion-related area in the multimodal image, such as edge, contour, color, saturation, and other lesion information, and then these feature information is analyzed, processed, and integrated to obtain the fused image features. The accuracy of lesion recognition on the fused features will be significantly higher than that on the original image directly.

Feature-level fusion compresses multimodal image information, only extracting and retaining useful information related to lesions. Then fuse high-frequency and low-frequency feature data to obtain features enhanced by frequency information. In addition, the information from the ultrasound image is used as a guide and supplement, so that the extracted feature information is comprehensive and rich, and the lesion can be identified and segmented. The federated learning strategy is also introduced into the technology, which will make the data acquisition and training process safe and efficient.

3.5 The Overall Realization Process and Preliminary Practice Results of Endoscopic Image Big Data Platform Serving Precision Medicine

The endoscopic image big data platform of this project can provide real-time high-definition endoscopic images, and can automatically identify the lesion area and mark relevant pathological features, thus assisting doctors in the diagnosis and prediction of diseases. The process of this platform serving the above-mentioned precision medicine can be divided into three stages.

The first stage is big data collection. First, the initial endoscopic images were collected using the miniaturized endoscopic imaging technology and multimodal endoscopic technology of this project. Then, the designed endoscopic image super-resolution convolutional neural network is used to convert some low-resolution endoscopic images into high-resolution endoscopic images to construct a high-definition endoscopic big data set.

The second stage is the training stage of the intelligent analysis network model. Using the constructed dataset, neural network training is performed on the designed multimodal data fusion network and cross-modal fusion data intelligent labeling network.

The third stage is clinical application. Deploying the above-trained network model on the device can realize an intelligent diagnosis and precise medical treatment, and assist doctors in disease diagnosis and treatment through a friendly human-computer interaction process.

Some preliminary practical results have been achieved so far. In the big data collection stage, using some basic data provided by the hospital, the super-resolution neural network suitable for the endoscope of this project has been trained. Moreover, the automatic conversion of low-resolution endoscopic images to high-resolution endoscopic images is realized, and the obtained high-definition data set can be used for further research related to data annotation and lesion identification.

4 Conclusion

Precision medicine and smart medicine are crucial for the development of medical and healthcare in China in the new era, and are essential for improving the health of people across the country. The construction of an endoscopic image big data platform is an important part of big data intelligent diagnosis technology, and is expected to accelerate the breakthrough and leap-forward development of key technologies such as endoscopic imaging hardware equipment, multi-modal data fusion, and high-precision intelligent diagnosis. As the big data model continues to deepen and optimize, the future endoscopic image big data platform will further develop in the direction of being large and refined. By establishing a new standard for rapid screening of optical endoscopy through an end-to-end correlation mechanism between the microscopic morphology of photons and cells to the macroscopic outcome of pathological features, precision medicine and smart medicine can be promoted to a more mature stage of development.

References

1. The Central People's Government of the People's Republic of China. The Communist Party of China Central Committee and the State Council Office issue the "Healthy China 2030 Planning Outline" [EB/OL]. http://www.xinhuanet.com//politics/2016-10/25/c_1119785867.htm. 25 October 2022

2. Achord JL, Muthusamy VR (2019) The history of gastrointestinal endoscopy. Elsevier
3. De Groen PC (2017) History of the endoscope [scanning our past]. Proc IEEE 105(10):1987–1995
4. Lv P, Liu F, Lü K, Qi Z (2002) History of endoscopy development. Chin J Med Hist 32(1):10–14
5. Elahi SF, Wang TD (2011) Future and advances in endoscopy. J Biophotonics 4(7–8):471–481
6. Zhou S (2002) Diagnosis and treatment with electronic laryngoscope and fiber laryngoscope, 1st edn. Shanghai Second Military Medical University Press, Shanghai
7. Zhang W, Zhou Z, Guan S, Yu Z (2017) Research status and development trend of electronic endoscope. Chin Med Equip 32(1):93–98
8. Chen J, Cai T (2015) Analysis of China's endoscope industry development in 2014. China Med Device Inf 10:16–21
9. Rosen II, Hall TC, Mettler F, Wicks J, Kelsey CA, Gustafson DE (1980) A computerized database system for medical diagnostic studies (DIASTU). Comput Programs Biomed 12(2):249–261
10. Zhu J (2008) Study on the database of image workstation in wireless endoscope system. Chin J Med Instrum 14(10):5
11. Ji H, Chang J, Zhu Q (2015) Discussion on the construction of breast ultrasound database based on medical image information management system. Chin J Gen Surg (Electron Ed) 9(5):3
12. Wu H, Weng X, Wang L et al (2016) Research on storage and mining technology of medical image big data. Chin Digit Med 11(2):5
13. Ortega S, Callico GM, Plaza ML et al (2016) Hyperspectral database of pathological in-vitro human brain samples to detect carcinogenic tissues. IEEE
14. Fabelo H, Ortega S, Szolna A et al (2019) In-vivo hyperspectral human brain image database for brain cancer detection. IEEE Access 7:39098–39116
15. Hirasawa T, Aoyama K, Tanimoto T et al (2018) Application of artificial intelligence using a convolutional neural network for detecting gastric cancer in endoscopic images. Gastric Cancer 21(4):653–660
16. Zhu Y, Wang QC, Xu MD et al (2019) Application of convolutional neural network in the diagnosis of the invasion depth of gastric cancer based on conventional endoscopy. Gastrointest Endosc 89(4):806–815
17. Misawa M, Kudo SE, Mori Y, Cho T, Kataoka S, Yamauchi A et al (2018) Artificial intelligence-assisted polyp detection for colonoscopy: initial experience. Gastroenterology 154(8):2027–2029
18. Urban G, Tripathi P, Alkayali T, Mittal M, Jalali F, Karnes W et al (2018) Deep learning localizes and identifies polyps in real time with 96% accuracy in screening colonoscopy. Gastroenterology 155(4):1069–1078
19. Wang P, Xiao X, Brown JG et al (2018) Development and validation of a deep-learning algorithm for the detection of polyps during colonoscopy. Nat Biomed Eng 2(10):741–748
20. Tencent Technology on Artificial Intelligence (2017) Tencent releases an AI tool that could solve the problem of early screening for esophageal cancer. Inf Comput (15):1
21. Ming LT, Hsiao HH, Cheng HC et al (2018) Metalenses: advances and applications. Adv Opt Mater 6(18):1800554
22. Liu W, Cheng H, Tian J et al (2020) Diffractive metalens: from fundamentals, practical applications to current trends. Adv Phys: X 5(1):1742584
23. Boominathan V, Robinson JT, Waller L et al (2022) Recent advances in lensless imaging. Optica 9(1):1–16
24. Edgar MP, Gibson GM, Padgett MJ (2019) Principles and prospects for single-pixel imaging. Nat Photonics 13(1):13–20
25. Agustsson E, Timofte R (2017) NTIRE 2017 challenge on single image super-resolution: dataset and study. IEEE
26. Piazza C, Cocco D, Benedetto LD et al (2010) Narrow band imaging and high definition television in the assessment of laryngeal cancer: a prospective study on 279 patients. Eur Arch Otorhinolaryngol 267(3):409–414

27. Collins T, Maktabi M, Barberio M, Bencteux V, Jansen-Winkeln B, Chalopin C, Marescaux J, Hostettler A, Diana M, Gockel I (2021) Automatic recognition of colon and esophagogastric cancer with machine learning and hyperspectral imaging. Diagnostics 11(10):1810
28. Fu Y, Huang Z, He H, Ma H, Wu J (2018) Flexible 3 x 3 Mueller matrix endoscope prototype for cancer detection. IEEE Trans Instrum Meas 67(7):1700–1712
29. Wang S, Cong Y, Fan H, Liu L, Li X, Yang Y, Tang Y, Zhao H, Yu H (2016) Computer-aided endoscopic diagnosis without human-specific labeling. IEEE Trans Biomed Eng 63(11):2347–2358
30. Najafabadi MM, Villanustre F, Khoshgoftaar TM, Seliya N, Wald R, Muharemagic E (2015) Deep learning applications and challenges in big data analytics. J Big Data 2(1):1–21
31. Yang H, Zhang R, Wang MiYe (2021) Development of a multimodal medical image annotation system based on image cloud. West China Med J 36(9):6
32. Lee TJ, Siau K, Esmaily S, Docherty J, Stebbing J, Brookes MJ, Broughton R, Rogers P, Dunckley P, Rutter MD (2019) Development of a national automated endoscopy database: the United Kingdom national endoscopy database (NED). United Eur Gastroenterol J 7(6):798–806
33. Luo X, Mori K, Peters TM (2018) Advanced endoscopic navigation: surgical big data, methodology, and applications. Annu Rev Biomed Eng 4(20):221–251
34. Gibson GM, Johnson SD, Padgett MJ (2020) Single-pixel imaging 12 years on: a review. Opt Express 28(19):28190–28208
35. Zhang J, Nie Y, Fu Q, Peng Y (2019) Optical–digital joint design of refractive telescope using chromatic priors. Chin Opt Lett 17(5):052201
36. Zhang J, Su R, Ren W, Fu Q, Nie Y (2021) Learnable reconstruction methods from RGB images to hyperspectral imaging: a survey

Jingang Zhang is an Associate Professor at the University of Chinese Academy of Sciences and the Executive Director of the Intelligent Imaging Center. He is also a Distinguished Professor at Xidian University, a member of the Expert Group in the Information field of the MinJingang Zhang is an Associate Professor at the University of Chinese Academy of Sciences and the Executive Director of the Intelligent Imaging Center. He is also a Distinguished Professor at Xidian University, a member of the Expert Group in the Information field of the Ministry of Science and Technology, an adjudicator of the State-owned Assets Supervision and Administration Commission of the State Council for Venture Capital Funds, and a review expert for the National Natural Science Foundation of China. His research is focused on cutting-edge interdisciplinary technologies, including optics and intelligent computing, clinical and medical optics, remote sensing monitoring and big data analysis. His research is primarily focused on the application of technology research and achievement transformation in the fields of medical health and intelligent monitoring. Dr. Zhang has led more than 20 national-level scientific research projects, such as the National Natural Science Foundation, the National Major Science and Technology Project, the Chinese Academy of Sciences Research Equipment Special Project, and the Air Force Equipment Special Project. He has been awarded more than 10 invention patents and 4 software copyrights, published over 30 academic papers, and co-authored 2 books.

Infrastructures

Concept, Landscape, and Progress of the Global Open Science Cloud

Jianhui Li, Lili Zhang, and Liangming Wen

Abstract To address the common challenges that human face in the context of globalization, the open sharing of research infrastructures has become vital important. However, it is still prevalent to see silos, gaps, and grey areas in the sharing of research resources across domains and disciplines in data centers, repositories, platforms, facilities, and initiatives. To help bridge the gaps among research infrastructures, we propose to co-design and co-build the Global Open Science Cloud (GOSC) during the CODATA 2019 Beijing Conference. Based on a lightweight review of global research infrastructures, this paper introduces the concept, landscape, and progress of the Global Open Science Cloud (GOSC) initiative. Besides, we report on the progress of the GOSC Initiative in terms of strategic planning, governance mechanisms, and international alignment. Looking into the future, GOSC will be enriched by the open, collaborative, trustworthy, and sustainable framework to provide tailored cloud-based services for typical intercontinental scientific research scenarios.

Keywords Open science · Research infrastructure · Cloud federation · Global Open Science Cloud

1 Introduction

Faced with major global challenges, the United Nations (UN) (2015) adopted the "*2030 Agenda for Sustainable Development*" [1]. To achieve these Sustainable Development Goals (SGDs), open science [2] offers robust solutions by supporting the active participation of diverse stakeholders, including the scientific community, the industry, and the general public. Open science refers to openness, transparency, and collaboration throughout the research process, where research conditions, services,

J. Li (✉) · L. Zhang · L. Wen
Computer Network Information Center, Chinese Academy of Sciences, Beijing, China
e-mail: lijh@cnic.cn

L. Wen
University of Chinese Academy of Sciences, Beijing, China

China's e-Science Blue Book 2023, https://doi.org/10.1007/978-981-99-8270-7_24

and resources are integrated and open for sharing. It emphasises the open research culture, collaborative research paradigms, and innovative IT-driven approaches to improve access to research resources, such as scientific publications, data, and others, while respecting to freedom, openness, collaboration, and sharing [3].

Open science has been demonstrated as a future-ended paradigm for addressing global challenges, such as during the COVID-19 pandemic and others, in particular by facilitating research collaboration and the flow of diverse research resources and knowledge across the globe. Many regional, national, and institutional organizations have taken efforts to initiate open science in initiatives, infrastructures, and other areas. For instance, the Finland government launched the "*Open Science and Research Initiative*" in 2014 following the "*Roadmap for Open Science*" across the European Union (EU), Poland, France, Canada, etc. The European Commission (EC) published the "*Open Innovation, Open Science, Open to the World,*" which outlines the EU's open science efforts, and launched the "*European Open Science Agenda* and *European Cloud Initiative*" support the full implementation of open science The "*Netherlands' Plan on Open Science*" (2017), the "*National Plan for Open Science in France*" (2018), the "*New Law on Science and Research*" in Serbia (2019), the "*Open Science by Design: Realizing a Vision for 21st Century Research*" in the US, and its "*Open Science Framework*" are all examples that highlight different actions in open science, including open access to publications, open access to research data, and open collaboration in initiatives. China is also playing an active role. For example, the national-level regulation "*Measures for Managing Scientific Data*" (2018) provides guidelines for sharing research data. Moreover, twenty national scientific data centers facilitate the sharing of research resources by infrastructure-level and service-level open science implementation. In addition, the Open Science Identity (OSID) Initiative also helps to link research resources for accessibility and reusability.

Open research infrastructures have become fundamental components of open science [4]. Thus, we see great potential in building enhanced information infrastructures to integrate existing research resources for better science. Currently, infrastructure development is flourishing at regional, national, and institutional levels, such as the practices in Europe (i.e., European Open Science Cloud), Asia (i.e., Malaysian Open Science Platform, China Science and Technology Cloud), Africa (i.e., African Open Science Platform), Oceania (i.e., Australian Research Data Commons), America (i.e., Open Science Grid in the US), etc. In light of this background, this paper will focus on the current development of global research infrastructures and present an open science initiative called the Global Open Science Cloud (GOSC).

2 Review of Research Infrastructures in the Globe

The efforts of building research e-infrastructures are carried out in different regions (see examples in Table 1), among which open science-oriented clouds, also called Open Science Clouds (OSCs), are gradually coming into the spotlight. In general,

Table 1 Representative initiatives to open research infrastructures

Nations or regions	Key actions (laws, initiatives, infrastructures)	Year
USA	*National information infrastructure act of 1993*	1993
European Union	Open access infrastructure research for Europe (OpenAIRE)	2009
United Kingdom	Common basic data set (CBDS)	2012
USA	Center for open science	2013
United Kingdom	National information infrastructure (NII)	2015
United Kingdom	*National infrastructure delivery plan 2016–2021*	2016
South Africa and others	African open science platform (AOSP)	2016
European Union	European open science cloud (EOSC)	2016
Australia	Australian research data commons (ARDC)	2018
France	European science cluster of astronomy and particle physics ESFRI (ESCAPE)	2019
Malaysia	Malaysia open science platform (MOSP) initiative	2019

OSCs aim to provide users with composite solutions of various resources and services, including computing resources, data storage, network connectivity, and hardware resources [5], following a "pay-as-you-go" approach. For example, USA took the lead in initiating the Open Science Data Cloud (OSDC) project (2009) [6]. Moreover, the EC initiated the European Cloud Initiative (2016) to integrate existing research information infrastructures, resources, and knowledge to jointly build a harmonized digital information environment.

Nevertheless, with the increasing and dynamic demand for cloud resources, centralized deployment approaches are facing limitations in scalability, performance, etc. [7]. Therefore, federated cloud services have become popular to tackle the challenges [8], especially featuring enhanced interconnectivity and interoperability. Furthermore, to help strengthen such interoperability across domains, the idea of co-design and co-build a global open science cloud was proposed at the Committee on Data of International Science Council (CODATA) Beijing Conference in October 2019. The idea is to promote the co-development of a common GOSC through a broad international collaborative network, to foster interconnectivity and interoperability among global open science research infrastructures [9].

3 Introduction to the Global Open Science Cloud Concept

GOSC is an ecosystem designed to connect multiple research infrastructures to meet the specific needs of scientific activities. As an inter-cloud collaborative environment that integrates technologies, infrastructures, standards, policies, and organizations [10], GOSC connects multiple research resources across various stakeholders to support scientific discovery in applicable scenarios. It provides solutions for different scenarios. For example, within the GOSC framework, federated strategies and customised middle-wares can support the sharing of distributed supercomputing resources, enabling sufficient and balanced workloads and feeding both high-performance and high-throughput computing needs [11, 12]. Federated learning also provides an innovative method to facilitate the sharing of research resources. For instance, in the pursuit of the SDGs, algorithms, software, data, and other resources from different member states should be brought together, however, there are also a number of challenges to overcome during this process. This way, federated learning can provide ways forward by keeping data locally but sharing trained models to support decision-making [13]. The GOSC framework also coutributes to offset constraints in a distributed research environment. For example, GOSC integrates various resources into a virtualized federation, supported by a trusted authentication and authorization infrastructure to share resources in a broader context [14] following the supply and demand rules [15].

Besides, GOSC adopts federated cloud service models with prominent resource orchestration, service quality control, governance mechanisms, and other aspects. Cloud service users can access and share resources provided by the federation, thus breaking the resources supply constrains. To guarantee a certain quality of service delivery, interoperable specifications, and protocols should be established and followed by all federation members to provide services smoothly across domains. In terms of governance mechanisms, the GOSC examines the rules for identity identification, resource operation, and resource transactions. Participants involved in federation governance are sublect to the dual constraints of local and federal rules. Therefore, GOSC enables dynamic, scalable, reliable and efficient management of distributed cloud resource services [16].

As a comprehensive ecosystem integrating multiple research resources, the design of the GOSC Initiative includes six layers framework (see Fig. 1). The research resource layer integrates open science resources in the global. The software and tools layer mainly integrates various cloud-based operational toolkits for handing open scientific resources and open source virtual academic exchange communities. The resource organization layer implements the standardized encapsulation and organization of heterogeneous open research resources and common software tools. Existing open research infrastructures (cloud service platforms) in the service platform layer connect with integrated resources, tools, policies, and knowledge. The application layer supports interoperability among researchers, scientific resources, software tools, service platforms, and scientific applications. The scientific application layer supports the application of GOSC in scenario-based science programmes. In addition, the open-sharing governance and sustainable development mechanism of research resources provides policy support for the communication and interconnectivity between the different layers. For example, the research resource layer interacts with the resource coordination layer through standardized protocols. The application coordination and scientific application layers implement service schedules through the application program interfaces. These interactions and services schedules should follow the agreed Service Level Agreement (SLA) and specific service quality requirements within the GOSC trusted ecosystem.

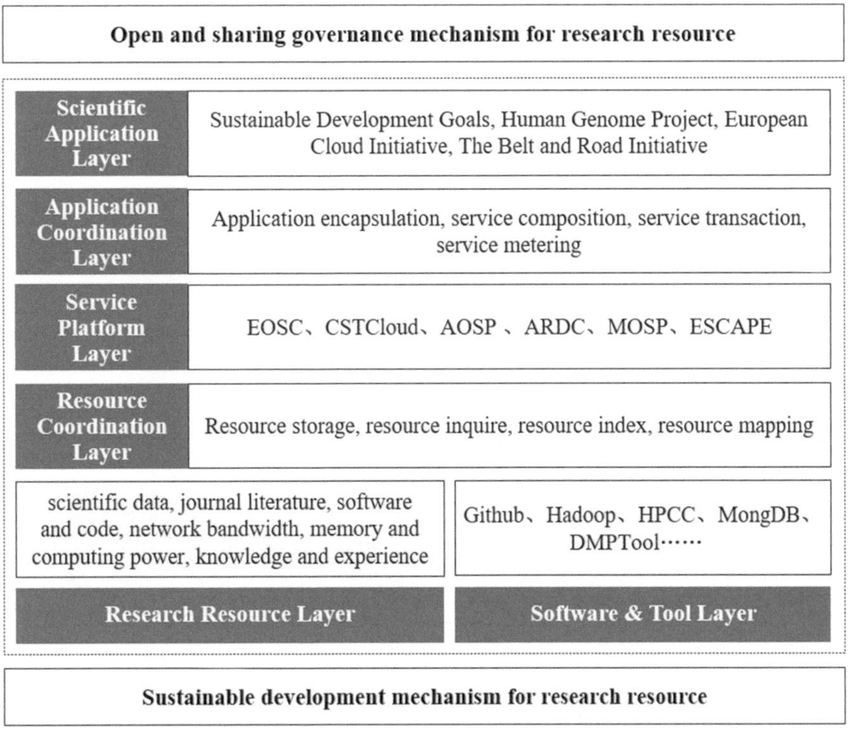

Fig. 1 Logical framework of GOSC

4 Landscape of the GOSC

To bridge the gaps among different research e-infrastructures, the GOSC plans to explore guiding policies, interoperable protocols, sustainable mechanisms, typical application scenarios, and others. In addition, stakeholders are present to jointly develop research frameworks, supporting software, and testbeds to connect various national and regional open science clouds. Intercontinental digital infrastructure and virtual research environment for global collaboration and open science will provide scientific researchers worldwide with an interconnected network, computing federation, FAIR (Findable, Accessible, Interoperable, Reusable) data, secured authorization and authentication services, etc. Figure 2 outlines an ideal form of the future-ended GOSC.

According to this landscape, the implementation of the GOSC may raise concerns, so we need to take action to dispel the doubts. First, we need to establish a broad international collaborative network across various disciplines to support the co-development of GOSC. Second, policies, standards, and governance mechanisms should be developed under mutual trust and collaboration for scientific research to co-build the open science infrastructure ecosystem. Third, we need to establish open

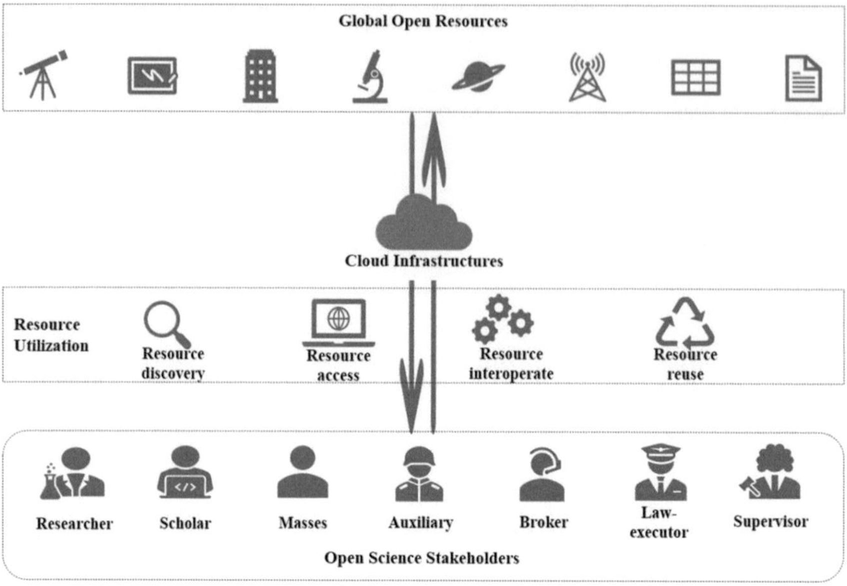

Fig. 2 Ideal form of GOSC

and interconnected systems under the guidance of FAIR principles to help validate the interoperability of intercontinental OSC resources and services. Fourth, practices in typical data-intensive areas for large-scale demonstrations of cloud service applications to help strengthen the robustness and resilience of GOSC technologies.

5 Progress of the GOSC Initiative

To be fully implemented, the GOSC Initiative has been jointly endorsed by CODATA and the Chinese Academy of Sciences (CAS) since early 2021, calling for global collaboration for open science engagement at the infrastructure-level. As a result, experts from all over the world are engaged in this initiative to jointly shape the future of global open science infrastructures for scientific discovery (work plan details, see Fig. 3).

Currently, the GOSC Initiative has established a Steering Group (SG), followed by four Working Groups (WGs) and five Case Studies (CS) to facilitate international engagement [17] (see details in Table 2). Besides, a global collaborative network is also emerging. The GOSC Initiative has attracted experts from over thirty countries and regions to participate in the GOSC dialogue. There is a series of conferences for community outreach [18], such as the conversations during the "Global Open Science Cloud Workshop" (November 2020), "International FAIR Convergence Symposium" (November 2020), "Mobilizing the Global Open Science Cloud (GOSC) Initiative:

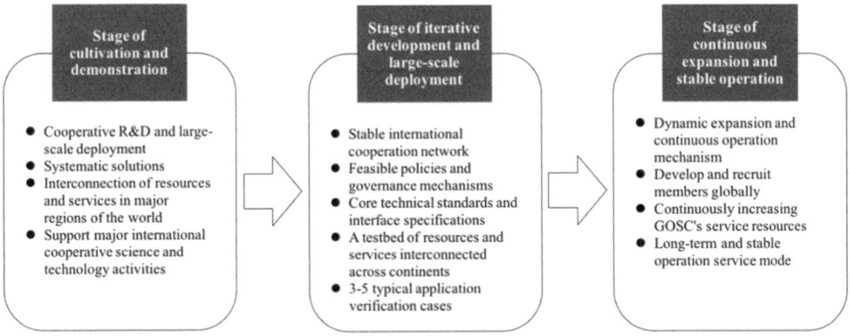

Fig. 3 Implementation work plan of the GOSC initiative

Priority, Progress, and Partnership (December 2020)", and the "Global Open Science Cloud Launch Event" (June 2021) [19]. A "Global Open Science Cloud Landscape" report has been released to report the latest status of open science clouds around the world.

Among all the engagements, CAS's Computer Network Information Center (CNIC) also plays an important role. Through strategic collaboration with the European Grid Infrastructure (EGI), a joint testbed linking China and Europe has been established to support disciplinary showcases in space physics, biology, and earth sciences [20, 21]. In addition, cloud federation schemes have been drawn in tailored research scenarios to fit into the complex inter-cloud environment.

6 Conclusions

The call to building collaborative research communities with a shared future for humanity is intertwined with the global changes unseen in a century. Open science has great potential and ability to highlight all these challenges, in which we see the indispensable role that GOSC can play. The implementation of the GOSC Initiative seeks to help bridge the gaps between different research clouds by exploring possible ways to strengthen the management and sharing of various OSC resources and the provision of services between clouds. However, the development of GOSC is a long-term, iterative, and continuous process. Therefore, composite solutions should include evolving policies, transparent resource flows and interoperable technologies [22] based on mutual trust and smooth mechanisms for OSC dialogue in the global context. Moreover, global stakeholders should work together to enrich this GOSC framework to facilitate the sharing of global research resources and the extensive interconnection of research infrastructures for scientific discovery in the infinite frontiers of science.

Table 2 Workgroups and case studies in GOSC initiative

Name	Research focuses
WG—strategy, governance, and sustainability	• GOSC governance models • GOSC's mission, landscape, and strategic initiatives • GOSC funding sources • Engagement mechanism of stakeholder
WG—policy and legal	• Policy guidelines • Policy interoperable solutions • GOSC guiding principles • Engagement with case studies
WG—technical infrastructure	• Technical interoperation between digital research infrastructures • Test templates of the GOSC framework • GOSC testbed and use case • International research collaborations
WG—data interoperability	• Cross-domain data models • Data interoperability solutions • Data interoperability standards and metric • Long-term preservation strategies for semantic tools
CS—incoherent scatter radar data fusion and computation	• Radar data and metadata federation model • Radar data collaborative production and sharing model • Radar data quality control in processing • Federated learning enhancement algorithm
CS—biodiversity and ecology information platform	• Consistency issues in data storage • Biosphere reserve data systems • Online cloud services for modular functions • Intelligent image analysis
CS—SDG-13 climate change and natural disasters	• Homogeneity and heterogeneity study in climate change and natural disasters • Causes and frequency of extreme regional climates • Cloud federation technologies maturity validation
CS—sensitive data federation analysis model in population health	• Distributed data analysis and learning system • Federated platform to access shielded data • A prototype system for the analysis of sensitive data
CS—open, reproducible raw diffraction data for access in pandemics	• Metrics of 'definitive reusability' • Semantic interoperability solutions • (Meta) data co-production and sharing model • Workflow specification to accommodate federated data learning

Note The data are collected from the CODATA GOSC web page at https://codata.org/initiatives/decadal-programme2/global-open-science-cloud/

Acknowledgements This work is supported by the National Key R&D Program of China (No. 2021YFE0111500), the CAS Program for fostering international mega-science (No. 241711KYSB20200023), and the National Natural Science Foundation of China (No. 72104229).

References

1. Bexell M, Jönsson K (2017) Responsibility and the United Nations' sustainable development goals. Forum for Dev Stud 44(1):13–29
2. Guo H (2021) Open data and science drive the realization of SDGs. Sci Technol Rev 39(2):68–69
3. Ramachandran R, Bugbee K, Murphy K (2021) From open data to open science. Earth Space Sci 8(5):e2020EA001562
4. Liao F, Hong X, Wang Y et al (2019) The data and computing platform is an important infrastructure which drives modern scientific research development. Front Data Comput 1(5):2–10
5. Manvi SS, Shyam GK (2014) Resource management for infrastructure as a service (IaaS) in cloud computing: a survey. J Netw Comput Appl 41(5):424–440
6. Grossman RL, Gu Y, Mambretti J et al (2010) An overview of the open science data cloud. In: Proceedings of the 19th ACM international symposium on high performance distributed computing. Chicago Illinois, USA, pp 377–384
7. Ghazi A, Bahaei H, Arani MG (2014) A new framework for the evaluation of QoS in cloud. Int J Comput Appl 107(1):44–49
8. Ray BK, Saha A, Khatua S et al (2019) Toward maximization of profit and quality of cloud federation: solution to cloud federation formation problem. J Supercomput 75:885–929
9. Computer Network Information Center, Chinese Academy of Sciences (2022) International initiative: global open science cloud [EB/OL]. http://english.cnic.cn/nws/202106/t20210608_271753.html. 24 January 2022
10. Chinese Academy of Sciences (2022) Global open science cloud initiated for dialogue and alignment between open science platforms [EB/OL]. https://english.cas.cn/newsroom/news/202106/t20210630_273264.shtml. 24 January 2022
11. Inkpen K (2009) Many computers, many people, and everything in between: supporting shared computing. In: Proceedings of the 13th international conference on computer supported cooperative work in design, Santiago Chile, USA, p 1
12. Cabellos L, Plasencia IC, Fernández-Del-Castillo E et al (2011) Scientific workflow orchestration interoperating HTC and HPC resources. Comput Phys Commun 182(4):890–897
13. Yang Q, Liu Y, Chen T et al (2019) Federated machine learning: concept and applications. ACM Trans Intell Syst Technol 10(2):1–19
14. Stergiou C, Psannis KE, Kim B et al (2018) Secure integration of IoT and cloud computing. Futur Gener Comput Syst (78):964–975
15. Singh J (2019) Exploring expansion and innovations in cloud computing. Int J R&D Innov Strat 1(1):46–59
16. Goiri I, Guitart J, Torres J (2010) Characterizing cloud federation for enhancing providers' profit. In: Proceedings of the 3rd IEEE international conference on cloud computing, Miami, USA, pp 123–130
17. Computer Network Information Center, Chinese Academy of Sciences (2022) Global open science cloud workshop connects policy, science and technology experts for open science infrastructure [EB/OL]. http://english.cnic.cn/rsearch/rp/202106/t20210608_271755.html. 26 January 2022
18. China Science and Technology Cloud (2022) Global open science cloud series discussion held successfully [EB/OL]. http://www.cstcloud.net/news/36.jhtml. 26 January 2022

19. Chinese Academy of Sciences (2022) CAS program for fostering international mega-science "Global Open Science Cloud" was launched [EB/OL]. https://www.cas.cn/yx/202106/t20210630_4795892.shtml. 26 January 2022
20. Chinese Academy of Sciences (2022) SCTCloud and EOSC launched strategic cooperation [EB/OL]. http://www.cas.cn/yx/202006/t20200624_4751116.shtml. 26 January 2022
21. Chinese Academy of Sciences (2022) CNIC and EGI to cooperate on open science cloud [EB/OL]. https://english.cas.cn/newsroom/news/202006/t20200624_239185.shtml. 28 January 2022
22. Computer Network Information Center, Chinese Academy of Sciences (2022) Review of the global open science cloud initiative in 2021 [EB/OL]. http://english.cnic.cn/coll/ic/202112/t20211223_295159.html. 29 January 2022

Jianhui Li is the professor, Ph.D. supervisor of UCAS, and director of the CSTCloud department of the Computer Network Information Center (CNIC), Chinese Academy of Sciences (CAS).

He is also the Deputy Director General of the International Research Center of Big Data for Sustainable Development Goals (CBAS). Jianhui Li is the PI of the CAS "Global Open Science Cloud" project and also in charge of several other national-level key projects in developing open science infrastructures. He specializes in Big Data and cloud computing, open data and open science, digital research infrastructures.

The Trusted System and International Service Capacity Construction of Science Data Bank (ScienceDB)

Yuanchun Zhou, Pengyao Wang, Chengzan Li, Zongwen Li, Lulu Jiang, Zeyu Zhang, and Jia Liu

Abstract The value of research data as an important driving factor for scientific research has become increasingly prominent. In recent years, the publication of papers associated data has performed a rapid growth trend internationally. While the number of international papers published by Chinese researchers continues to grow, China is more embracing open science and actively constructing open science infrastructure and other related work. However, the culture and atmosphere of open data sharing in China is still immature, and there is a lack of credible research data publishing platform infrastructure that has international influence and is recommended by international publishers for data storage. To this end, the Chinese Academy of Sciences has built a general research data publishing platform-Science Data Bank (ScienceDB for short). At present, the platform has been recommended by many international mainstream publishing groups and indexed by the authoritative research data retrieval platforms. On the basis of analyzing the development situation of open data sharing at home and abroad, this paper systematically introduces the overall goal, technical structure, credible system, construction achievements and service effects of Science Data Bank from the demand of building an international data publishing platform that serves the global scientific community. This paper also looks forward to the next construction ideas and development direction of Science Data Bank.

Keywords Science data bank · Data publishing · Trustworthy architecture · International standard

1 Introduction

Scientific data is the input and output of scientific research activities and the core driving factor of scientific and technological innovation. In the era of big data, the open sharing of scientific data has become the common demand of all stakeholders

Y. Zhou (✉) · P. Wang · C. Li · Z. Li · L. Jiang · Z. Zhang · J. Liu
Computer Network Information Center (CNIC) of the Chinese Academy of Sciences (CAS), Beijing, China

China's e-Science Blue Book 2023, https://doi.org/10.1007/978-981-99-8270-7_25

in the scientific community. Through the open sharing of scientific data, we can promote scientific progress, reduce duplication of labor and improve productivity, promote more effective cooperation between science and policy, promote the long-term progress of scientific research and education, bring new solutions to social problems, shorten the incubation cycle of new products and meet the public's information demands, etc. [1].

Open access to scientific data comes from the World Data Center system created for the international geophysical year held from 1957 to 1958. It is required that the data must be stored in a machine-readable format and enhance the accessibility of the data as much as possible [2]. As the value and significance of data openness and sharing are gradually recognized, a large number of data openness initiatives, policies and principles have been formulated and published. In 1996, scientists and funding agencies of the international human genome project jointly put forward Bermuda principles, requiring data disclosure and resource sharing in accordance with the principle of common ownership, sharing and common action, and requiring the data to be submitted to the international gene database within 24 h [3]. In 2014, members of Force11 (Future of Research Communications and e-Scholarship) jointly proposed the principle of FAIR (Findable, Accessible, Interoperable and Reusable), which provides guidance for data producers and publishers to improve the reusability of scientific data from the four aspects of data discoverability, accessibility, interoperability and reusability [4]. In 2019, members of Research Data Alliance put forward the TRUST (Transparency, Responsibility, User focus, Sustainability and Technology) principles for the credibility of data repository, which was officially published through *Scientific Data* journal in 2020. This principle provides guidance on how to build a trusted data repository from the dimensions of transparency, responsibility, user orientation, sustainability and technical capability [5, 6].

During this period, some journals and publishing institutions, such as National Academies Press, PLoS, Nature, BMC and many other academic publishers and journals, put forward clear policies on the storage and dissemination of scientific research data, and issued policies one after another to require authors to open the data and materials related to publications according to the data requirements of journals and store them in appropriate and standard electronic forms, ensure open access to publications and supplementary data and materials, and allow all readers to obtain, copy and disseminate them free and permanently [7].

These open sharing initiatives, policies and principles have played a key guiding role in the in-depth development and promotion of data open sharing activities. In the past few decades, more and more countries, institutions, universities and international organizations have established scientific data repositories to allow permanent access and use of data in a reliable environment.

In 2001, China started the construction of scientific data sharing project. In 2004, a special project for the construction of national science and technology infrastructure platform was established to comprehensively promote relevant work [8]. Over the years of practice and exploration, China has accumulated and established an increasingly perfect scientific data policy system. In March 2018, the general office of the State Council issued *the Measures for the Management of Scientific Data*, which made it clear that the scientific data formed by government funds should be shared and utilized in accordance with the principle of openness as the norm and non-openness as the exception [9]. The Ministry of Science and Technology and the Ministry of Finance released the list of optimization and adjustment of the national science and technology resource sharing service platform in June 2019, and established 20 national science data centers and 30 national biological germplasm and experimental material resource databases to provide high-quality science and technology resource sharing services for scientific research, technological progress and social development.

After years of efforts, China has made great progress in the open sharing of scientific data, and has built a number of infrastructures such as open sharing service platforms of scientific data. However, with the increasingly prominent value of scientific data as an important driving factor of scientific research, the construction, operation and maintenance services of supporting infrastructure such as scientific data repository in China still face many problems and challenges. At present, the scope of disciplines covered by China's scientific data repository and other infrastructure is still limited, and mainly concentrated in traditional disciplines such as biomedicine and geoscience with open and shared advantages. Infrastructure such as open and shared service platforms in other disciplines are relatively scarce, and infrastructure construction and open sharing culture need to be continuously strengthened. In addition, by benchmarking international mainstream data repositories, such as Figshare, Zenodo, Dryad, etc., we can find that international mainstream data repositories are effective in many aspects, such as data policies, international standards and norms, platform credibility capacity building, data resource scale, ecological construction, influence and so on [10]. In recent years, driven by international mainstream publishers, the publication of paper related data has formed a rapid growth trend in the world. The number of international papers published by Chinese researchers is the largest in the world every year, and a large number of paper related data are published according to the publishing rules. However, at present, the scale, quality, credibility and influence of China's open sharing data service platform have not been fully established, the storage platform of paper related data and publishing services are insufficient, and the cooperation with the publishing industry at home and abroad is insufficient. As a result, the phenomenon of overseas storage of scientific data in China is common, and the continuous aggregation ability of international high-quality resources of domestic platforms is insufficient.

Therefore, the data repository should give full play to its platform and technical advantages, play a good coordinating role in the creation, circulation and utilization of international data, meet the infrastructure construction challenges of global data publishing and data sharing, strengthen international cooperation and interconnection, build a global data resource network, and play an important role in improving the accessibility and interoperability of shared data. Especially in the era of big data, adopting reasonable infrastructure and technical route to support the efficient transmission, safe and long-term storage, stable and reliable service of massive data, and following the international principles and standards such as the TRUST Principles, building an independent, controllable and internationally influential trusted scientific data repository service platform are an important infrastructure and current main challenge in the process of scientific research informatization construction in China. Adhering to the intensive construction of data repository infrastructure can effectively avoid problems such as scattered resources, repeated construction and inability to form scale effect, comprehensively improve the safe reliable operation and maintenance service capacity of repository infrastructure.

2 Science Data Bank Trusted Architecture

2.1 Technology Architecture

Science Data Bank (ScienceDB) is a public, general-purpose data repository aiming to provide data services (e.g. data acquisition, long-term preservation, publishing, sharing and access) for researchers, research projects/teams, journals, institutions, universities, etc. and ScienceDB supports a variety of data acquisition and data licenses. ScienceDB is dedicated to promoting data findable, citable and reusable on the prerequisite of protecting the rights and interests of data owners. The goal of ScienceDB is to establish an international, safe and reliable general scientific data repository. Focusing on this goal, ScienceDB makes comprehensive use of cloud native technologies to establish its service system in an intensive way. ScienceDB supports the rapid construction of data community and promotes the data sharing ecology. Combined with big data analysis technology, ScienceDB provides a series of value-added services such as data community construction, institutional portrait analysis and influence tracking. ScienceDB follows the open sharing principles such as FAIR, TRUST, DOI and the international norms of academic publishing, supports the international dissemination and utilization of published scientific data resources, and realizes the international standardization and open sharing of scientific data related to papers. Meanwhile, ScienceDB comprehensively strengthened the construction of safe and reliable service capacity of data and platform to improve the credibility of the platform.

The trusted technical architecture of ScienceDB is consisted of three specific systems (Fig. 1): cloud service system, standardization system and security protection system.

2.2 Cloud Service System

Cloud service system is composed of three layers from bottom to top: foundation layer, service layer and presentation layer. The foundation layer provides the network environment, storage and computing resources of the whole cloud platform. The service layer comprehensively utilizes the resources of the foundation layer to form a

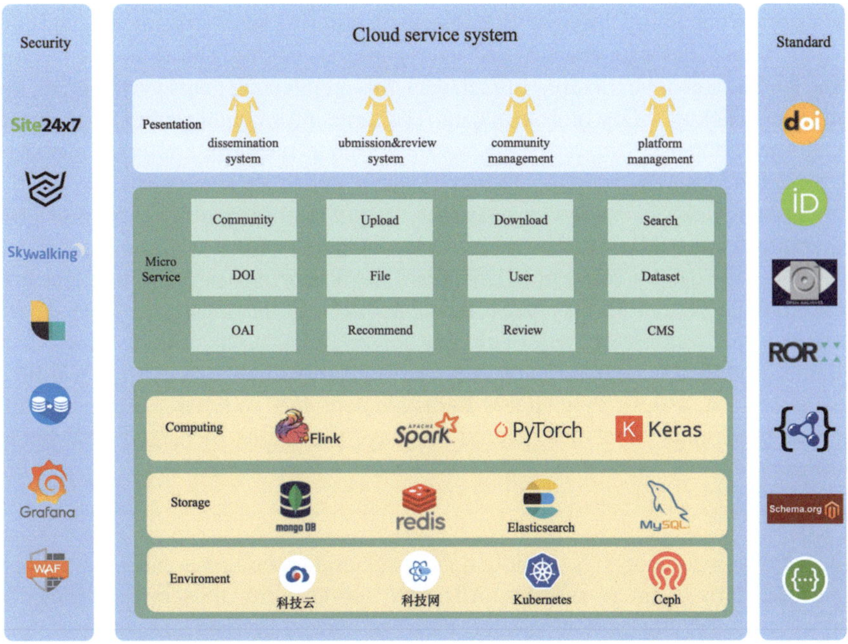

Fig. 1 ScienceDB technical architecture

micro service cluster of ScienceDB. The presentation layer uses the interface of micro service cluster to provide users with dataset submission, publishing, downloading and other services.

Foundation Layer

The foundation layer is divided into three parts: environment, database and computing. The environment part provides basic software and hardware resources, the storage part provides structured, semi-structured and unstructured data access capabilities, and the computing part provides real-time and offline big data computing platform and artificial intelligence analysis and processing capabilities.

The environment of whole ScienceDB system is built based on infrastructure of China Science and Technology Cloud. In order to ensure the stability and reliability of global services, the platform accelerates the global access to ScienceDB through CDN. Kuberntes provides a virtualized running environment to support the management and scheduling of software containers. Ceph provides file and object storage capability in the cloud system.

ScienceDB's data consists of structured information (such as user information), semi-structured information (such as dataset metadata information and log information) and unstructured information (such as dataset description text). Therefore, the database architecture selects various types of database software to support ScienceDB's data service. Specifically, MySQL is chosen for providing structured data storage capability, MongoDB is selected for storing semi-structured data, ElasticSearch provides data retrieval and aggregation capability, and Redis caches hot data. At the same time, in terms of metadata, this architecture helps to strengthen the redundant backup ability of system data. The same information elements are stored in different databases in various forms at the same time (for example, metadata information is stored in MongoDB and ElasticSearch at the same time).

The computing layer provides a variety of different computing capabilities for the entire ScienceDB cloud service system, including offline analysis capabilities for large-scale data, timely and reliable real-time data analysis capabilities, and artificial intelligence algorithm processing capabilities. In the computing layer, Spark is used to process offline data (including log analysis, batch data cleaning, etc.), Flink is used to provide real-time processing capability (including user log real-time analysis and statistics), Pytorch and Keras are combined to provide artificial intelligence processing capability (natural language processing of dataset description information).

Service Layer

The service layer adopts the front-end and back-end separation architecture, which is divided into 12 micro services. All micro services provide services through a unified gateway. The gateway has the ability of current limiting and authentication to ensure the stable operation of the system. All cloud-native micro services run in the Kubernetes cluster provided by the foundation layer in the form of container.

Based on the cloud native capacity expansion technology and automatic scheduling capability [11], all service upgrades of ScienceDB adopt the rolling upgrade mode, which ensures the continuity and stability of services in the process of software upgrade. On the premise of ensuring multiple service replicas, load balancing and container isolation, the service layer shares node resources as much as possible based on the cluster performance monitoring results provided by Promethus, effectively reduces the number of physical machines and makes full use of the intensive deployment of host performance.

Micro services interact with each other by SideCar agent. Combined with SideCar agent and gateway configuration, the whole micro service system supports A/B test and provides authorization and encryption of calls between services. Micro service implements automatic load balancing of microservice calls to ensure fair distribution of resources and realizes the tracking and observation of the whole link of micro service invocation.

Presentation Layer

The presentation layer is divided into four systems: dissemination system, submission and review system, community management system and platform management system.

The dissemination system serves public users. Users can search and retrieve datasets published in ScienceDB. At the same time, with the help of the ScienceDB's standardization system, the dissemination system promotes the dataset to a series of international index platforms including Data Citation Index, Google Dataset Search,

Mendeley Data, Scopus and so on. It can promote the international dissemination of dataset and enhance their influence.

The submission and review system is oriented to dataset authors, peer reviewers and editors. Its main business is the submission, review and editing of datasets. The system cleans and standardizes the organization, funder, paper information through automatic methods, and standardizes the original information of datasets. The system supports the private link access of the dataset during the paper review process to meet the access of the third-party editor to the dataset before the dataset is published.

The community maintenance system maintains and manages the data community in the ScienceDB platform. The data community is the data governance unit in ScienceDB. Through the community model, journals, conferences and other organizations can manage their paper related data and publish data on ScienceDB. The system mainly maintains the display template, basic information, home page content and link information in the community.

The platform management system is the background management system of ScienceDB, which provides the maintenance and supervision functions of the whole platform. The main functions include: the management of platform users, the monitoring and management of Kubernetes cluster, the tracking of DOI registration status, the management of platform display content, etc.

2.3 Standardization System

ScienceDB is committed to building a scientific data publishing platform that complies with international standards and regulations and can effectively integrate its published content into the international academic resource communication ecology. To this end, ScienceDB continues to improve the internationalization and standardization capabilities of the platform, and is committed to implementing the FAIR principle of data sharing, and has initially formed a standardization construction system of the platform. The system mainly includes three aspects: data format standardization, data publishing process standardization and data interoperation standardization.

Data Format Standardization

The ScienceDB metadata standard ensures the standardization and standardization of data description in element design and element description standard, and improves the international communication ability of data. Among them, the element design is based on Dublin Core (DC), which is compatible with DOI and Chinese Science and Technology Resource (CSTR) in the support of identification standards, effectively enhancing the interconnection of platform resources with other domestic and foreign science and technology resources. The element description process integrates international identification systems such as personnel identification information (such as ORCID), organization identification information (such as ROR, GRID) and fund identification information (such as CrossRef Funder), effectively enhancing the operability and interoperability of data. In addition, the standardized construction of data for the data service provides a good foundation and it's compatible with various identification service system. On the one hand, it can standardize the user registration and metadata to submit the related important elements in the process of standardization, on the other hand, the majority of users can quickly locate research body, found belongs to the scientific dataset provides accurate and convenient way. The standardized construction of data could better display scientific data and research institutions of the affiliation and highlight the user's scientific data research strength.

Standardization of data files, a set of standards on data file format has been developed. There are more than 10 types of file types, including plain text, markup language, text documents and spreadsheet etc., and their corresponding preferred recommended format and non-preferred format are more than 50 types, which can effectively improve the reusability and long-term accessibility of data released on the platform.

Standardization of Data Publishing Process

In terms of data publishing process standardization, ScienceDB has established strict data publishing process, including dataset metadata filling, data file uploading, data file security scanning, automatic detection of data format and metadata format, data review, revision, publishing and other complete publishing process.

In terms of standardizing the data submission process, ScienceDB has established a data submission process that follows international standard metadata to support the standardized description and publication of scientific data. In addition, the overall platform adopts i18N multi-language architecture design, supporting at least two languages, Chinese and English, to ensure the consistency of Chinese and English bilingual content entities.

In terms of file collection standardization, ScienceDB supports file upload with directory structure, supports Http and Ftp protocols to upload, and provides functions such as resumable upload and reupload of large files. For all files uploaded to the platform, the system automatically performs digital digest MD5 calculation for them. The result of MD5 calculation is written to the system to ensure the consistency of the whole life cycle of files in the platform environment.

In terms of the standardization of tiered storage after data submission, it also refers to the implementation of OAIS (Open Archival Information System) [12] to classify data storage space according to different stages of publication of data files. This includes the cache space after submission, the review space during the review stage, and the archiving and publishing space after official publication. Multi-copy storage of data files and remote disaster recovery are realized to ensure the safe and reliable storage and service of data.

Data Interoperability Standardization

ScienceDB follows the Open Archives Initiative, Protocol for Metadata Harvesting (OAI-PMH) and is built to provide standardized open access metadata services (Fig. 2). OAI-PMH is a general metadata open harvesting protocol, which effectively solves the interoperability barrier at the metadata level of the repository and plays an important role in data sharing and dissemination. ScienceDB ensures

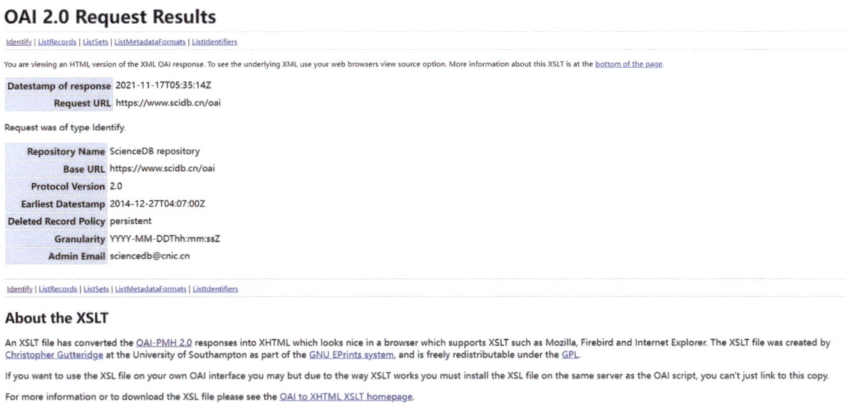

Fig. 2 ScienceDB integrates the OAI-PMH service

good machine readability by providing metadata harvesting services based on the OAI-PMH protocol and providing dataset metadata in XML format.

In addition, ScienceDB construction provides ScienceDB API Doc services based on the Swagger framework. Any third-party program can access and call the Open API of ScienceDB to access and query the metadata of the site. ScienceDB open metadata supports Dublin Core, Schema.org and other metadata standards, and supports multiple different data return formats (XML, JSON-LD, etc.) for the same dataset in the design and implementation of the open interface.

2.4 Persistent Identification System

The persistent identifier is one of the basic information infrastructure for science data to be FAIR (Findable, Accessible, Interoperable, Reusable), it is the unique identification of the scientific data. Persistent identifier is well known as the essential part of the scientific data repository, for example, figshare adopts DOI as the persistent identifier, NCBI using PMID, EUDAT choosing Handle system, GBIF using its own identifier systems. In China, the national standard CSTR (Common Science and Technology Resource Identification, GB/T 32843—2016) is released in 2016. The main different with other identifiers is that it defined various science and technology resources which are distinguished using the two digital type code. CSTR has the ability to support the traceability through the full life cycle of scientific data and its related items, including numbering mechanism, registration, resolution and management.

ScienceDB built its own scientific data identification system based on CSTR, and it is the concrete practice of CSTR in the scientific e data field. The technical architecture of CSTR is a compatible interworking. It provides the compatible resolution service for DOI, Handle, ORCID, ROR and so on. The CSTR system includes three

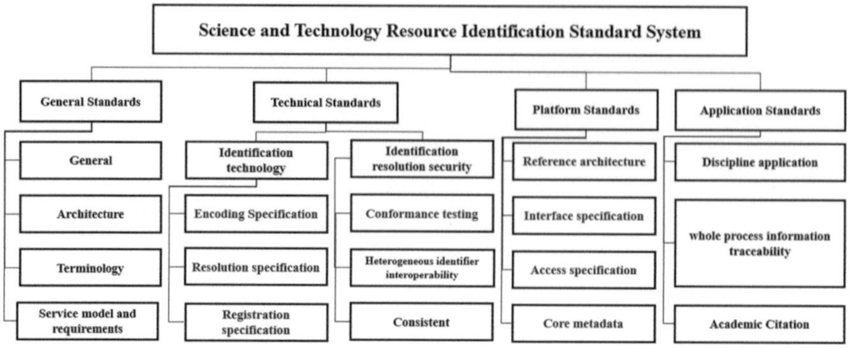

Fig. 3 The CSTR standards

aspects: the standardization system, the principles and requirements and the service platform.

2.4.1 CSTR Standards

The standard system of identification of scientific and technological resources (Fig. 3) consists of four categories: the general standards, the technical standards, the platform standards and the application standards. The "Scientific data numbering standard" is undergoing national standard, besides, the "Dissertation numbering standard", "Scientific research instruments and equipment numbering standard", "patent numbering standard" had the specific application cases. The CSTR standard system is adopted in space science, remote sensing, plant germplasm and other disciplines. The prefix of ScienceDB is "31253". ScienceDB has developed a set of unified numbering specification for scientific data with different granularity and core metadata standards that can interact with the international communities.

2.4.2 CSTR Principle and Mechanism

The CSTR adheres to "Persistent principle" which is the Internationally universal for identifiers. Once the identifier is assigned, it could not be deleted. The CSTR standard defines numbering mechanism, each CSTR identifier is composed of three parts, the first five digits is the registration authority code, the next two digits is the type code for different science and technology resources, and the last part is defined by the submitting organization expandability. The resolution service of CSTR (https://www.cstr.cn) provides the compatible resolution not only for CSTR, but also supports DOI, Handle and so on. The CSTR defines a set of core metadata systems, and provides extensible metadata standards for different scientific and technological resources, such as geographical location data.

2.4.3 CSTR Identification Platform

The CSTR Identification Platform (https://www.cstr.cn/en) is initialed in 2016 based on the National Standard GB/T 32843—2016 "Science and technology resource identification", providing unique identification service for global scientific data, papers, scientific institutions, researchers, scientific instruments, patents and other scientific and technological resources. CSTR is also accepted as the international standards, such as IANA, etc.

The CSTR Identification Platform aims to promote establishing a global science and technology resource identification architecture, which is able to tracing the global influence, cross-disciplinary, cross-regional and cross-platform, to realize the rapid positioning and acquisition worldwide, and be used as the digital base for open science, contributed to the scientific research innovation.

The CSTR identifier provides four key services. First one is the Registration Service. Through the registration of science and technology resource identifiers and metadata, the published scientific and technical resources are made more discoverable and accessible. Second one is the Resolution Service. The convenient enquiry and search services are provided by using scientific and technical resource identifiers, keywords and metadata as inputs. Third one is the Associated Services. It links papers to authors, papers to resources, research processes and results data in a vast network of associations through scientific and technical resource markers. Fourth one is the Data Service. It provides multi-dimensional statistical analysis of the total amount of resources, resolution volume and query volume, forming a comprehensive data service.

2.5 Security System

Building a safe and credible security system is the basis for ScienceDB to provide sustainable services. ScienceDB's security system focuses on strengthening security in the following three directions.

2.5.1 Disaster Recovery

In terms of file system disaster recovery, ScienceDB uses two file systems to back up and store the entity files submitted by users, namely open-source CEPH file storage and commercial GPFS storage. The two sets of storage are deployed in different locations to ensure the safety and reliability of data files through remote mirror backup.

In terms of database disaster recovery, firstly, all databases of ScienceDB are built in cluster mode, and all databases have real-time hot backup capability. Secondly, the core database nodes are deployed in Kubernetes. In case of abnormal nodes, the self-healing of the database can be realized through the automatic scheduling ability of Kubernetes [13]. In addition, ScienceDB adopts a regular backup strategy to regularly back up the database content to remote nodes.

2.5.2 Network Protection

In order to protect the user data in the network and ensure the stability and security of the system's services worldwide, ScienceDB has adopted a variety of means to standardize and strengthen network security.

On the one hand, the core business of ScienceDB provides external services through CDN to ensure the stability of users around the world when accessing the core business of ScienceDB. At the same time, WAF firewall is enabled at the exit of CDN to filter attack requests.

On the other hand, with the help of Kubernets SideCar mode, the technical team installed RASP (application runtime self-protection) [14] on all ScienceDB self-developed software. RASP is directly injected into services to provide function level real-time protection, and detect unknown vulnerabilities without updating policies and upgrading protected application code.

2.5.3 Monitorability and Observability

In addition to the security at the file and network levels, another important security indicator in ScienceDB security system is the monitorability and observability of system and service applications to ensure early detection and prevention of abnormal problems. The constructability of ScienceDB monitoring platform is mainly from the following two aspects:

In terms of network, ScienceDB uses Site 24×7 service to realize the automatic monitoring ability of platform services from more than 30 nodes around the world, and gives the availability and loading speed reports of each node in real time. Within the micro service cluster, ScienceDB has built a service call link tracking system, which samples and analyzes the whole life cycle information of service calls in real time, and alerts abnormal calls.

In terms of platform operation monitoring, all service software in ScienceDB has completed cloud native deployment, and the whole platform software can be monitored through Kubernets dashboard. Meanwhile, ScienceDB has installed Promethus database in its Kubernetes cluster, which collects metrics and displays it through Grafana, so as to visually monitor CPU, Memory, Network and other main indicators.

3 Effectiveness of ScienceDB Services

3.1 Effectiveness of ScienceDB Capacity Building

Data Publishing Services

ScienceDB dataset submission interface is friendly and interactive, such as supporting Chinese and English bilingual metadata filling, filling interface display prompt information, organization name query and other functions, to assist submitters to accurately fill in the metadata of the dataset. All datasets are subject to a rigorous data review process, and eventually accepted published datasets are automatically assigned CSTR and DOI identifiers and provided with standard dataset reference formats and so on.

Data International Access Service

ScienceDB ensures smooth international access through CDN services. The platform adopts dynamic CDN for content distribution and user access acceleration, with 2800+ CDN global nodes and 500+ overseas nodes, supporting five major operators. Using DCDN WAF firewall to protect data information, to provide global users with fast, secure and stable dataset access network.

Data Usage Tracking Service

ScienceDB tracks the impact of the global use of the dataset. The page of the dataset (Fig. 4) provides a real-time visual representation of global visits and downloads, as well as timely tracking of academic papers referencing the data, providing an effective reference for the impact of the dataset.

Below in *A High-resolution Air Quality Reanalysis Dataset over China (CAQRA)* as an example (Fig. 5), the dataset published details page visualize the following information: total view of the dataset, file downloads, citation downloads and so on. At the same time, it also shows the number of global visits and downloads of the dataset in the form of the distribution of the number of users in the world map.

ScienceDB tracks the global citation influence of the dataset in real time, and displays the latest citation information of the dataset in real time on the home page (Fig. 6), as well as the ranking according to the number of visits and citations. The latest citation information of the data will also be sent to the data author in the form of email.

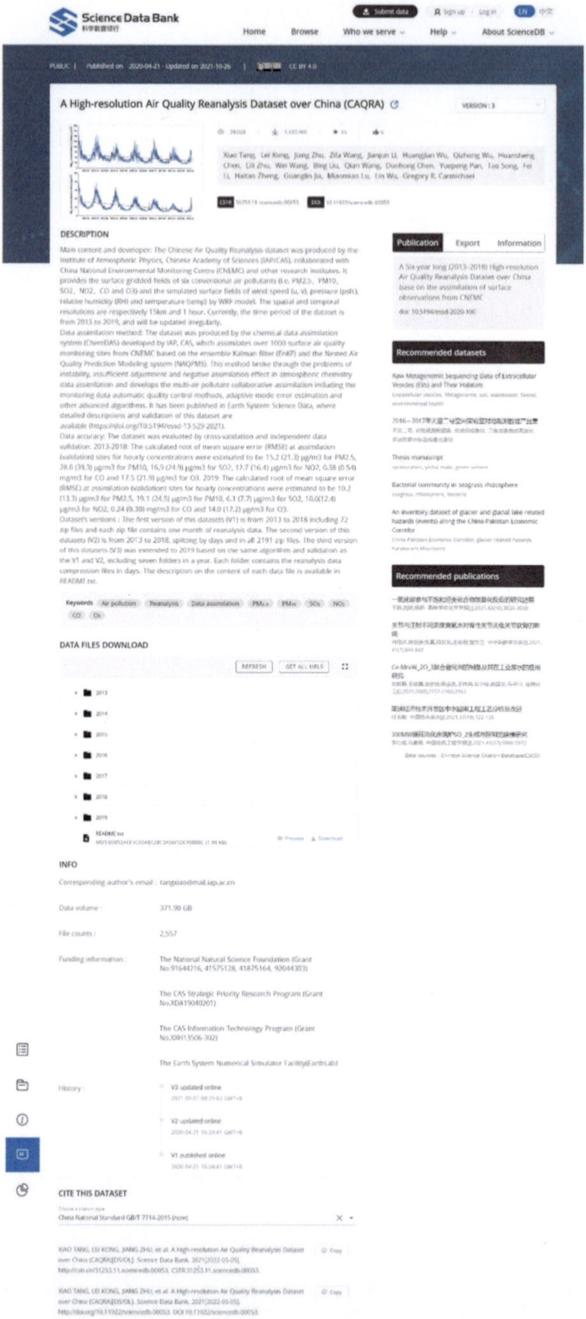

Fig. 4 Dataset publishing interface

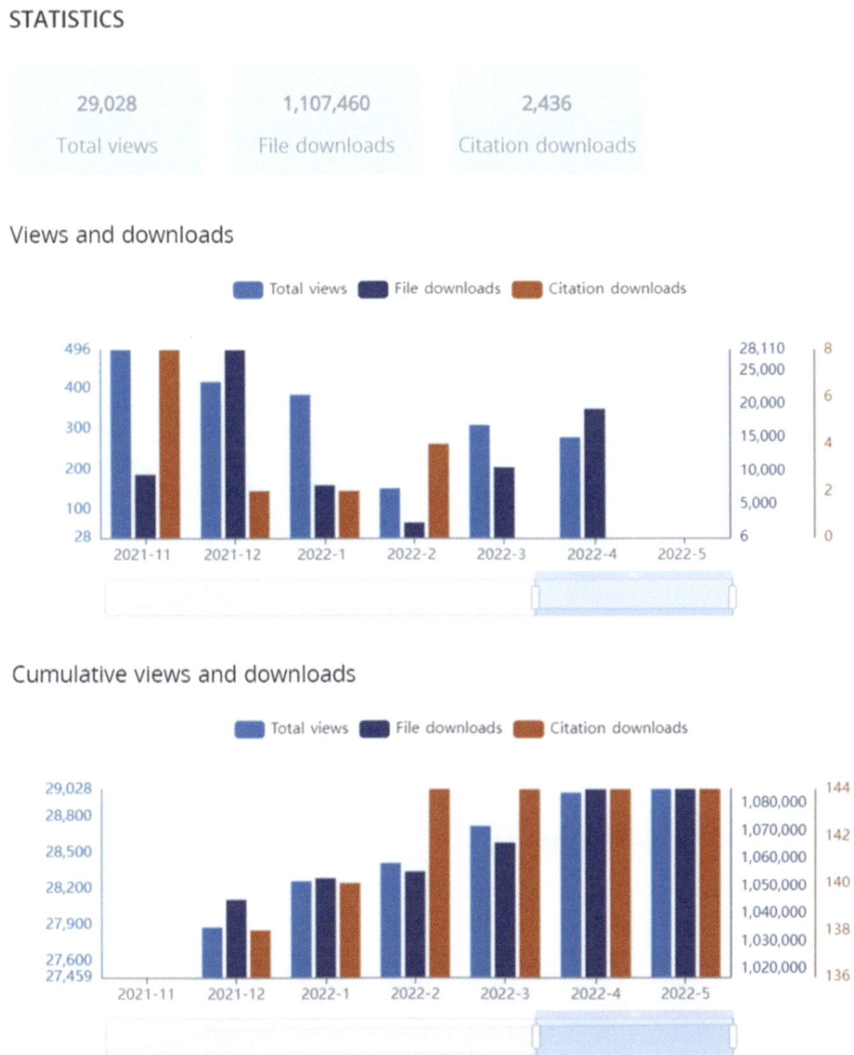

Fig. 5 Influence statistics of *A High-resolution Air Quality Reanalysis Dataset over China (CAQRA)*

Fig. 6 Ranking of visits and citations

3.2 *The Supporting Achievement of Science Data Bank*

For now, ScienceDB has been included as the recommended general repository by Springer Nature, Taylor & Francis, Cell Press, Elsevier, AGU journals and has also been indexed by Data Citation Index, Google Dataset Search, Scopus and Mendeley Data.

In addition, ScienceDB is open for free registration and use by the global public. By January 2022, the platform had received 4,174,548 user visits, 14,034,354 file downloads, and 1,724,875 publicly shared datasets (Fig. 7). The authors of the datasets came from more than 30 countries and regions around the world, including the Netherlands, the United States, the United Kingdom, India, Brazil, Spain, South Africa, Iran, Italy, Vietnam, France, and visiting users in nearly 160 countries and regions around the world. In recent years, ScienceDB releases have continued to increase, with the number of releases in 2021 increasing by 3.14 times compared to 2020. The number of international datasets has also increased year by year, especially since ScienceDB was designated as the recommended common repository by Springer Nature, which has increased by 7.84 times compared to before the recommendation, and the total proportion of international datasets has reached 14.3%. In terms of ecological construction of data community, ScienceDB has successively built more than 150 data communities of journals, institutions and societies, and continues to promote the construction of open sharing ecology of scientific data in the way of data community.

ScienceDB supports data publishing in several national scientific data centers, including the National Basic Science Data Center and the National Space Science Data Center. ScienceDB also supports the data submission of the CAS Big Earth Data Science Engineering Program (CASEarth). In addition, *China Scientific Data* supported by ScienceDB was awarded the "The fifth China Publishing Government Award for Advanced Publishing Entities".

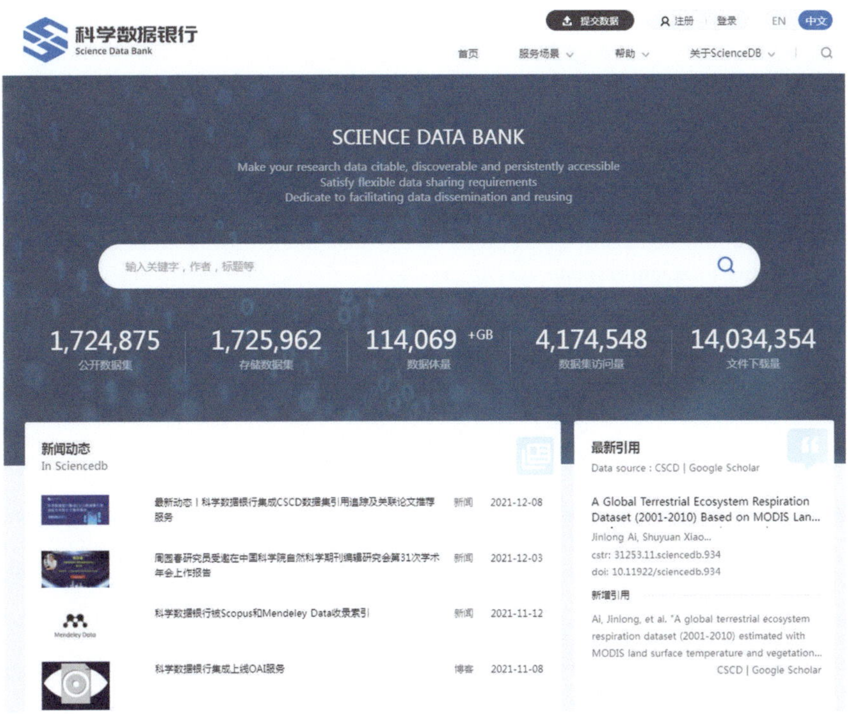

Fig. 7 ScienceDB datasets and access statistics

4 Summary and Prospect

The value of scientific data as an important driving factor of scientific research has become increasingly prominent. With the joint efforts of national governments, international organizations and academia, the open sharing of scientific data has made considerable progress. In recent years, with the rapid growth of the publication of paper related data in the world, China's scientific data are facing the problems of immature open sharing culture and ecology, lack of authoritative and trusted repository with international influence and so on.

Science Data Bank is committed to building a safe and reliable general scientific data publishing and storage platform, actively exploring the construction of open sharing ecosystem based on intensive cloud services of data community from the perspective of technical architecture, strengthening the construction of open sharing principles and international standards of the platform, constantly strengthening the safe and reliable service capacity of the platform and data, and improving the credibility of the platform. Through efforts, ScienceDB has achieved certain construction and service results. In the future, ScienceDB will continue to provide good platform services, continuously gather high-quality scientific data resources, and carry out application research of scientific data fusion analysis technology in combination with the development idea of big data, artificial intelligence and digital computing integration, so as to serve scientific researchers and scientific research innovation. Our team will continue to promote the ecological and cultural construction of China's open sharing of scientific data, explore an open data sharing model suitable for China's national conditions, and promote the sustainable development of China's open sharing of scientific data.

References

1. Zhang L, Wen L, Shi L, Zheng X, Li J (2018) Progress in scientific data management and sharing. Bull Chin Acad Sci 33(8):774–782
2. Li H, Ma J, Wang N et al (2013) Research review on scientific data organization and management at home and abroad. Libr Inf Serv 57(23):130–136
3. Huang R (2009) New treads of international organizations in open access. Libr Inf 2009(5):63–69
4. Wilkinson MD, Dumontier M, Aalbersberg IJJ et al (2016) The FAIR guiding principles for scientific data management and stewardship. Sci Data 3(1):1–9
5. Lin D, Crabtree J et al (2020) The TRUST principles for digital repositories. Sci Data 7:144. https://doi.org/10.1038/s41597-020-0486-7
6. Delgado J, Romero CA et al (2021) Approaches to the integration of TRUST and FAIR principles. In: Short papers proceedings of the First SWForum workshop on trustworthy software and open source 2021: virtual conference, March 23–25, 2021. CEUR-WS.org, pp 1–5
7. Piwowar HA, Day RS, Fridsma DB (2007) Sharing detailed research data is associated with increased citation rate. PLoS ONE 2(3):e308
8. Su J, Shi L, Wang Z et al (2015) Thoughts and countermeasure to promote management and sharing of scientific data and information resources. China Sci Technol Resour Rev 5:45–49

9. Xing W, Hong C (2019) Opening is normal—interpretation of items about scientific data sharing and utilization in the scientific data man. Libr Trib 39(1):117–124
10. Zhang T, Liu F (2018) Progress in data publishing. Chin J Sci Tech Period 29(5):453–459
11. Balalaie A, Heydarnoori A, Jamshidi P (2016) Microservices architecture enables devops: migration to a cloud-native architecture. IEEE Softw 33(3):42–52
12. Lee CA (2010) Open archival information system (OAIS) reference model. Encycl Libr Inf Sci 3:4020–4030
13. Li F (2019) Cloud-native database systems at Alibaba: opportunities and challenges. Proc VLDB Endow 12(12):2263–2272
14. Huang YU, Shuai W, Huamin JIN (2020) RASP based web security detection method. Telecommun Sci 36(11):113

Yuanchun Zhou Ph.D., researcher, doctoral supervisor, distinguished researcher of the Chinese Academy of Sciences, his research field is scientific big data and knowledge graph. He is currently the deputy director of the Computer Network Information Center of the Chinese Academy of Sciences, the director of the Science and Technology Committee, the director of the General Center for Scientific Data of the Chinese Academy of Sciences, and the vice chairman of the Scientific Data Committee of the China Information Association. He has successively undertaken the key projects of the National Natural Science Foundation of China, the National Key R&D Program, the Ministry of Science and Technology's Innovative Method Special Project, and the Industrial Internet Innovation and Development Project, and has led the construction of the Chinese Academy of Sciences' scientific data center system. He has obtained 5 national standards; published 2 monographs; more than 20 authorized patents, and some achievements have been transferred; relevant achievements have been published in internationally renowned journals and conferences such as IJCAI, AAAI, ACM TKDD, ACM TIST, Nucleic Acids Research, etc.

Construction and Services of Chinese Science and Technology Resource Portal Under a New Innovation Paradigm of Science and Technology

Hui Zhang and Bo Xu

Abstract Sharing of scientific and technological (S&T) resources implements the national S&T development strategy and comprehensively promotes S&T progress and socioeconomic development, a vital measure to improve Chinese innovation capability and competitiveness. To integrate, share, and service of S&T resources, the construction of the Chinese Science and Technology Resource Portal (CSTRP) focuses on developing new paradigm and spurring new demands of S&T innovation. CSTRP builds standard specification and resource identification systems to realize the aggregation and sharing service of multi-category S&T resources in China, which promotes the optimal allocation and efficient utilization of S&T resources and accelerates innovative construction in China. This paper comprehensively introduces the system architecture, key technologies, function modules, and important achievements of CSTRP. It provides guidelines on Chinese S&T resources information aggregation, service sharing, and decision making.

Keywords CSTRP · S · T resource opening and sharing · S · T innovation

1 The Significance of Constructing Chinese S&T Resources Portal (CSTRP) in the New Era

In recent years, with the rapid development of big data, artificial intelligence, and other emerging information technologies, scientific data, scientific facilities and instruments, biological germplasm and experimental materials, and other science and technology (S&T) resources have been generated, circulated, and used with increasing frequency. As a result, the amount of S&T innovation resources and resource information has grown exponentially. The characteristics of S&T resources,

H. Zhang (✉)
Beihang University, Beijing 100191, China
e-mail: hzhang@buaa.edu.cn

B. Xu
The National R&D Infrastructure Center of China, Beijing 100038, China

China's e-Science Blue Book 2023, https://doi.org/10.1007/978-981-99-8270-7_26

such as their massive amount, heterogeneity, and scattered distribution, have made acquiring and utilizing the resources more challenging, resulting in high cost and low innovation efficiency. To address the difficulty in sharing S&T resources, a new paradigm of S&T innovation has been constructed to managee, open, and share S&T resource information. Big data and artificial intelligence have been applied to achieve deep mining and precise matching among the massive amount of S&T resource information, thus discovering the underlying knowledge and wisdom. These measures are the new goal for the digitization of S&T resources in the new era [1]. The promotion of open sharing of S&T resources serves as a bridge that connects the vast number of S&T resources and the demand for them. Therefore, it is urgent to create a more open, transparent, secure, and credible innovation ecosystem.

The Chinese S&T Resources Portal (CSTRP) is a platform constructed by the National S&T Infrastructure Platform Center which was jointly supported by the Ministry of S&T and the Ministry of Finance. The CSTRP was officially launched in September 2009 and made a major revision in 2019. The current version has integrated a vast amount of S&T resources information, including scientific data, biological germplasm and experimental materials, major scientific research infrastructures, and large scientific instruments and equipment. In addition, it is a logically unified and physically distributed network system designed for the management and services of S&T resources. The CSTRP has become a central hub for gathering China's S&T resources information, providing an open window for information release and achievements display. It has also established itself as a support system for decision-making in S&T resources management and a hub for the exchange of S&T resources between China and foreign countries [2].

According to *The Notice of the General Office of the State Council (of China) on Regulations of Scientific Data Management (GBF [2018] No. 17)* and *The Notice of the Ministry of S&T and the Ministry of Finance on Regulations of National S&T Resources Sharing Service Platform Management (GKFJ [2018] No. 48)*, the latest version of the CSTRP should serve as a national platform for the network management of scientific data, as well as a national platform for the information release and network management of S&T resources sharing services. As a multifunctional platform, the CSTRP serves as the portal system for S&T resources sharing services in China. Specifically, it can release the catalogs of scientific data and other S&T resources according to unified standards and provide diversified information services in terms of S&T resources. Furthermore, it also supports the establishment, operation, management, and evaluation of national platforms for S&T resources sharing services (hereinafter as "national platforms"), such as the National Scientific Data Center and National Biological Germplasm and Experimental Materials Center. CSTRP takes charge of the construction and operation of these national platforms, and interconnects with various departments, localities and service platforms related to S&T resources, such as research institutes and universities.

The new version of CSTRP has made a comprehensive upgrade and optimization in service function, service mode, and service content. The integration of advanced information technologies such as Cloud Computing, Artificial Intelligence and Blockchain has led to the development of a series of standards, including metadata standards for S&T resources, data collection and transfer schema, unified user authentication norms, registration, and resolution of S&T resources identifiers, etc. Combining personalized services, accuracy information retrieval, resource identification, property rights protection, and other scenarios, CSTRP develops innovative R&D and applications to promote the open sharing and utilization of S&T resources comprehensively. The home page of CSTRP is shown in Fig. 1.

2 The Main Functions of CSTRP

As the portal of S&T resources sharing services, CSTRP serves as the access entrance of national S&T resources catalog, and it is also the platform for the related researchers to obtain S&T resources services [3]. The goal of CSTRP is to be a safe, convenient, flexible, authoritative, reliable, and interconnected platform. It consists of modules for S&T resources information management, identification management, S&T resources information retrieval and classification, information service, performance evaluation, and user unified identity authentication.

2.1 Information Management of S&T Resources

The information management function of S&T resources includes collection, archives, analysis, and release. It plays an important role in optimizing the utilization of S&T resources [4]. According to the standards of S&T resources metadata, the CSTRP collects and integrates the information, such as the name, identifier, keyword, description, and property, forming the S&T resources catalog. Based on the S&T resources catalog, this module can accomplish tasks such as resource classification, resource release, resource evaluation, identifier registration, information update, and data quality evaluation. This not only ensures the quality of information exchange, supports the management and service of S&T resources, but also keeps the resource entities (such as scientific research instruments, biological germplasm, etc.) in the S&T resources owner or service provider. The CSTRP forms a distributed service architecture in which resource information can be uniformly retrieved and found, and each platform serves the resource entities independently. The CSTRP not only improves the efficiency of sharing S&T resources, but also protects intellectual property rights.

At present, the CSTRP divides the S&T resources information into four categories: scientific data, biological germplasm and experimental materials, major scientific research infrastructure, and large scientific research facilities. To obtain resources quickly and accurately, S&T resources information is further divided into secondary and tertiary categories, such as genomic science data, space science data, ecological science data, etc.

Fig. 1 The home page of CSTRP

2.2 Scientific and Technological Resources Identification Management

Science and Technology Resources Identification (CSTR-ID) is a unified encoding regulation for S&T resources. Every S&T resource has a unique ID for management, service, circulation, evaluation, ownership confirmation, and source tracing.

CSTR-ID classification regulation is constructed according to the national standard *Science and technology Resources Identification*. CSTR-ID system is designed and accomplished on the CSTRP, solving the problems of uniqueness and permanence in S&T resources discovery, search, and statistics, providing support for the ownership confirmation and source tracing of S&T resources after the registration of S&T identifier. CSTR is not only an important part of the open sharing framework of S&T resource, but also one of the important infrastructures and service to promote the open service and stable management of S&T resources, which lays a solid foundation for the standardized management and unified identification of S&T resources [5].

Relying on the CSTRP, we have designed and formed a protocol for synchronizing data in the distributed S&T resources identification system. A series of software including the distributed S&T resources identification management system, the data synchronization system, and the metadata tracing system have been developed. The main functions of CSTR-ID system are S&T resources online registration, identification allocation, statistical analysis, ID resolving service, metadata tracing, and other features.

2.3 Information Retrieval of S&T Resources

The information retrieval of S&T resources consists of two services, including keyword-based retrieval and category-based navigation provided by the CSTRP. Users can search for resources by entering keywords in the search box or obtain the resources through the classification system. The process of resources retrieval includes:

1. The owner or service provider of the S&T resources collects and delivers the S&T resources catalog (metadata) to the corresponding national platform.
2. The national platform of each research field integrates and shares resource information to CSTRP according to the metadata standard.
3. CSTRP establishes the resources catalog index and classification tree.
4. Once the users find the desired resources through information retrieval service, the online service system of the national platform to which the resource belongs will provide the resource service. The data can be browsed and downloaded through an online service system. For non-data resources, resource acquisition services can be provided directly by the relative national platforms.

2.4 Information Service of S&T Resources

The type of information includes S&T news, service cases and thematic resources relating to the sharing of S&T resources, which are reprinted from the national platforms in various field. Through CSTRP, users can conveniently browse the profiles, news, service cases, and standard specifications of various national platform.

Thematic resources information is one of the key contents of CSTRP. It is a collection of resources on a certain topic which is formed after data mining, such as the COVID-19 S&T Data Resources Service and the Summer Flooding Scientific Data Resources Service. This module provides users with a thematic, knowledge-based, and non-profitable S&T resources service.

2.5 Performance Evaluation of S&T Resources Service

CSTRP offers online management functions for platform construction, operation management, and performance assessment. Using a range of self-developed evaluation and monitoring tools, CSTRP provides statistical analysis of the operation status, usage, and resource information increasing or decreasing for national platforms. This data can be used for service evaluation and decision support by the administration department of S&T resources.

The evaluation of CSTRP consists of multiple factors. To accurately measure the performance of S&T resources sharing services, the new version of CSTRP summarizes the experience of sharing service performance evaluation over the years. Based on the specificity, objectivity, comprehensiveness, accessibility, simplicity, and flexibility of the assessment indicators, CSTRP continuously collects and evaluates indicators such as user visits, page views, pages error rate, website connectivity rate, URL-link alive rate of S&T resources catalog, and website average response time for each national platform. At the same time, users' satisfaction with the services of the national platform is investigated from resource quality and service quality through questionnaires and other means. The statistical results are presented in various visual forms, such as graphics and statements, to report the operation and service situation of each national platform in a more comprehensive way. It can also support the performance assessment of each national platform.

2.6 User Unified Identity Authentication

The S&T resources management and service process includes four user roles: resource administrator, service provider, owner, and user. The resource administration role is responsible for specific management tasks such as resource statistics, user access statistics, data quality evaluation, user satisfaction surveys, and service performance evaluation. The service provider role is the organization that provides various S&T resources services, including 20 established national scientific data centers, 31 national biological germplasm and experimental material repositories, and others. The resource owner role refers to the holder of property rights or other similar rights in S&T resources. They can provide S&R resources service independently or authorize other service parties after registering the resources. The user role refers to the end-users of S&T resources, including government agencies, scientific research institutions, enterprises, individuals, and others needing services through CSTRP.

To ensure the security of user information, CSTRP has established a completely trust and authorized management mechanism, a distributed user registration system, a user authentication system, a resource authorization management system, and a unified framework of S&T resources authorization strategy. CSTRP also implements secure sharing of user profiles between the online service systems of national platforms and CSTRP. Users can single sign on CSTRP to access multiple national platforms conveniently and quickly, which greatly improves the convenience of accessing resources.

3 Achievements of CSTRP Service

Based on the *Specifications for S&T Resource Information Integration on CSTRP,* CSTRP has aggregated vast amounts of S&T resources information and provides various information services, such as the S&T resources information navigation, resources retrieval, thematic resources, and service cases, supporting the S&T resources open sharing and efficient utilization.

By October 2021, the S&T resources directory of CSTRP has integrated 51 national platform resources, which include 20 national scientific data centers and 31 national germplasm resource centers, major scientific research infrastructure facilities, and large scientific research instruments, containing more than 2.83 million items of resource metadata in total. Among them, there are over 100,000 S&T datasets and more than 2.63 million biological germplasm and experimental materials, as well as 82 major scientific research infrastructure facilities and more than 100,000 items of large scientific research instruments. The information integration enables the administrative department to keep track of the progress of research activities, promoting the S&T resources open sharing and efficient utilization further and further.

CSTRP has more than 2.73 million S&T resource identifiers and 54 identification registration organizations till October 2021. There are three S&T resources identification registration agencies, which are *Beihang University, Institute of Scientific and Technical Information of China, and Computer Network Information Center of Chinese Academy of Sciences.* The construction of CSTR-ID System standardizes identifier coding rules and provides a foundational guarantee for the resource location, traceability, citation, counting and evaluation of S&T resources. The CSTR-ID system improves S&T resources management and standardization ability, enhances the efficiency of S&T resources management and services sharing, protects S&T resources copyrights and makes them traceable.

CSTRP information services mainly include news, thematic resources, service cases, etc. Thematic resources are divided into important thematic resources and general thematic resources based on their subject selection and contents. CSTRP has released 48 thematic resources according to the characteristics of platform resources and national R&D priorities till October 2021.

CSTRP has released two important resource themes, one is COVID-19 S&T Resources Service and another is Summer Flooding Scientific Data Resources Service. "COVID-19 S&T Resources Service Subject" was released by 6 national platforms: the National Gene Organization Science Data Center, the National Microbial Science Data Center, the National Pathogenic Microorganism Resource Center, the National Earth System Science Data Center, the National Population Health Science Data Center, and the National Standard Material Resource Center. "Summer Flooding Scientific Data Resources Service Subject" was published by the National Earth Observation Data Center, the National Qinghai-Tibet Plateau Data Center, the National Earth System Data Center, and the National Meteorological Science Data Center.

Besides, CSTRP has published 246 typical service cases of S&T resources from different national platforms, covering the major national strategies, major scientific research, social livelihood, and enterprise innovation, etc. The service case includes the service title, time, target and scope, background, significance, performance, content and main approaches, as well as publishing agent.

Based on the requirements of *Specifications for S&T Resource Information Integration on CSTRP* and *Specifications for National S&T Resources Sharing Service Platform Online System Construction and Operation Management,* CSTRP monitors and evaluates online of 51 national platforms' operation and service performance. The evaluation factors include resource catalogs volume, number of visiting users, hot search words, items of news, information quality, etc. The report of the national platforms' operation and services is regularly generated based on the evaluation statistical analysis results, which effectively supports the administrative departments in evaluating the service performance of national platforms.

4 Future Prospects

The CSTRP continues to collect various S&T data resources from the whole society, and integrate advanced technologies, such as Big Data, Artificial Intelligence, and Blockchain. It provides effective scientific data, resources and information services to support scientific and technological research and development. As the development of the fast data growing and technologies, the CSTRP needs to meet users' new requirements and enhance its core position and leading role in S&T resources sharing with cutting-edge technologies, high quality of resource and services.

In the future, the CSTRP will continue to optimize its performance according to *Regulations of Scientific Data Management,* and *Regulations of National S&T Resources Sharing Service Platform Management*, and provide more efficient S&T management services as an important national platform and make effective promotions in S&T resources exchange and cooperation.

References

1. Junjie Z, Xiaomeng Z, Shaoliang F et al (2019) Research on user portrait construction in the sharing service of science and technology resource. China Sci &Technol Resour Rev 51(4):11
2. Shaoli Z, Xiaoqi Z, Hui Z et al (2018) Innovation and practice of technology resource sharing network model: taking China's science and technology resources sharing network as an example. Sci Technol Manag Res 38(13):10
3. National Science and Technology Infrastructure Management Method [EB/OL]. http://www.most.gov.cn/xxgk/xinxifenlei/fdzdgknr/fgzc/gfxwj/gfxwj2018/201802/t20180224_138207.html
4. Lei S, Wei Y (2012) Thoughts on establish a long-term mechanism for the convergence of science and technology planning resources. China Sci & Technol Resour Rev 44(4):2–5
5. GB/T 32843-2016, Science and technology resource identification

Hui Zhang, Ph.D., professor, School of computer science at Beihang University, deputy director of the National Engineering Research Center for Science and Technology Resources Sharing Service of China. He is in charge of technology R&D and system operation for CSTRP. He has undertaken several national R&D projects, including National Key Research and Development Program of China "Research on Distributed S&T Resources Hierarchy and Service Evaluation Technology", the National S&T infrastructure platform construction project "Operation and Service of National S&T infrastructure platform Portal System", and the National Natural Science Foundation of China (NSFC) project "Research on Key Technologies of Data Sharing Service of NSFC projects", etc. His main research interests are big data, S&T resources sharing, computer network and information security. He has published more than 60 SCI and EI papers and won the second prize of the Ministry S&T Progress Award.

A Reference Software Architecture for Managing, Processing and Serving Multi-field Scientific Data

Huajin Wang, Zhihong Shen, Changfa Lu, Xiaojie Zhu, Yuanchun Zhou, Weihua Feng, and Yongsheng Wang

Abstract Scientific data resources typically have the characteristics of rich data types, diverse fields, and close semantic links. Traditional e-scientific technologies cannot simultaneously support the fusion management, processing and service of multi-field scientific data resources simultaneously, leading to the repeated development of scientific data software. This article proposes a reference software architecture for the managing, processing and serving of multi-field scientific data. This architecture consists of function modules like hybrid cloud management, integrated elastic deployment, heterogeneous data fusion store, data pipeline, and data middle platform. It can provide cross-domain heterogeneous infrastructure integration, heterogeneous data fusion management, data pipeline processing, and public master data services. This can improve the technical level of multi-field scientific data governance, service and utilization.

Keywords Scientific data · Open science · Heterogeneous fusion · Data middle platform · Data pipeline

1 Introduction

Scientific data are generally characterized by rich types, diverse fields and strong semantic links. On the one hand, traditional e-science technology struggles to simultaneously support the fusion management, processing, and service of scientific data resources in different fields, leading to the repeated development of scientific data software. On the other hand, when conducting problem-oriented scientific research, data resources in all relevant fields should be fully gathered and cross-analyzed to ensure that the research work is carried out comprehensively. However, scientific data

H. Wang (✉) · Z. Shen · C. Lu · X. Zhu · Y. Zhou
Computer Network Information Center, Chinese Academy of Sciences, Beijing, China
e-mail: wanghj@cnic.cn

W. Feng · Y. Wang
Zhengzhou Tobacco Research Institute, China National Tobacco Corporation, Beijing, China

China's e-Science Blue Book 2023, https://doi.org/10.1007/978-981-99-8270-7_27

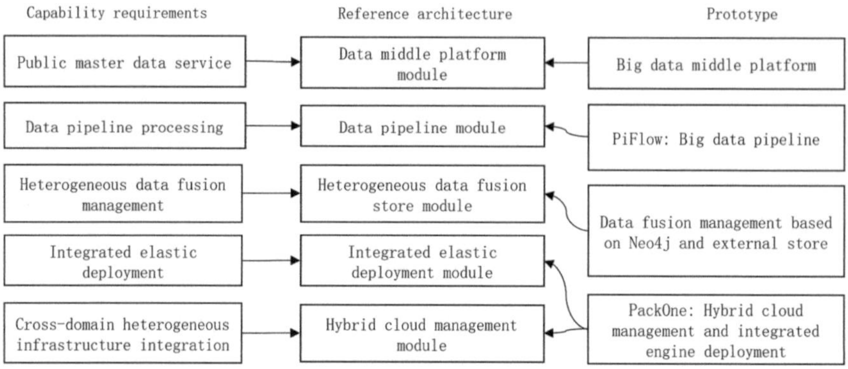

Fig. 1 The research roadmap and related contents for this work

face barriers to software reuse due to the heterogeneity of storage environment, data types and field semantics. This hinders the connection and operation of the scientific data from different fields, and leads to the scattered management and isolated analysis of scientific data in research practices. As a result, semantic links in multi-field data resources cannot be fully utilized. It is urgent to use cloud computing and big data technologies to build a reference software architecture that can provide multi-field scientific data gathering, storage, processing, analysis and services at one stop. This will guide the construction of e-science infrastructure.

The existing cloud computing and big data related theories [1, 2] and tools [3–6] are very mature. When applying them to the reference architecture, we should focus on how to provide a collection of functions that are specifically required by multi-field scientific data based on the standard features of cloud computing and big data. To this end, this paper first analyzes the requirements for scientific data management, processing, and service capabilities. Then it proposes the functional module design of the reference architecture and finally gives the prototype implementation of the reference architecture. The research roadmap and related contents are shown in Figure 1.

2 Capability Requirements for the Reference Architecture

Cross-domain integration capability of heterogeneous infrastructure. Since data resources in different disciplines often belong to different institutions, cross-field data fusion also means cross-domain data resource gathering. The heterogeneous IT infrastructures of different institutions, the tight coupling between the storage system and underlying hardware, and the isolated intranet together hinder cross-domain data transfer. The heterogeneity of the storage environment results in cross-domain data gathering that can only be done by manually searching and downloading data. Therefore, the reference architecture must be able to integrate cross-domain heterogeneous infrastructure and open up data transmission links between different institutions to allow programmatic lookups and downloads of cross-domain data resources.

Integrated elastic deployment capability of the software stack. The software stack related to scientific data governance, service, and analysis is a complex software system. It consists of underlying big data store/processing software, middle-level big data warehouse software, and upper-level algorithm models. Besides, the reference architecture must be able to integrate such a software stack and deploy it to form a scientific big data cluster that can be managed, configured and monitored in a unified manner. To keep pace with the growing scale of scientific data, the computing/storage resource of the cluster should be elastically scalable. To meet ease of use and maintainability, users should be allowed to selectively deploy the software set they need to reduce the clustering complexity as much as possible.

Fusion management capability of heterogeneous data. Problem-oriented scientific research must first discover and obtain scientific data in related fields and then conduct fusion analysis. There are often rich semantic links between scientific data. A data store that can preserve data links can facilitate data discovery, acquisition, and analysis in an efficient manner. However, scientific data has significant heterogeneity in data format. For example, species metadata is structured and usually stored in relational databases; Patents and papers are text-structured and usually stored in document databases; Graphic data are usually stored in distributed file systems. The reference architecture should be able to manage heterogeneous data in a fusion way, that is, to support linked storage and linked retrieval of various scientific dataset. Due to the data structure differences, heterogeneous data is stored on heterogeneous systems. In this case, linked storage can only be realized by a series of linking ETL tasks like reading data from various data stores, processing heterogeneous data into data of the same intermediate format, and identifying the links between data intermediates. Traditionally, the linking ETL pipeline needs to be manually written and has poor reusability, resulting in low overall efficiency of the linked query. Therefore, the reference architecture should support the automatic generation and execution of the linking pipeline, and support associated reading/writing and linked retrieval of heterogeneous data.

Data pipeline processing capability. After years of research and practice, scientific institutions have collected abundant data resources on intellectual property rights and entities such as the genome, remote sensing, radar, etc. However, data processing still faces challenges like repeated collection, difficult data finding and messy version management. It is urgent to use big data technology to standardize and streamline the process of data collection, transmission, processing, storage, classification, indexing and utilization. This will realize automatic data collection and pipeline processing. In addition, the reference architecture should provide typical data processing operators and pipeline templates for the target research field to support the customization and reuse of operators and pipelines.

Serviceability of public master data. Master data refers to the data models of entities encountered in research activities. For example, the master data of biological species can be described by metadata of its habitat, morphology, etc. Unified master data is a prerequisite for cross-domain data sharing and is a basic means to integrate scientific data from different fields. The reference architecture should absorb the latest progress in big data technology to form a master data service suitable for various scientific applications.

3 Functional Module Design of the Reference Architecture

The reference software architecture for managing, processing and serving multi-field scientific data is shown in Fig. 2. This section introduces the design of each functional module from the bottom up.

3.1 Hybrid Cloud Management Module

This module manages the IT infrastructure underlying the software architecture based on the hybrid cloud scheme and enables the reference architecture to integrate cross-domain heterogeneous infrastructures. It uses virtualization technology to pool

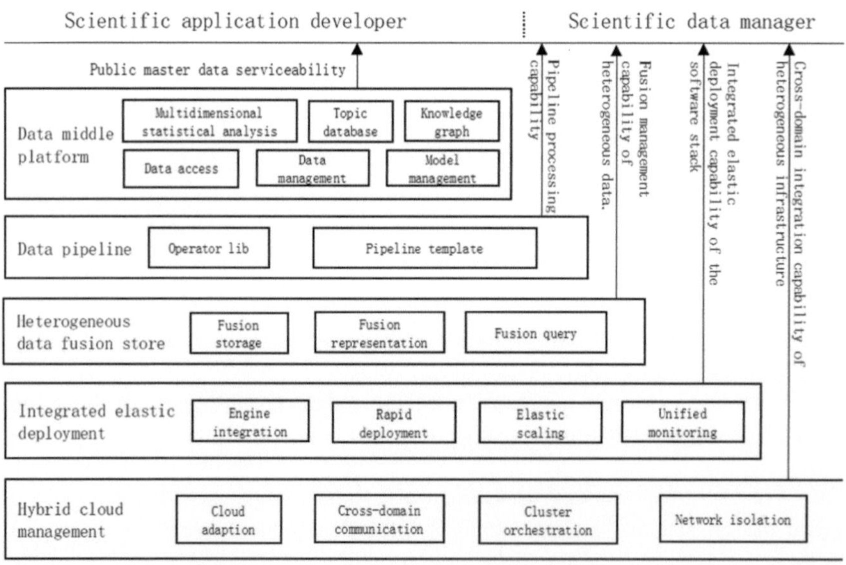

Fig. 2 The reference software architecture for managing, processing and serving multi-field scientific data

computing and storage resources and provides the same APIs to add, delete, modify and query virtual machines on different clouds. The main functional components are:

(1) **Cloud adaptation**. At present, the hybrid cloud solution is dominated by public cloud vendors. They virtualize the customer's local data center into a private cloud and connect it to the vendor's public cloud service through VPN technology. This way locks that customer to a specific public cloud vendor, resulting in cost pressure and management risk. To overcome this problem, a set of standard open APIs should be developed to shield the cloud API differences on resource management and provide consistent APIs for adding, deleting, modifying and checking virtual machines as well as storage volumes.
(2) **Cross-domain communication**. This component is set to provide cross-domain infrastructure integration capability by supporting mutual communication between virtual machines located in different data centers.
(3) **Cluster arrangement**. This component should support the one-click creation or destruction of clusters. A cluster refers to a collection of virtual machines and storage volumes running a software stack. With the cooperation of the cross-domain communication component, a cross-domain cluster can be created similarly.
(4) **Network isolation**. This component provides resource isolation capability by blocking network communication between different clusters by default.

3.2 *Integrated Elastic Deployment Module*

This module is responsible for software integration, rapid deployment and elastic cluster scaling.

(1) **Engine integration**. This component should be able to seamlessly integrate scientific software with the big data software stack. The integration protocol should be able to express the typical master-slave-client architecture of big data software and their dependencies on each other in the form of configuration files. A typical engine integration scheme is depicted in Fig. 3.

(2) **Rapid deployment**. This component should support the automatic deployment of software stack into the hybrid cloud to form a scientific big data cluster. The deployment efficiency should be controlled at the minute level to quickly respond to users' ad-hoc demands. Before deployment, users should be allowed to select the software set and the cluster size. After deployment, they can start and stop scientific data software in batches. In addition, network communication between different clusters is blocked by default to meet resource isolation requirements.
(3) **Elastic scaling**. This component should be able to dynamically scale scientific big data clusters according to computing/storage resources usage. At present, elastic scaling is divided into two schemes: Progressive and quantitative. The

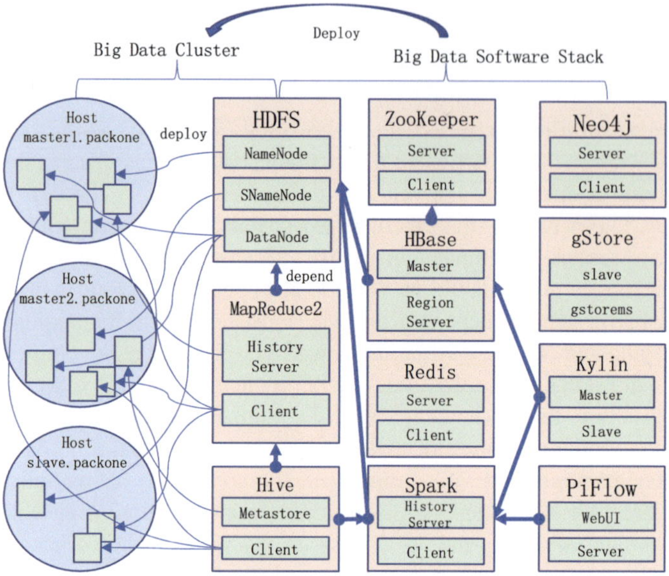

Fig. 3 A typical big data integration scheme

progressive scaling method monitors the competition of the upper application for computing and storage resources and dynamically increases or decreases the allocated resources. The quantitative scaling method can allocate corresponding computing/storage resource quotas for applications in advance by estimating the resource requirements of target applications. For simplicity, this component should give priority to supporting the progressive scaling method.

3.3 Heterogeneous Data Fusion Store Module

This module provides the fusion management capability of structured, semi-structured and unstructured scientific data by optimizing the storage structure, processing operator and query language to express the links of heterogeneous scientific data. It includes the underlying fusion storage that can preserve multiple formats of data with their links, abstract data manipulating operators that can be adapted by heterogeneous data store systems, and the fusion query language that can express common link analysis methods.

(1) **Fusion query**. This component allows using a SQL-like query language to retrieve heterogeneous data interactively. To meet the low delay requirement, the backend execution engine should be able to generate in-situ data processing pipelines automatically, avoiding manually writing data ETLs.
(2) **Fusion representation**. When the SQL-like language is parsed into intermediate data and query plans, the operators in the plan, especially the join operator, need

their operands to be expressed with a unified data model. Therefore, we need to design data models that can represent heterogeneous data of the same entity in a fusion way.

(3) **Fusion storage**. To support the fusion representation of heterogeneous data of the same entity, the data should be saved in the same physical storage structure by default to reduce the conversion cost between different data models. When the physical storage structure encounters unfitting data, this component should be able to access these data directly through APIs of these data's holding system.

3.4 Data Pipeline Module

This module is oriented to the streaming processing requirements of scientific data, and provides typical data processing operators and typical pipeline templates in the target research field.

(1) **Data processing operators**. Scientific data processing procedures include data collection, cleaning, fusion and storage, which have typical pipeline characteristics. Considering the massive, multi-source and heterogeneous characteristics of scientific data, this module must be able to adopt big data technology to realize multi-source and multi-type data gathering/cleaning/linking/sinking in the form of plug-in data processing operator.

(2) **Pipeline templates**. This component assembles data processing pipelines with multiple operators mentioned above. Considering the reusability and operation/maintenance cost, the pipelines should be saved as templates to support customization.

3.5 Data Middle Platform Module

This module provides data model management, topic database, multidimensional statistical analysis and knowledge graph services for upper-layer applications.

(1) **Data model management**. This component serves upper application developers with standardized data models, which solves the schema inconsistency between applications and databases. Its functions include data standardization, master data model creation/configuration/retrieval, model change monitoring and model completion.

(2) **Topic database**. This component organizes data with the same label or class into a logic database and generates public data access APIs that can meet the master data standard and the application requirements at the same time. These APIs serve for data publishing, data registration and resource discovery, etc.

(3) **Knowledge Graph**. This component serves the knowledge graph construction by human–computer interfaces, such as data selection, node modeling and relation generating. Moreover, it should be able to analyze data links and do topic mining of large-scale knowledge graphs.

4 The Prototype of the Reference Architecture

4.1 PackOne: Hybrid Cloud Management and Integrated Elastic Deployment

PackOne is a cloud-integrated deployment tool for big data software stack, based on which the prototype can manage OpenStack/H3CloudOS/EVCloud and deploy various big data software into the hybrid cloud automatically and elastically by the joint call to the cloud APIs and Apache Ambari APIs.

To decouple from specific cloud vendors, PackOne has developed a set of standard APIs for cloud resource management, and provided adaptation drivers for all supported cloud architectures. A driver is responsible for translating the standard APIs into the native call interfaces of the target cloud architecture. To realize cross-domain communication, each cloud member is equipped with an edge gateway based on the Wireguard protocol.

On the integrated elastic deployment, PackOne adopts Apache Ambari's component definition framework and software dependency description protocol, which abstracts the big data software stack into a series of interdependent services, such as HDFS, MapReduce, Spark, Hive, etc. Moreover, PackOne uses *scale* to express the cluster's creating specification. A scale formulizes the image/template/storage/services configurations of each node in the cluster. As depicted in Fig. 4, PackOne provides Web UI for cluster management.

4.2 Data Fusion Management Based on Neo4j and External store

This software employs Cypher, Neo4j, ElasticSearch and HBase to provide heterogeneous scientific data fusion management.

As depicted in Fig. 5, this software has a three-tier structure:

(1) **Fusion storage**. This tier adopts an entity-oriented data store schema, which organizes graph-structured, document-structured and unstructured data into linked data by the same entity ID, and uses Neo4j, ElasticSearch and HBase to hold them respectively. To preserve data links, the entity IDs and their relationship are stored locally in Neo4j. Besides, data stored in ElasticSearch and HBase are all accompanied by their entity IDs.

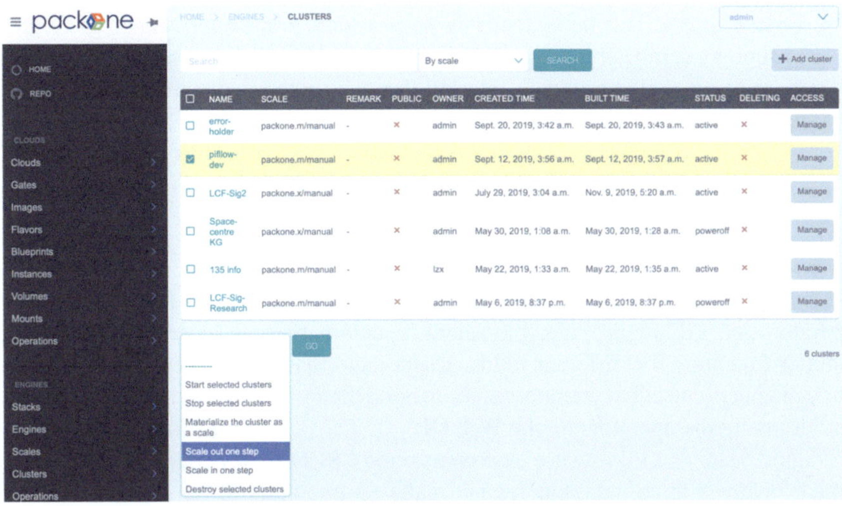

Fig. 4 PackOne's cluster management interface

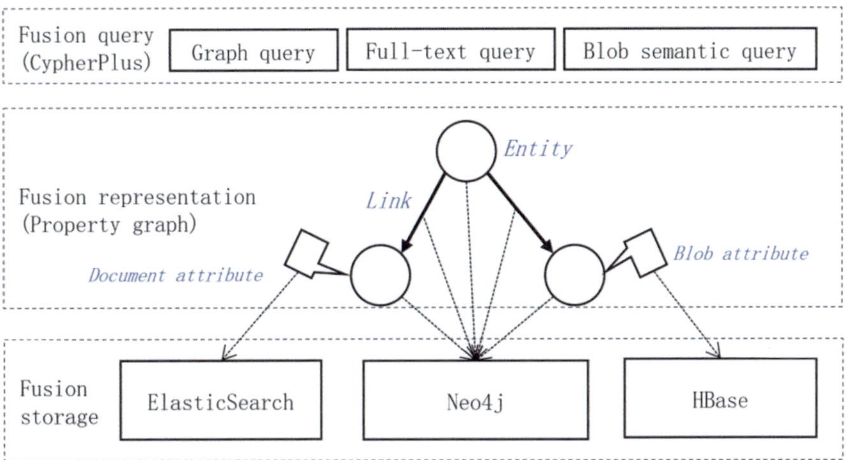

Fig. 5 Heterogeneous data fusion management architecture

(2) **Fusion representation**. This tier achieves fusion representation of heterogeneous data with the help of the attribute graph model. Specifically, the entity and its document/binary data are mapped into the vertex and its attributes. The links between entities are mapped into edges. For example, an image and a document belonging to the same biological specie are linked through a shared vertex, and genetic relationships between different species are expressed by edges between vertexes.

(3) **Fusion query**. This tier extends the open-source graph query language Cypher to support vertices filtering according to their IDs/attributes/relationships, and document/unstructured data filtering through semantic operators such as full-text retrieving and similarity computing.

4.3 PiFlow: Big Data Pipeline

PiFlow [7] is a big data pipeline execution engine. Based on the customizable operators set, PiFlow provides the scientific data gathering, cleaning, linking and sinking functions for different fields. Users can create their operators simply by customizing preinstalled operators and can build their pipeline templates by dragging and dropping the operators on the Web UI.

Figure 6 shows a PiFlow use case on tobacco CSCD achievement data gathering, which involves three data sources and calls 87 operators to do incremental data collecting and warehousing. These operators are responsible for the detection and correlation of talents, institutions, keywords and other information from papers, standards, patents, achievements and books. The collected heterogeneous data and their relationships are stored in the data fusion management software.

4.4 Big Data Middle Platform

The big data middle platform enables upper applications to access all kinds of data assets, and enables the application developer to create/publish topic databases and knowledge graphs. Specifically, the platform includes the following functional components:

(1) **Metadata management**. This component stores and manages metadata information of database/table/field, data linkage and data change history, etc. Initial configurations and automatic detection strategies are adopted to keep metadata naming to be standard.
(2) **Data asset management**. This component provides a layered data model and unified development specifications including data standardization, and model creation/configuration/retrieval functions, to solve the inconsistency between architecture design and application development.
(3) **Data service**. This component creates, publishes, retrieves and visualizes topic database services and knowledge graph services.

Taking tobacco scientific research as an example, the big data middle platform has realized the unified access, governance and service of various data such as papers, achievements, projects, institutions and persons. At present, it has accumulated 10^7 entities and 3×10^7 relationships. Figure 7 shows the tobacco science & technology knowledge graph service created by the platform.

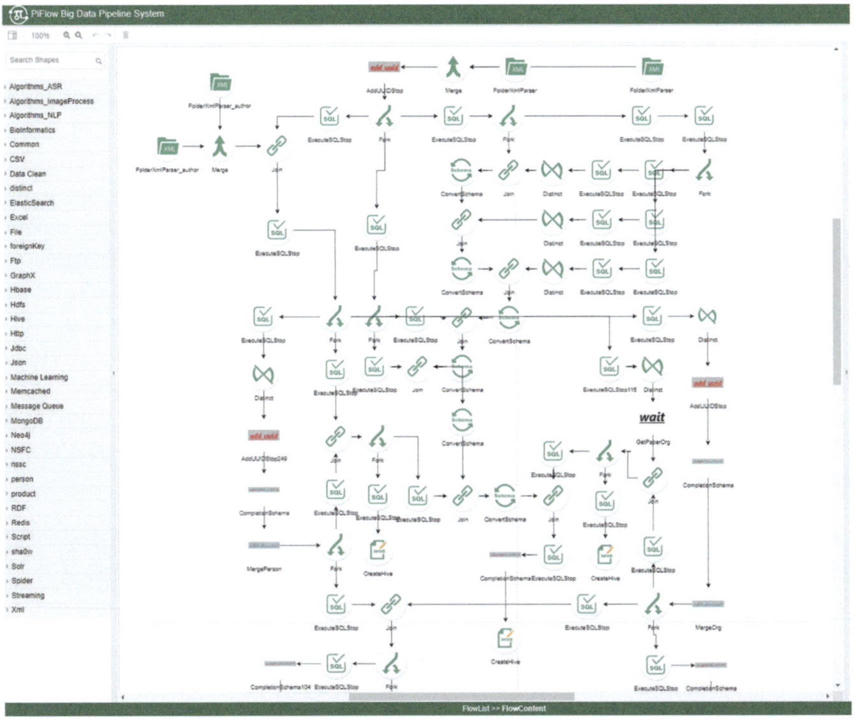

Fig. 6 Scientific data pipeline use case

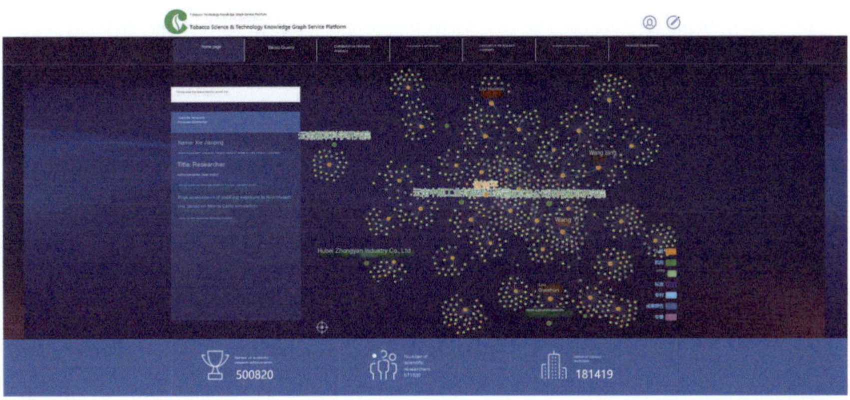

Fig. 7 Big data middle platform use case: Tobacco science & technology knowledge graph service

5 Summary

Scientific data in different fields are rich in both formats and semantic links. We lack effective methods to do the fusion management, processing and service of multi-field data. To solve these problems, this paper proposes a reference software architecture composed of function modules like hybrid cloud management, integrated elastic deployment, heterogeneous data fusion store, data pipeline and data middle platform, which can provide cross-domain integration capability of heterogeneous infrastructure, integrated elastic deployment capability of the software stack, fusion management capability of heterogeneous data, data pipeline processing capability and serviceability of public master data respectively. Based on the existing works of PackOne and PiFlow, we built a prototype implementation of the reference architecture. The reference architecture design is inevitably idealized. When applying it to practice, we should choose the most relevant functions according to the actual requirements, and make tradeoffs between the costs and benefits.

Acknowledgements This paper was funded by China National Tobacco Corporation (Project No. 110202101030 SJ-01).

References

1. Shi W, Zhang X, Wang Y et al (2019) Edge computing: state-of-the-art and future directions. J Comput Res Dev 56(1):69–89
2. Li J, Shen Z, Meng X (2017) Scientific big data management: concepts, technologies and system. J Comput Res Dev 54(2):235–247
3. Zaharia M, Chowdhury M, Franklin M J et al (2010) Spark: cluster computing with working sets. In: Proceedings of the 2nd USENIX conference on Hot topics in cloud computing: USENIX, pp 1–7
4. Zaharia M, Chowdhury M, Das T et al (2012) Resilient distributed datasets: a fault-tolerant abstraction for in-memory cluster computing. In: Proceedings of USENIX symposium on networked systems design and implementation—NSDI, pp 1–14
5. Carbone P, Katsifodimos A, Ewen S et al (2015) Apache Flink: stream and batch processing in a single engine. Bull IEEE Comput Soc Tech Comm Data Eng 36(4):28–38
6. Thusoo A, Sarma JS, Jain N et al (2009) Hive: a warehousing solution over a map-reduce framework. Proc VLDB Endow 2(2):1626–1629
7. Xiaojie ZHU, Zihao ZHAO, Yi DU (2020) PiFlow: a model driven big data pipeline framework. J Comput Appl 1:1–12

Huajin Wang was born in 1987. He graduated from University of Chinese Academy of Sciences with a doctor's degree in June 2018. At the present, he is the assistant professor at Computer Network Information Center, Chinese Academy of Sciences. His main research field is big data management and processing technology. He has published several papers in important journals and conferences such as the *Journal of software and CCGrid*. He participated in the development of open-source software such as PackOne and PandaDB (a heterogeneous data fusion management system).

MIX
Papier aus verantwortungsvollen Quellen
Paper from responsible sources
FSC® C105338
FSC
www.fsc.org

If you have any concerns about our products,
you can contact us on
ProductSafety@springernature.com

In case Publisher is established outside the EU,
the EU authorized representative is:
Springer Nature Customer Service Center GmbH
Europaplatz 3, 69115 Heidelberg, Germany

Printed by Libri Plureos GmbH
in Hamburg, Germany